ECO-HYDROLOGY

In the last two decades the importance of hydrological processes in ecosystems and the effects of plants on hydrological processes have become increasingly apparent. These relationships can be grouped under the term 'eco-hydrology'.

Eco-Hydrology introduces and explores diverse plant–water interactions in a range of environments. Leading ecologists and hydrologists present reviews of the eco-hydrology of drylands, wetlands, temperate and tropical rain forests, streams and rivers, and lakes. The authors provide background information on the water relations of plants, from individual cells to stands together with an in-depth review of scale issues and the role of mathematical models in eco-hydrology.

Eco-Hydrology is the first book to offer an overview of the complex relationships between plants and water across a range of terrestrial and aquatic environments. The text will prove invaluable as an introduction to hydrologists, ecologists, conservationists and others studying ecosystems, plant-life and hydrological processes.

Andrew Baird is lecturer in Physical Geography at the University of Sheffield, and **Robert Wilby** is senior lecturer in Physical Geography at the University of Derby, and a project scientist at the National Center for Atmospheric Research, Boulder, Colorado.

A VOLUME IN THE
ROUTLEDGE PHYSICAL ENVIRONMENT SERIES
Edited by Keith Richards
University of Cambridge

The Routledge Physical Environment series presents authoritative reviews of significant issues in physical geography and the environmental sciences. The series aims to become a complete text library, covering physical themes, specific environments, environmental change, policy and management as well as developments in methodology, techniques and philosophy.

ECO-HYDROLOGY

Plants and water in terrestrial and aquatic
environments

*Edited by Andrew J. Baird and
Robert L. Wilby*

London and New York

First published 1999
by Routledge
2 Park Square, Milton Park, Abingdon, Oxon, OX14 4RN

Simultaneously published in the USA and Canada
by Routledge
270 Madison Ave, New York NY 10016

Routledge is an imprint of the Taylor & Francis Group

Transferred to Digital Printing 2007

Typeset in Garamond by Keystroke, Jacaranda Lodge, Wolverhampton

British Library Cataloguing in Publication Data
A catalogue record for this book is available from the British Library

Library of Congress Cataloging-in-Publication Data
A catalog record is available on request

ISBN 0–415–16272–6 (hbk)
ISBN 0–415–16273–4 (pbk)

Publisher's Note
The publisher has gone to great lengths to ensure the quality of this reprint
but points out that some imperfections in the original may be apparent

CONTENTS

CONTENTS

FIGURES

FIGURES

FIGURES

TABLES

CONTRIBUTORS

Andrew R.G. Large, Department of Geography, University of Newcastle Upon Tyne, UK.

Mark Mulligan, Department of Geography, King's College London (University of London), UK.

Karel Prach, Faculty of Biological Sciences, University of South Bohemia, Czech Republic.

John M. Roberts, Hydrological Processes Division, Institute of Hydrology, Wallingford, Oxfordshire, UK.

David S. Schimel, National Center for Atmospheric Research, Boulder, Colorado, US.

John B. Thornes, Department of Geography, King's College London (University of London), UK.

Melvin T. Tyree, Aiken Forestry Services Laboratory, Burlington, Vermont, US.

John Wainwright, Department of Geography, King's College London (University of London), UK.

Robert G. Wetzel, Department of Biological Sciences, University of Alabama, College of Arts and Sciences, Tuscaloosa, Alabama, US.

Bryan D. Wheeler, Department of Animal and Plant Sciences, University of Sheffield, Sheffield, UK.

PREFACE

We live in a time when the boundaries between academic disciplines are becoming increasingly blurred. Many contemporary environmental problems and important research questions can only be addressed by collaboration between allied disciplines. For example, the management of wetland ecosystems for nature conservation may involve ecologists working closely with hydrologists in order to understand the water level and water quality requirements of a given plant community in a particular wetland. Equally, the management of a wetland for nature conservation also requires an appreciation of which species and habitats to conserve, and here we enter the realms of social science, environmental ethics and environmental law.

This volume is an attempt to formalise an area of overlap between two disciplines, namely ecology and hydrology. In the last two decades, ecologists have become increasingly aware of the importance of hydrological processes to ecosystem functions. Although many workers accept simple conceptual models such as the hydrosere (which shows a relationship between plants and the position of the water table), there is still a lack of detailed understanding about how individual plants or assemblages of plants are affected by and, in turn, influence hydrological processes. Hydrologists have also become more aware of the effects of plants on hydrological processes. For example, in arid and semi-arid environments, hydrologists have found that the distribution of plants has a profound effect on overland flow processes and erosion. Equally, the distribution and cover of plants is critically controlled by water availability.

Such relationships may be grouped under the term *eco-hydrology*. However, this term should not be confused with 'hydro-ecology', which is used in a narrower sense to describe the study of ecological and hydrological processes in rivers and floodplains. Although the term 'eco-hydrology' has been frequently coined by ecologists to describe interactions between water tables and plant distributions in wetlands, it can be used to describe plant–water interactions in other environments. In dealing with eco-hydrology, this volume therefore covers a wider range of hydrological environments than is implied by 'hydro-ecology'. Although this book aims to introduce the reader to some of the themes within eco-hydrology, it is not an introductory text in the usual sense. It is assumed that readers have a good basic understanding of hydrology and ecology, and accordingly the volume is pitched at advanced undergraduates and researchers coming to the subject for the first time. Although this book is an edited collection we have adopted a slightly unconventiontal approach to its preparation. We felt that it was impossible for one, two or even three authors to provide a sufficiently expert overview of the different areas within eco-hydrology. This necessitated adopting a truly multi-author approach. In contrast to multi-author conference proceedings, which are now so common, the detailed structure and content of the book were designed in advance and the authors carefully matched to the proposed chapters. However, one positive feature of conferences is that authors can take on board constructive comments made by other delegates before publishing their work in the proceedings. In essence the refereeing base is widened. Recognising this advantage, the authors of each chapter were invited by the editors to an open workshop on eco-hydrology held at the University of Sheffield on 18 October 1997 and sponsored by the British Ecological Society and the British Hydrological Society. The workshop was attended by over fifty academics, professional hydrologists and practising ecologists. The papers presented at the meeting stimulated much useful discussion and feedback, and we are indebted to all those who attended the workshop for making it a success.

In addition to the comments received at the workshop, each chapter was also independently reviewed, and we would like to thank Dr Jo Bullard, Dr Tim Davie, Professor George Hornberger, Dr Richard John Huggett, Dr Hugh Ingram, Patricia Rice, Professor Neil Roberts, Dr Peter Smithson, Professor David Thomas and Professor Ian Woodward for their refereeing. We would also like to thank Sarah Carty from Routledge for her patience with the senior editor, who delivered the final maunscript five months later than predicted. Her e-mails were the model of firm politeness! Sarah Lloyd encouraged the senior editor to pursue the idea of the volume in the first place, while three anonymous reviewers of the original book outline made valuable comments concerning the proposed content. Many of the figures in the book were expertly redrawn to a consistent style by Paul Coles and Graham Allsop of Cartographic Services, Department of Geography, University of Sheffield. We are also indebted to Oliver Tomlinson, Andy Barrett, Simon Birkett and Sally Edwards at the University of Derby for assisting in the preparation of maps and figures. Finally, we are grateful to all the authors for agreeing to contribute their respective chapters at a time when there is increasing pressure on academics to write internationally refereed research papers at the expense of book chapters.

We would like to dedicate this book to our families: to Laura, Aileen and Hester (AJB), and to Dawn and Samuel (RLW).

Sheffield and Boulder
August 1998

1

INTRODUCTION

Andrew J. Baird

WHAT IS ECO-HYDROLOGY?

Eco-hydrology, as its names implies, involves the study of both hydrology and ecology. However, before attempting a definition of eco-hydrology, and therefore a detailed description of what this book is about, it is useful to consider briefly how hydrology and ecology have been studied separately and together in the past. As a *discipline*, hydrology has a long history, which has been described in detail by Biswas (1972) and Wilby (1997a). As a *science*, its origins are more recent. Bras (1990: p. 1) defines modern hydrology as 'the study of water in all its forms and from all its origins to all its destinations on the earth.' Implied in this definition is the need to understand how water cycles and cascades through the physical and biological environment. Also implied in this definition is the principle of continuity or the balance equation; that is, in order to understand the hydrology of a system we must be able to account for all inputs and outputs of water to and from the system as well as all stores of water within the system. As noted by Baird (1997) after Dooge (1988), the principle of continuity can be considered as the fundamental theorem or equation in hydrology. The dominance of the principle is relatively recent, and it is surprising to discover that the earliest hydrologists were unable to perceive that rainfall was the only ultimate source of stream and river flow. Remarkably, it was not until the studies of Palissy (1510–1590) that it was realised that rain falling on a catchment was sufficient to sustain stream discharge (Biswas, 1972: p. 152). Much of the history of hydrology is dominated by the necessity to secure and distribute potable water supplies (Wilby, 1997a). The discipline, therefore, has an engineering background.

Today, hydrology is still in large part an engineering discipline concerned with water supply, waste water disposal and flood prediction and, as studied by engineers, has close links with pipe and channel hydraulics. This link between hydrology and hydraulics together with the practical application of hydrological knowledge often forms the focus of standard hydrology and hydraulics textbooks written for engineering students, such as Chow (1959),

1

Linsley *et al.* (1988) and Shaw (1994). Notwithstanding the seminal work of Palissy and others, hydrology has emerged as a science only in the last few decades. The publication of Ward's (1967) *Principles of Hydrology* represents one of the first attempts to set down the fundamental principles of hydrology as a science for a non-specialist audience. There are now many texts and journals that deal with scientific hydrology; examples of the former include Parsons and Abrahams (1992), Hughes and Heathwaite (1995) and Wilby (1997b), and examples of the latter include *Journal of Hydrology*, *Water Resources Research* and *Hydrological Processes*. However, since about the early 1980s hydrologists have paid increasing attention to the relationship between water in the landscape and ecological processes. For example, in the past a hydrologist might have thought of plants in a river channel merely as representing a particular roughness coefficient for use in Manning's 1889 Universal Discharge Formula. Now an increasing number of hydrologists are concerned with how flow velocities affect plant growth in channels and the relationship between river flow regimes and ecological processes in riparian habitats (see, for example, Petts and Bradley, 1997; Petts *et al.*, 1995). This has spawned the so-called sub-discipline of *hydro-ecology*, which has as its focus the study of hydrological and ecological processes in rivers and floodplains and the development of models to simulate these interactions (for example, the Physical HABitat SIMulation model or PHABSIM – see Bovee, 1982). Plants and ecological processes are no longer regarded by hydrologists as static parts of the hydrological landscape. Another striking example of hydrologists studying the role of plants in the hydrological landscape has been in studies of evapotranspiration and rainfall interception (for a recent example, see Davie and Durocher, 1997a and b). In a similar fashion, ecologists have become much more sophisticated in their appreciation of water storage and transfer processes in ecosystems (see below). Finally, the 1990s, in particular, have seen a blurring of the distinction between engineering hydrology and scientific hydrology. Engineers are increasingly concerned with the impact of engineering works on ecological processes and 'naturalisation' of previously engineered rivers (Kondolf and Downs, 1996). Equally, engineers are centrally involved in scientific advances in the discipline. For example, engineers as well as scientists are studying complex flow structures and sediment movement in natural and semi-natural channels using sophisticated field and laboratory equipment and computer models (for a collection of relevant papers see Ashworth *et al.*, 1996; see also Hodskinson and Ferguson, 1998; Niño and García, 1998; Tchamen and Kahawita, 1998).

The history of ecology is described in reasonable detail in a number of standard ecology textbooks, such as Brewer (1994), Stirling (1992) and Colinvaux (1993). According to Brewer (1994), the term 'ecology' (German *Ökologie*) was first coined in 1866 by the German zoologist Ernst Haeckel, who based it on the Greek *oikos*, meaning 'house'. Although Haeckel used the term to describe the relations of an animal with its physical and biological environment, the term has almost always been used to describe the study of the relationship of any organism or group of organisms with their environment *and* one another. As noted by Brewer (1994, pp. 1–2), 'Ecology is a study of interactions. A list of plants and animals of a forest is only a first step in ecology. Ecology is knowing who eats whom, or what plants fail to grow in the forest because they can't stand the shade or because, when they do grow there, they get (*sic*) eaten.' Ecology developed from natural history and until the middle of the century was concerned primarily with describing communities and the 'evolution' of communities through successional processes. Ecological succession *sensu* Clements (1916) refers to ordered change in a community to a final stable state called the climax. The essential ideas of succession are well illustrated by two systems: the coastal sand dune system and the lakeside system. In the former, the successional system is called the psammosere, in the latter the hydrosere. Interestingly, succession in both was seen to be the product of a close linkage between hydrological processes and ecological processes. In the hydrosere, for example, it was assumed that plants further from the water's edge were less tolerant of waterlogging. Thus, both successions are early examples of eco-hydrological models in which there was explicit recognition of the role of water in plant growth and survival. It is now known that the simple progressions assumed in the classic descriptions of the psammosere and hydrosere rarely, if ever, occur and that the links between plants and their physical environment are more complicated than was once thought. For example, in wetlands research, Wheeler and Shaw (1995: p. 63) note that:

> Conservationists would generally welcome a clear understanding of the inter-relationships between mire vegetation and hydrology, to help them predict the likely effects of hydrological change upon vegetation. . . . However, despite quite a large number of studies . . . the hydrology of fens and the composition of their vegetation is not at all well understood, except in gross terms.

This theme is developed in more detail in Chapter 5 of this volume (see also below).

The two-way linkage between plant growth and survival and hydrological processes has been studied in a range of environments, not just wetlands. For example, in recent studies Veenendaal *et al.* (1996) looked at seedling survival in relation to moisture stress under forest canopies and in forest gaps in tropical rain forest in West Africa, while Pigott and Pigott (1993) investigated the role of water availability as a determinant of the distribution of trees at the boundary of the mediterranean climatic zone in southern France. There are many other examples of non-wetland eco-hydrological study in the ecological literature, and these include Berninger (1997),

Jonasson *et al.* (1997), Stocker *et al.* (1997), Bruijnzeel and Veneklaas (1998) and Hall and Harcombe (1998). However, the term 'eco-hydrology', which is used to describe the study of these links, seems to have been coined originally to describe only research in wetlands (see, for example, Ingram, 1987) and appears to have been in use by wetland ecologists for at least two decades (G. van Wirdum and H.A.P. Ingram, personal communication). In an editorial of a collection of papers on eco-hydrological processes in wetlands in a special issue of the journal *Vegetatio*, Wassen and Grootjans (1996: p. 1) define eco-hydrology purely in terms of processes occurring in wetlands:

> Ecohydrology is an application driven interdisciplin [*sic*] and aims at a better understanding of hydrological factors determining the natural development of wet ecosystems, especially in regard of their functional value for nature protection and restoration.

It is instructive to read the papers in the special issue of *Vegetatio*, because they give an insight into how users of the term 'eco-hydrology' conduct their research. A recurrent theme of the papers is that hydrological, hydrochemical and vegetation patterns in the studied ecosystems are measured separately and *then* related to each other. This is particularly evident in the papers of Wassen *et al.* (1996), who compare fens in natural and artificial landscapes in terms of their hydrological behaviour, water quality and vegetation composition, and Grootjans *et al.* (1996a), who investigate vegetation change in dune slacks in relation to ground water quality and quantity. Somewhat surprisingly, manipulative experiments, whether in the laboratory or in the field, on factors affecting growth of plants do not figure highly in the research programmes presented. A second theme in the papers is that of the practical application of scientific knowledge to ecosystem management, especially for nature conservation. The compass of eco-hydrology as defined by these papers would, therefore, appear to be somewhat narrow. First, the term is used to describe wetlands research. Second, it describes predominantly field-based research where links between hydrological and ecological variables are sought; it does not appear to include manipulative experiments. Third, the term is associated with practical application of scientific ideas, especially in nature conservation. Interestingly, and somewhat confusingly, Grootjans *et al.* (1996b; cited in Wassen and Grootjans, 1996) appear to recognise that the term can be applied more widely. To them, 'ecohydrology is the science of the hydrological aspects of ecology; the overlap between ecology, studied in view of eco-logical problems.' Although broader, this definition still includes an emphasis on solving ecological 'problems', where presumably these are similar to the conservation problems mentioned above.

Very recently, Hatton *et al.* (1997) have used the term to describe plant–water interactions in general and suggest that Eagleson's (1978a–g) theory of an *ecohydrological equilibrium* should form the focus of eco-hydrological

4

research. The ideas of Hatton *et al.* on eco-hydrological modelling are discussed briefly later in this volume (Baird, Chapter 9; see also below).

THE SCOPE OF THIS TEXT

In this book, a wider definition than either (Wassen and Grootjans, 1996; Grootjans *et al.*, 1996b) given above is used. In recognising that *eco* is a modifier of *hydrology* it could be argued that eco-hydrology should be more about hydrology than ecology. However, it is undesirable and probably impossible to consider the links between plants and water solely in terms of how one affects the other. Thus, while this book tends to focus on hydrological processes, it also considers how these processes affect plant growth. Additionally, as noted above, there is no intrinsic reason why eco-hydrology should be solely concerned with processes in wetlands. Eco-hydrological relations are important in many, indeed probably all, ecosystems. Although such linkages are very important in wetlands, they are arguably of equal importance in forest and dryland ecosystems, for example. Therefore, an attempt is made in this volume to review eco-hydrological processes in a *range* of environments, thus following the broader definition of eco-hydrology implied by Hatton *et al.* (1997). Eco-hydrological processes are considered in drylands, wetlands, forests, streams and rivers, and lakes. However, it is probably impossible to compile a volume which looks at every aspect of eco-hydrology. In recognition of this, the book focuses on plant–water relations in terrestrial and aquatic ecosystems. Thus the role of marine ecosystems in the global hydrosystem, although extremely important and acknowledged in the conclusion, is not considered in any detail. Full consideration of the topic would require a volume in its own right. For similar reasons, the role of water as an environmental factor controlling animal populations is not dealt with. Even with the focus on terrestrial and aquatic ecosystems it has been impossible to provide a comprehensive overview of all ecosystems. Thus, eco-hydrological processes in tundra and mid-latitude grasslands, for example, are not discussed. Despite this selective focus, it is hoped that the material presented herein will still reveal the key research themes and perspectives within eco-hydrology. Finally, in this volume a range of eco-hydrological research methods, including manipulative experiments, are reviewed.

Each of the five chosen environments or ecosystem types mentioned above are dealt with in Chapters 4 to 8. In Chapter 4, John Wainwright, Mark Mulligan and John Thornes consider eco-hydrological processes in drylands. The authors look at how dryland plants cope with generally low and highly variable amounts of rainfall. They consider how plants affect runoff and water-mediated erosion in drylands. The issue of scale is also dealt with. Spatially, the role of small-scale process interactions on large-scale soil–vegetation–atmosphere transfers (SVAT) in drylands is considered.

Temporally, vegetation evolution in the Mediterranean is considered in terms of climate change.

In Chapter 5, Bryan Wheeler looks at water and plants in wetlands. The bulk of the chapter considers how water levels in wetlands exert a control on plant growth and survival. Wheeler notes that it is vital to consider more than just bulk amounts of water; water level regime can exert as much control on plant growth as can absolute level. However, despite ample evidence for water levels affecting plant growth, it is clear that simple deterministic relationships between level or regime and vegetation species composition have remained elusive. Complicating factors in the relationship between vegetation composition and water levels include the source of the water, which affects its nutrient status, and the effect of water on other processes within wetland soils such as oxidation–reduction (redox) reactions. Wheeler also considers the role of plants in affecting wetland water levels, principally through their control of the structure of the organic substrate found in many wetlands.

In Chapter 6, John Roberts explores the relationships between plants and water in forests. The emphasis here is on how trees affect the delivery of water to the ground surface and how they affect soil moisture regimes through the processes of evapotranspiration. Emphasis is placed on methods of measuring and modelling the various processes considered, especially canopy interception and transpiration. Roberts focuses on processes in temperate forests and tropical rain forests, and draws comparisons between water transfer processes in the two. He acknowledges that relatively little is known about plant–water relations in boreal and mediterranean forests and that these ecosystems should form the focus of future studies of forest hydrology.

Chapter 7, written by Andrew Large and Karel Prach, considers plants and water in streams and rivers. Plants exert a considerable influence on the hydraulic properties of channels, principally through their effect on channel roughness or friction to flow. The exact effect will depend on river stage and the species composition and density of in-channel vegetation. Flow regime, in turn, has a profound effect on plant growth and survival in channels. If the flow is too powerful, plants can become uprooted from the channel substrate or cannot become established. Equally, the stability of the substrate, itself related to flow regime, can affect rooting of in-channel plants. As in wetlands, hydrological regime is as important as mean conditions, and floods and droughts can pose special problems for plants growing in channels.

In Chapter 8, Robert Wetzel examines plants and water in lakes. The role of plants in controlling lake water levels is considerable. Here it is the littoral vegetation that is most important. Although planktonic organisms can modify thermal conditions and influence evaporation, stratification and related mixing processes in open water, they are of relatively minor importance in the hydrology of lake basins as a whole. Indeed, evapotranspiration from lake-edge vegetation can be several times greater than open-water evaporation, and can exceed other forms of water loss from lake systems. The drawdown of lakeside water tables can in turn change patterns of groundwater flow to and from a lake. Wind-generated water movements within lake waters transport sediment and nutrients and can, therefore, affect where plants are able to become established within the lake system. These water movements are also greatly affected (reduced) by the presence of littoral emergent vegetation.

To underpin and augment the thematic material on particular environments, three 'generic' chapters have been included. In the first of these, Chapter 2, Melvyn Tyree looks at the water relations of plants. The purpose of this chapter is to provide the reader with basic information on how plants use water. Many existing hydrology textbooks consider transpiration briefly, but few look in detail at the movement of water from roots to leaves and the effect of waterlogging and drought on plant growth and survival. Thus Chapter 2 provides a basic foundation for all the other chapters, but in particular Chapters 4 to 8. One theme within Chapter 2 is scale: water relations of plants are considered at the scale of the cell, whole plants and stands of plants. The theme is continued in Chapter 3, where Robert Wilby and David Schimel demonstrate how eco-hydrological relations can be considered at different spatial and temporal scales. Scale is an increasingly important theme in both hydrology and ecology, and it is important to understand how different eco-hydrological processes can be represented at different scales, from the small research plot to the regional and global scale. It is also important to consider how we can link observations made at different scales and how we can evaluate the effect of processes occurring at one scale on larger- and smaller-scale processes. In Chapter 9, Andrew Baird considers the role of models in eco-hydrology. Models are frequently used to formalise understanding of environmental processes and for theory testing. The growing importance of, and need for, eco-hydrological models is stressed. However, it is important that models are sensibly designed in order to improve scientific understanding of eco-hydrological processes. There is the persistent danger that models can become over-complex abstractions of reality that are difficult to understand and use.

The volume concludes with Chapter 10, in which Robert Wilby suggests future directions in eco-hydrological research. He returns to the theme of scale and considers the prospects for a planetary-scale eco-hydrology. He also examines the role of an experimental approach within eco-hydrology and looks at how an understanding of eco-hydrological processes can be used in the management of human-impacted ecosystems.

REFERENCES

Ashworth, P.J., Bennett, S.J., Best, J.L. and McLelland, S.J. (eds) (1996) *Coherent Flow Structures in Open Channels*, Chichester: Wiley, 733 pp.

Baird, A.J. (1997) Continuity in hydrological systems, in Wilby, R.L. (ed.) *Contemporary Hydrology: Towards Holistic Environmental Science*, Chichester: Wiley, pp. 25–58.

Berninger, F. (1997) Effects of drought and phenology on GPP in *Pinus sylvestris*: a simulation study along a geographical gradient, *Functional Ecology* 11: 33–42.

Biswas, A.K. (1972) *History of Hydrology*, Amsterdam: North Holland.

Bovee, K.D. (1982) *A Guide to Stream Habitat Analysis Using the Instream Flow Incremental Methodology*. Instream Flow Information Paper 12, FWS/OBS-82/26. Office of Biological Sciences, US Fish and Wildlife Service, Fort Collins.

Bras, R.L. (1990) *Hydrology: an Introduction to Hydrologic Science*, Reading, Massachusetts: Addison-Wesley.

Brewer, R. (1994) *The Science of Ecology* (second edition), Fort Worth: Harcourt Brace.

Bruijnzeel, L.A. and Veneklaas, E.J. (1998) Climatic conditions and tropical montane forest productivity: the fog has not lifted yet, *Ecology* 79: 3–9.

Chow, V.T. (1959) *Open Channel Hydraulics*, New York: McGraw-Hill, 680 pp.

Clements, F.E. (1916) *Plant Succession*, New York: Carnegie Institute of Washington.

Colinvaux, P. (1993) *Ecology 2*, New York: Wiley.

Davie, T.J.A. and Durocher, M.G. (1997a) A model to consider the spatial variability of rainfall partitioning within deciduous canopy. 1. Model description, *Hydrological Processes* 11: 1509–1523.

Davie, T.J.A. and Durocher, M.G. (1997b) A model to consider the spatial variability of rainfall partitioning within deciduous canopy. 2. Model parameterization and testing, *Hydrological Processes* 11: 1525–1540.

Dooge, J.C.I. (1988) Hydrology in perspective, *Hydrological Sciences Journal* 33: 61–85.

Eagleson, P.S. (1978a) Climate, soil, and vegetation. 1. Introduction to water balance dynamics, *Water Resources Research* 14: 705–712.

Eagleson, P.S. (1978b) Climate, soil, and vegetation. 2. The distribution of annual precipitation derived from observed storm sequences, *Water Resources Research* 14: 713–721.

Eagleson, P.S. (1978c) Climate, soil, and vegetation. 3. A simplified model of soil water movement in the liquid phase, *Water Resources Research* 14: 722–730.

Eagleson, P.S. (1978d) Climate, soil, and vegetation. 4. The expected value of annual evapotranspiration, *Water Resources Research* 14: 731–739.

Eagleson, P.S. (1978e) Climate, soil, and vegetation. 5. A derived distribution of storm surface runoff, *Water Resources Research* 14: 741–748.

Eagleson, P.S. (1978f) Climate, soil, and vegetation. 6. Dynamics of the annual water balance, *Water Resources Research* 14: 749–764.

Eagleson, P.S. (1978g) Climate, soil, and vegetation. 7. A derived distribution of annual water yield, *Water Resources Research* 14: 765–776.

Grootjans, A.P., Sival, F.P. and Stuyfzand, P.J. (1996a) Hydro-chemical analysis of a degraded dune slack, *Vegetatio* 126: 27–38.

Grootjans, A.P., Van Wirdum, G., Kemmers, R.H. and Van Diggelen, R. (1996b) Ecohydrology in the Netherlands: principles of an application-driven interdisciplin [*sic*], *Acta Botanica Neerlandica* (in press) (cited in Wassen and Grootjans, 1996).

Hall, R.B.W. and Harcombe, P.A. (1998) Flooding alters apparent position of floodplain saplings on a light gradient, *Ecology* 79: 847–855.

Hatton, T.J., Salvucci, G.D. and Wu, H.I. (1997) Eagleson's optimality theory of an ecohydrological equilibrium: quo vadis? *Functional Ecology* 11: 665–674.

Hodskinson, A. and Ferguson, R.I. (1998) Numerical modelling of separated flow in river bends: model testing and experimental investigation of geometric controls on the extent of flow separation at the concave bank, *Hydrological Processes* 12: 1323–1338.

Hughes, J.M.R. and Heathwaite, A.L. (eds) (1995) *Hydrology and Hydrochemistry of British Wetlands*, Chichester: Wiley.

Ingram, H.A.P. (1987) Ecohydrology of Scottish peatlands, *Transactions of the Royal Society of Edinburgh: Earth Sciences* 78: 287–296.

Jonasson, S., Medrano, H. and Flexas, J. (1997) Variation in leaf longevity of *Pistacia lentiscus* and its relationship to sex and drought stress inferred from $\delta^{13}C$, *Functional Ecology* 11: 282–289.

Kondolf, G.M. and Downs, P. (1996) Catchment approach to planning channel restoration, in Brooks, A. and Shields, F.D., Jr (eds) *River Channel Restoration: Guiding Principles for Sustainable Projects*, Chichester: Wiley, pp. 129–148.

Linsley, R.K., Jr, Kohler, M.A. and Paulhus, J.L.H. (1988) *Hydrology for Engineers* (SI metric edition), London: McGraw-Hill, 492 pp.

Niño, Y. and García, M. (1998) Using Lagrangian particle saltation observations for bedload sediment transport modelling, *Hydrological Processes* 12, 1197–1218.

Parsons, A.J. and Abrahams, A.D. (eds) (1992) *Overland Flow: Hydraulics and Erosion Mechanics*, London: UCL Press, 438 pp.

Petts, G.E. and Bradley, C. (1997) Hydrological and ecological interactions within river corridors, in Wilby, R.L. (ed.) *Contemporary Hydrology*, Chichester: Wiley, pp. 241–271.

Petts, G.E., Maddock, I., Bickerton, M. and Ferguson, A.J.D. (1995) Linking hydrology and ecology: the scientific basis for river management, in Harper, D.M. and Ferguson, A.J.D. (eds) *The Ecological Basis for River Management*, Chichester: Wiley, pp. 1–16.

Pigott, C.D. and Pigott, S. (1993) Water as a determinant of the distribution of trees at the boundary of the Mediterranean zone. *Journal of Ecology* 81: 557–566.

Shaw, E.M. (1994) *Hydrology in Practice* (third edition), London: Chapman & Hall, 569 pp.

Stirling, P.D. (1994) *Introductory Ecology*, Englewood Cliffs, New Jersey: Prentice-Hall.

Stocker, R., Leadley, P.W. and Körner, Ch. (1997) Carbon and water fluxes in a calcareous grassland under elevated CO_2, *Functional Ecology* 11: 222–230.

Tchamen, G.W. and Kahawita, R.A. (1998) Modelling wetting and drying effects over complex topography, *Hydrological Processes* 12: 1151–1182.

Veenendaal, E.M., Swaine, M.D., Agyeman, V.K., Blay, D., Abebrese, I.K. and Mullins, C.E. (1996) Differences in plant and soil water relations in and around a forest gap in West Africa during the dry season may influence seedling establishment and survival, *Journal of Ecology* 84: 83–90.

Ward, R.C. (1967) *Principles of Hydrology*, London: McGraw-Hill.

Wassen, M.J. and Grootjans, A.P. (1996) Ecohydrology: an interdisciplinary approach for wetland management and restoration, *Vegetatio* 126: 1–4.

Wassen, M.J., van Diggelen, R., Wolejko, L. and Verhoeven, J.T.A. (1996) A comparison of fens in natural and artificial landscapes, *Vegetatio* 126: 5–26.

Wheeler, B.D., and Shaw, S.C. (1995) Plants as hydrologists? An assessment of the value of plants as indicators of water conditions in fens, in Hughes, J.M.R. and Heathwaite, A.L. (eds) *Hydrology and Hydrochemistry of British Wetlands*, Chichester: Wiley, pp. 63–82.

Wilby, R.L. (1997a) The changing roles of hydrology, in Wilby, R.L. (ed.) *Contemporary Hydrology: Towards Holistic Environmental Science*, Chichester: Wiley, pp. 1–24.

Wilby, R.L. (ed.) (1997b) *Contemporary Hydrology: Towards Holistic Environmental Science*, Chichester: Wiley, 354 pp.

2

WATER RELATIONS OF PLANTS

Melvin T. Tyree

INTRODUCTION

Water relations of plants is a large and diverse subject. This chapter is confined to some basic concepts needed for a better understanding of the role of plants in eco-hydrology, and readers seeking more details should consult Slatyer (1967) and Kramer (1983).

First and foremost, it must be recognised that water movement in plants is purely 'passive'. In contrast, plants are frequently involved in 'active' transport of substances; for example, membrane-bound proteins (enzymes) actively move K^+ from outside cells through the plasmalemma membrane to the inside of cells. Such movement is against the force on K^+ tending to move it outwards, and such movement requires the addition of energy to the system to move the K^+. Energy for active K^+ transport is derived from ATP (adenosine triphosphate). While there have been claims of active water movement in the past, no claim of active water transport has ever been proved.

Passive movement of water (like passive movement of other substances or objects) still involves forces, but passive movement is defined as spontaneous movement in a system that is already out of equilibrium in such a way that the system tends towards equilibrium. Active movement, by contrast, requires the input of biological energy and moves the system further away from equilibrium or keeps it out of equilibrium in spite of continuous passive movement in the counter-direction. The basic equation that describes passive movement is Newton's law of motion on Earth where there is friction:

$$v = (1/f)F \qquad (2.1)$$

where v is velocity of movement (m s^{-1}), F is the force causing the movement (N) and f is the coefficient of friction (N s m^{-1}).

In the context of passive water or solute movement in plants, it is more convenient to measure moles moved per s per unit area, which is a unit of measure called a flux density (J). Fortunately, there is a simple relationship

11

between J, v and concentration (C, mol m^{-3}) of the substance moving: $J = Cv$. Also, in a chemical/biological context, it is easier to measure the energy of a substance, and how the energy changes as it moves, than it is to measure the force acting on the substance. Passive movement of water or a substance occurs when it moves from a location where it has high energy to where it has lower energy. The appropriate energy to measure is called the chemical potential, μ, and it has units of energy per mol (J mol^{-1}). The force acting on the water or solute is the rate of change of energy with distance, hence $F = -(d\mu/dx)$, which has units of J m^{-1} mol^{-1} or N mol^{-1} (because J = N m). So replacing F with $-(d\mu/dx)$ and v with J we have:

$$J = - K \, (d\mu/dx) \tag{2.2}$$

where K is a constant $= C/f$. Equation (2.2) or some variation of it is used to describe water movement in soils (Darcy's law – see Chapter 9) and plants. The variations on equation (2.2) generally involve measuring J in kg or m^3 of water rather than moles and measuring μ in pressure units rather than energy units.

WATER RELATIONS OF PLANT CELLS

The water relations of plant cells can be described by the equation that gives the energy state of water in cells and how this energy state changes with water content, which can be understood through the Höfler diagram. First let us consider the factors that determine the energy state of water in a cell.

The energy content of water depends on temperature, height in the Earth's gravitational field, pressure and mole fraction of water (X_w) in a solution. For practical purposes, we evaluate the chemical potential of water in a plant cell in terms of how much it differs from pure water at ground level and at the same temperature as the water in the cell; i.e., we measure

$$\Delta\mu = \mu - \mu_o \tag{2.3}$$

where μ_o is the chemical potential of water at ground level at the same temperature as the cell. It has become customary for plant physiologists to report $\Delta\mu$ (J mol^{-1}) in units of J per m^3 of water because this has dimensions equal to a unit of pressure (i.e. Pa = N m^{-2}), and this new quantity is called water potential (ψ). The conversion involves dividing $\Delta\mu$ by the partial molal volume of water ($\overline{V}_w$), i.e.,

$$\psi = \frac{1}{\overline{V}_w}\Delta\mu \tag{2.4}$$

So, in general, ψ is given by

$$\psi = P + \pi + \rho g h \tag{2.5}$$

where P is the pressure potential (the hydrostatic pressure),

$$\pi = \frac{RT}{V_w} \ln X_w$$

is called the osmotic potential (sometimes called osmotic pressure) and is approximated by $\pi = -RTC$, where C is the osmolal concentration of the solution, R is the gas constant and T is the kelvin temperature, ρ is the density of water, g is the acceleration due to gravity, and h is the height above ground level.

Water will flow into a cell whenever the water potential outside the cell (ψ_o) is greater than the water potential inside the cell (ψ_c). Let us consider the water relations of a cell at ground level, i.e., how water moves in and out of the cell in the course of a day. Water flows through plants in xylem conduits (vessels or tracheids), which are non-living pipes, and the walls of the pipes are made of cellulose. Water can freely pass in and out of the conduits through the cellulose walls (Figure 2.1A). The water potential of the water in the xylem conduit is given by:

$$\psi_x = P_x + \pi_x \tag{2.6}$$

During the course of a day, P_x might change from slightly negative values at sunrise (e.g. −0.05 MPa) to more negative values by early afternoon (let us say −1.3 MPa) and then might return again by the next morning, as explained in 'the cohesion–tension (C–T) theory and xylem dysfunction', below. The concentration of solutes in xylem fluid is usually very low (i.e., plants transport nearly pure water in xylem) so π_x is not very negative (let us say −0.05 MPa). So ψ_x might change from −0.1 to −1.35 MPa in the example given in Figure 2.1B. This daily change in ψ_x will cause a daily change in ψ_c as water flows out of the cell as ψ_x falls and into the cell as ψ_x rises.

The two primary factors that determine the water potential of a cell at ground level (ψ_c) are turgor pressure P_t (i.e. P inside the cell) and π of the cell sap (π_c):

$$\psi_c = P_t + \pi_c \tag{2.7}$$

Plant cells generally have π_c values in the range of −1 to −3 MPa; consequently P_t is often a large positive value whenever $\psi_c = \psi_x$. There are two reasons for cells having $P_t > 0$:

13

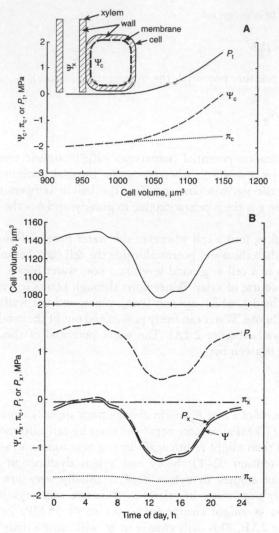

Figure 2.1 A drawing of a living cell with plasmalemma membrane adjacent to a xylem conduit (shown in longitudinal section). The living cell is surrounded by a soft cell wall composed of cellulose, which retards expansion of the cell and prevents collapse of the cell as P_t falls. The xylem conduit is surrounded by a woody cell wall (cellulose plus lignin), which strongly retards both expansion and collapse when P_x becomes negative. A: A Höfler diagram showing how the cell water potential (ψ_c), the cell osmotic potential (π_c) and the turgor pressure (P_t) change with cell volume. B: A representative daily time course of how cell or xylem water potential (ψ) changes during the course of a day. The component water potentials shown are xylem pressure potential (P_x), xylem osmotic potential (π_x), cell turgor pressure (P_t) and cell osmotic potential (π_c). Also shown is how cell volume changes with time. See text for more details.

1   The protoplasm of living cells is enclosed inside a semi-permeable membrane (the plasmalemma membrane) that permits relatively rapid trans-membrane movement of water and relatively slow trans-membrane movement of solutes, so the solutes inside the cell making π_c negative cannot move out to the xylem to make π_x more negative.

2   The membrane-bound protoplasm is itself surrounded by a relatively rigid-elastic cell wall, so the cell wall must expand as water flows into the cell to accommodate the extra volume; the stretch of the elastic wall places the cell contents under a positive pressure (much like a tyre pumped up by air puts the air in the tyre under pressure). The rise in P_t raises ψ_c until it reaches a value equal to ψ_x, at which point water flow stops.

The effect of water movement into or out of a cell is described by a Höfler diagram (Figure 2.1A). Entry of water into the cell has two effects: it causes a dilution of cell contents, hence π_c becomes slightly less negative; and P_t rises very rapidly with cell volume. The net effect of the increase in π_c and P_t is an increase in ψ_c towards zero. Conversely, a loss of water makes ψ_c, π_c and P_t fall to increasingly negative values. If ψ_c falls low enough, then P_t falls to and remains at zero in cells in soft tissue. In woody tissues, i.e. in cells with lignified cell walls, the lignification will prevent cell collapse and P_t can fall to negative values. The information in the Höfler diagram can be used to understand the water relations of cells in the course of a day.

A representative time course is shown in Figure 2.1B. Suppose the sun rises at 06:00 and sets at 18:00. Radiant energy falling on leaves will enhance the rate of evaporation above the rate at which roots can replace evaporated water, hence both ψ_c and ψ_x will fall to the most negative values in early afternoon (indicated by the solid line marked by ψ in Figure 2.1B). As the afternoon progresses the light intensity diminishes and the rate of water loss from the leaves falls below the rate of uptake of water from the roots; hence ψ increases. Overnight, the value of ψ will return to a value near zero in wet soils or more negative values in drier soils; in either case ψ reaches a maximum value just prior to dawn. The value of ψ prior to sunrise (called the pre-dawn water potential) is often taken as a valid measure of soil dryness (= ψ_{soil}) in the rooting zone of the plant. In the xylem, the osmotic potential (π_x) remains more or less constant and only slightly negative during the day. Therefore, all the change in ψ_x is brought about by a large change in P_x, which closely parallels changes in ψ. Similarly, in living cells, changes in ψ_c are brought about by large changes in P_t, while the cell osmotic potential (π_c) changes only slightly and remains a large negative value.

WATER RELATIONS OF WHOLE PLANTS

The water relations of a whole plant can be understood in terms of the fundamental physiological role of the leaf. The leaf is an organ designed to permit CO_2 uptake at a rate needed for photosynthesis while keeping water evaporation from leaves to a reasonably low rate. The roots have the function of extracting water from the soil to replace water evaporated from leaves.

The leaf structure of a typical plant is illustrated in Figure 2.2. The upper and lower epidermis of leaves are covered with a waxy cuticle, which reduces water loss from the leaf to negligible levels. All gas exchange into and out of the leaf is via stomata. The guard cells of the stomata are capable of opening and closing air passages that provide pathways for diffusion of CO_2 into the leaf and for loss of water vapour from the leaf. Photosynthesis occurs primarily in the palisade and mesophyll cells of the leaf (Figure 2.2B). Leaves are thin enough (about 0.1 mm thick) to permit photon penetration to chloroplasts (shown as dots in Figure 2.2B). The chloroplasts absorb the energy of the photon and, through a photochemical process, use the energy to convert CO_2 and water into sugar. The consumption of CO_2 in the chloroplasts lowers the concentration of CO_2 in the cells and thus sets up a concentration gradient for the diffusion of more CO_2 into the cells via the stomatal pores and mesophyll air spaces. Since mesophyll and palisade cell surfaces are wet, water will evaporate continuously from these surfaces, and water vapour will diffuse continuously out of the leaf by the same pathway taken by CO_2. Evaporated water is replaced continuously by water flow through veins in the leaves. The veins contain xylem for water uptake and phloem for sugar export.

The physiology of guard cells has evolved to optimise photosynthesis when conditions are right for the photochemical reaction, and to minimise water loss when conditions are wrong for photosynthesis. Stomata remain closed or only partly open when soils are dry; this is an advantage to the plant because roots are unable to extract water fast enough from dry soils to keep up with evaporation from leaves. Roots send a chemical signal to leaves, abscisic acid (ABA), which mediates stomatal closure (Davies and Zhang, 1991). Stomata open in sunlight when soil water is not limited and when the internal CO_2 concentrations fall below ambient levels in the atmosphere. In about half the known species, stomata have the additional capability of sensing the relative humidity (RH) of the ambient air and tend to close progressively as RH in the air adjacent to the leaves falls. Rapid stomatal opening is mediated by movement of K^+ from the epidermal cells to the guard cells (Figure 2.2C to F). The movement of K^+ makes π_c less negative in the epidermal cells and more negative in the guard cells. Consequently, water flows from the epidermal cells to the guard cells, P_t falls in the epidermal cells and rises in the guard cells. The mechanical effect of this water movement and change in P_t causes the guard cells to swell into the epidermal region and open a stomatal pore.

16

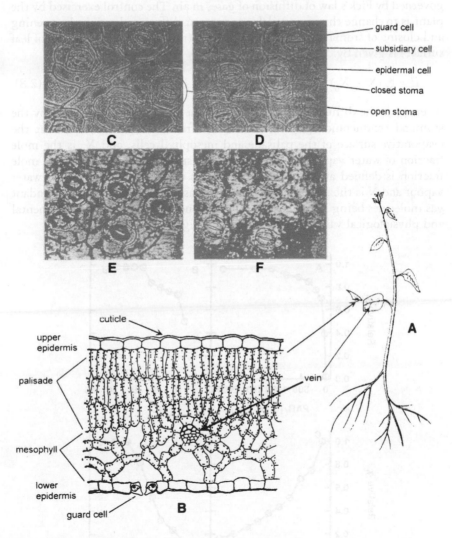

guard cell
subsidiary cell
epidermal cell
closed stoma
open stoma

C   D

E   F

cuticle

upper
epidermis

palisade

vein

mesophyll

lower
epidermis

guard cell

A

B

Figure 2.2 A: A typical plant. B: The lower leaf has been cut, and the cross-section is shown in enlargement at about 750×. C to F: Photographs of the lower leaf surface enlarged about 1000×. E and F have been stained to show location of K⁺. C and E illustrate the open state of stomata when the leaf is exposed to light. D and F are similar leaves in light but also exposed to abscisic acid (ABA), causing stomata to remain closed in light. Note that in open stomata K⁺ is concentrated in the guard cells (E), and the guard cells have moved apart to form an air passage for gas exchange (the stomatal pore). Note that the K⁺ is located in the epidermal cells when the stomata are closed (F). Adapted from Kramer (1983) and Bidwell (1979).

The evaporative flux density (E) of water vapour from leaves is ultimately governed by Fick's law of diffusion of gases in air. The control exercised by the plant is to change the area available for vapour diffusion through the opening and closing of stomatal pores. The value of E (mmol water per s per m^2 of leaf surface) is given by:

$$E = g_L (X_i - X_o) \qquad (2.8)$$

where g_L is the diffusional conductance of the leaf (largely controlled by the stomatal conductance, g_s), X_i is the mole fraction of water vapour at the evaporative surface of the palisade and mesophyll cells, and X_o is the mole fraction of water vapour in the ambient air surrounding the leaf. The mole fraction is defined as $X = n_w/N$, where n_w is the number of moles of water vapour and N is the number of moles of all gas molecules, the most abundant gas molecules being N_2 and O_2. The dependence of g_L on some environmental and physiological variables is illustrated in Figure 2.3.

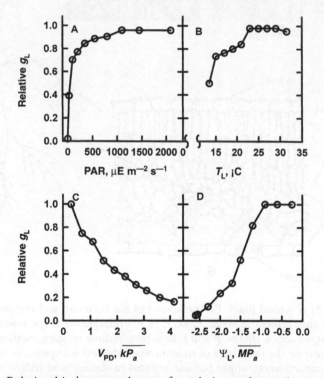

Figure 2.3 Relationship between change of g_L relative to the maximum value, i.e. g_L/g_{Lmax} and various environmental factors measured on *Acer saccharum* leaves. The environmental factors are A: photosynthetically active radiation (PAR); B: leaf temperature (T_L); C: vapour pressure deficit (V_{PD}); and D: leaf water potential (ψ_L). Adapted from Yang *et al.* (1998).

The maximum value of X occurs when RH is 100 percent, i.e. when the air is saturated with water vapour, and its value increases exponentially with the kelvin temperature of the air. The air at the evaporating surface of leaves is at the temperature of the leaf (T_L), and X_i is taken as the maximum value of X at saturation, which we can symbolize as $X_i = X(T_L)$. The value of X_o depends on the microclimate near the leaf, i.e. the air temperature and relative humidity. However, the microclimate of the leaf is strongly influenced by the behaviour of the plant community surrounding the leaf, so even though the leaf has direct control over the values of g_s and g_L, it has less control over E than might appear from equation (2.8).

The qualitative aspects of how leaves influence their own microclimate is easily explained. When the sun rises in the morning, the radiant energy load on the leaf increases. This has two effects. T_L rises as the sun warms the leaves; hence X_i rises and g_L increases as stomata open. But the increased evaporation from the leaves causes X_o to rise as water vapour from the leaves is added to the ambient air. Even changes of g_L under constant radiant energy load causes less change in E than might be expected from equation (2.8). When g_L doubles E also doubles, but only temporarily. The increased E lowers T_L because of increased evaporative cooling, and the increased E from all the leaves in a stand eventually increases X_o; hence $X_i - X_o$ declines, causing a decline in E. Consequently, equation (2.8) is not very useful in predicting the value of E at the level of plant communities. Leaf-level behaviour can be extrapolated to the community level if we take into account leaf-level solar energy budgets using equations that describe light absorption by leaves at all wavelengths and the conversion of this energy to temperature and heat fluxes. Studies of solar energy budgets have been made at both the leaf and stand level (Slatyer, 1967; Chang, 1968).

Eco-hydrologists are primarily concerned with E expressed as kg or m^3 of water per s per m^2 of ground. Fortunately, most of the changes in E at the leaf level can be explained in terms of net radiation absorbed at the stand level, and net radiation absorption is relatively easy to measure. This relationship is illustrated in Figure 2.4, where daily values of E at the leaf level (kg per m^2 per day) are correlated with daily values of net radiation (MJ per m^2 of ground per day) measured with an Eppley-type net radiometer. Equations presented elsewhere in this volume (Chapter 6) to describe stand-level E in terms of net radiation and other factors are validated by relationships like that in Figure 2.4. The other factors most commonly included account for plant control over E through leaf area index (leaf area per unit ground area), the effect of drought on g_L, and ambient temperature (see 'Factors controlling the rate of water uptake and movement', p. 29).

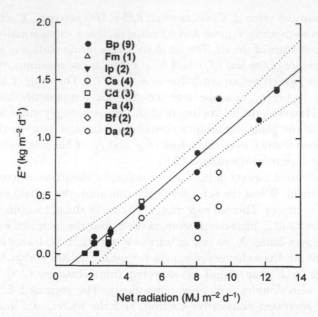

Figure 2.4 Correlation of daily water use of leaves (E^*) and net radiation (NR). Data are from potometer experiments with nine woody cloud forest species in Panama. The number of days per species is given in parentheses. The regression (with 95 percent confidence intervals) is for *Baccharis pendunculata* (Bp) ($E^* = -0.20 + 1.29\,NR$, $r^2 = 0.92$). Even when the data for all species are pooled, NR proved to be a very good predicter of daily water use ($E^* = -0.2 + 1.04\,NR$, $r^2 = 0.72$). Other species names are *Croton draco* (Cd), *Ficus macbridei* (Fm), *Inga punctata* (Ip), *Parathesis amplifolia* (Pa), *Blakea foliacea* (Bf), *Clusia stenophylla* (Cs) and *Dendropanax arboreus* (Da). From Zotz *et al*. (in press).

WATER ABSORPTION BY PLANT ROOTS

The primary factor affecting the pattern of water extraction by plants from soils is the rooting depth. Rooting depths can be extremely variable, depending on soil conditions and species of plant producing the roots (Figure 2.5A). Many of the early studies of rooting depth and the branching pattern of roots were performed in the 1920s and 1930s in deep, well-aerated prairie soils, where roots penetrate to great depths. At the extreme, roots have been traced to depths of 10 to 25 m, e.g. alfalfa (10 m), longleaf pine (17 m) (Kramer, 1983), and drought-evading species in the California chaparral (25 m) (Stephen Davis, personal communication). The situation is very different for plants growing in heavy soils, where 90 percent of the roots can be found in the upper 0.5 to 1.0 m.

In seasonally dry regions, e.g. central Panama, the majority of the roots may be located in the upper 0.5 m, but it is far from clear if the majority of

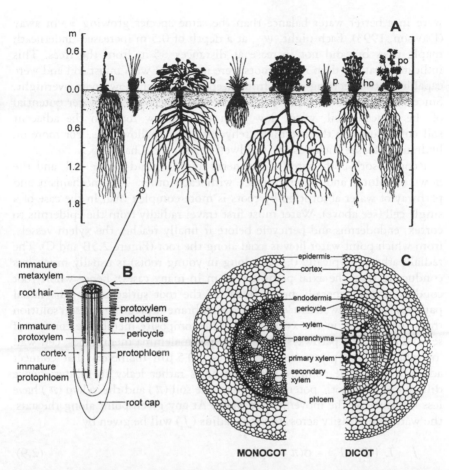

Figure 2.5 A: Differences in root morphology and depth of root systems of various species of prairie plant growing in a deep, well-aerated soil. Species shown are *Hieracium scouleri* (h), *Loeleria cristata* (k), *Balsamina sgittata* (b), *Festuca ovina ingrata* (f), *Geranium viscosissimum* (g), *Poa sandbergii* (p), *Hoorebekia racemosa* (ho) and *Potentilla blaschkeana* (po). B: A dicot root tip enlarged about 50×. C: Cross-section of monocot and dicot roots enlarged about 400×. Adapted from Kramer (1983), Steward (1964) and Bidwell (1979).

water absorption occurs in the upper 0.5 m during the dry season. Water use by many evergreen trees is higher in the dry season than in the wet season, even though the upper 1 m of the soils has much lower potential than the leaves of the trees ($\psi_{soil} << \psi_{leaf}$) (personal observation). Hence the role of shallow versus deep roots of woody species deserves more study. In addition, deeply rooted species may contribute to the water supply of shallow-rooted species through a process called 'hydraulic lift'. In one study, it was found that shallow-rooted species growing within 1 to 5 m of the base of maple trees were in a better water balance than the same species growing >5 m away (Dawson, 1993). Each night, ψ_{soil} at a depth of 0.5 m increased underneath maple trees but did not increase at distances >5 m from the trees. This indicated that the deep maple roots were in contact with moist soil and were capable of transporting water from deep roots to shallow roots overnight. Since the water potential of the shallow roots exceeded the water potential of the adjacent soil, water flow from the shallow roots to the adjacent soil contributed to the overnight rehydration of shallow soils. For more on hydraulic lift, see Richards and Caldwell (1987) and Chapter 6.

Roots absorb both water and mineral solutes found in the soil, and the flows of solutes and water interact with each other. The mechanism and pathway of water absorption by roots is more complex than in the case of a single cell (see above). Water must first travel radially from the epidermis to cortex, endodermis and pericycle before it finally reaches the xylem vessels, from which point water flow is axial along the root (Figure 2.5B and C). The radial pathway (typically 0.3 mm long in young roots) is usually much less conductive than the axial pathway (>1 m in many cases); hence whole-root conductance is generally proportional to the root surface area. The radial pathway can be viewed as a 'composite membrane' separating the soil solution from the solution in the xylem fluid. The composite membrane consists of serial and parallel pathways made up of plasmalemma membranes, cell wall 'membranes' and plasmodesmata (pores <0.5 μm diameter) that connect adjacent cells. The composite membrane is rather leaky to solutes; hence differences in osmotic potential between the soil (π_s) and the xylem (π_x) have less influence on the movement of water. At any given point along the axis, the water flux density across the root radius (J_r) will be given by

$$J_r = L_r [(P_s - P_x) + \sigma(\pi_s - \pi_x)] \qquad (2.9)$$

where L_r is the radial root conductance to water and σ is the solute reflection coefficient. For an 'ideal' membrane, in which water but not solutes may pass, $\sigma = 1$. But for the composite membrane of roots, σ is usually between 0.1 and 0.8. The system of equations that describes water transport in roots is quite complex when, for example, axial and radial conductances, the fact that each solute has a different σ, and the influence of solute loading rate (J_s) on water flow are taken into account. Water and solute flow in roots can be described by a standing gradient osmotic flow model, and readers interested in the details may consult Tyree et al. (1994b) and Steudle (1992).

Fortunately, the equations describing water flow become quite simple when the rate of water flow is high. The concentration of solutes in the xylem fluid is determined by the ratio of solute flux to water flux (J_s/J_w). Solute flux tends to be more or less constant with time, but water flux increases with increasing transpiration. When water flow is high, the concentration of solutes in the xylem fluid becomes quite small and approaches values comparable with that in the soil solution, and pressure differences become quite large, hence $(P_s - P_x) >> \sigma(\pi_s - \pi_x)$. Only at night or during rainy periods can values of $(P_s - P_x)$ approach those of $\sigma(\pi_s - \pi_x)$. So water flow $(J_w, \text{kg s}^{-1})$ through a whole root system during the day can be approximated by

$$J_w \cong K_r (P_s - P_{x,b}) \qquad (2.10)$$

where $P_{x,b}$ is the xylem pressure at the base of the plant and K_r is the total root conductance (combined radial and axial conductances).

THE PATHWAY OF WATER MOVEMENT (HYDRAULIC ARCHITECTURE)

Van den Honert (1948) quantified water flow in plants in a classical paper in which he viewed the flow of water in the plant as a catenary process, where each catena element is viewed as a hydraulic conductance (analogous to an electrical conductance) across which water (analogous to electrical current) flows. Thus, van den Honert proposed an Ohm's law analogue for water flow in plants. The Ohm's law analogue leads to the following predictions: (1) the driving force of sap ascent is a continuous decrease in P_x in the direction of sap flow; and (2) evaporative flux density from leaves (E) is proportional to the negative of the pressure gradient $(-dP_x/dx)$ at any given 'point' (cross-section) along the transpiration stream. Thus at any given point of a root, stem or leaf vein we have:

$$-dP_x/dx = AE/K_h + \rho g \, dh/dx \qquad (2.11)$$

where A is leaf area supplied by a stem segment with hydraulic conductivity K_h, and $\rho g \, dh/dx$ is the gravitational potential gradient, where ρ is the density of water, g is acceleration due to gravity, and h is vertical distance and x actual distance travelled by water in the stem segment.

In the context of stem segments of length (L) with finite pressure drops across ends of the segment we have:

$$\Delta P_x = LAE/K_h + \rho g \Delta h \qquad (2.12)$$

Figure 2.6 illustrates water flow through a plant represented by a linear catena of conductance elements near the centre and a branched catena of conductance elements on the left. The number and arrangement of catena elements is dictated primarily by the spatial precision desired in the representation of water flow through a plant; a plant can be represented by anything from one to thousands of conductance elements. A relatively recent review of the hydraulic architecture of woody plants can be found in Tyree and Ewers (1991).

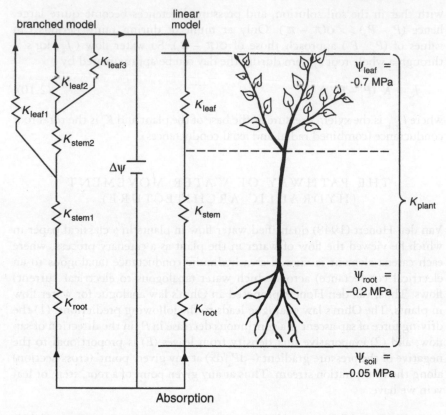

Figure 2.6 The Ohm's law analogy. The total conductance is seen as the resultant conductance (K) of the root, stem and leaf in series and parallel. Water flow is driven by the differences in water potential between the soil, ψ_{soil}, and the water potential at the evaporating surface, ψ_{evap}. On the right is the simplest Ohm's law analogy with conductances in series. On the left is a more complex conductance catena where some conductance elements are in series and some in parallel.

Values of K_{h} are measured on excised stem segments by connecting the segments to water-filled tubing and measuring the water flow rate, w (kg s^{-1}), resulting from an imposed pressure drop of ΔP (MPa) across the segment of length L (m); K_{h} is equated to $Lw/\Delta P$. Obviously, much more water will flow across a large-diameter stem segment than a small-diameter segment at any given ΔP; hence K_{h} changes over several orders of magnitude for stems 1 mm to 1 m in diameter. A useful way of comparing the hydraulic conductivity of stems of very different diameter is to compute specific conductivity (K_{s}), which is equivalent to the hydraulic conductivity of Darcy's law (see Chapter 9) and is given by $K_{\text{h}}/A_{\text{w}}$, where A_{w} is the cross-sectional area of conducting stem. Values of K_{s} provide information on the relative efficiency of stem segment. Stem segments with large-diameter vessels and many vessels per unit cross-section have larger K_s values than segments with small-diameter vessels. Another useful measure is obtained by dividing K_h by leaf area dependent on the segment to get leaf specific conductivity, $K_L = K_h/A$; K_L values give the hydraulic sufficiency of stems to supply water to leaves and permit the quick computation of pressure gradients, since $-dP_x/dx = E/K_L$ (ignoring gravity). Values of K_L and K_s are not uniform within large trees; values decline with declining diameter because small branches tend to have smaller-diameter vessels than large branches. There are also dramatic differences in K_L between species.

Values of K_L have been applied to complex 'hydraulic maps', where the above-ground portion of trees were represented by hundreds to thousands of conductance elements (Figure 2.6). Using known values of K_L and E, it is possible to calculate P_x versus path length from the base of a tree to selected branch tips. This has been done in Figure 2.7. Some species are so conductive (large K_L values) that the predicted drop in P_x is little more than is needed to lift water against gravity (see *Schefflera*). In other species, gradients in P_x become very steep near branch tips (see *Thuja*).

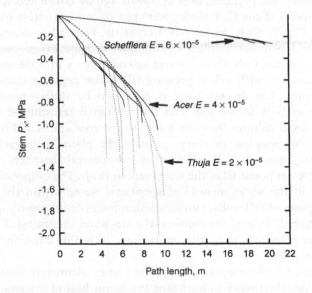

Figure 2.7 Pressure profiles in three large trees, i.e. computed change in xylem pressure (P_x) versus path length. P_x values were computed from the base of each tree to a few randomly selected branch tips. The 'drooping' nature of these plots near the apices of the branches is caused by the decline of leaf specific conductivity (K_L) from base to apex of the trees. The pressure profiles in this figure do not include pressure drops across roots or leaves. In some species, pressure drop across leaves can be more than shown in this figure. The pressure drop across roots is generally equal to that across the shoots (including leaves). E = evaporative flux density in kg s^{-1} m^{-2}.

The major resistance (i.e. inverse conductance) to liquid water flow in plants resides in the non-vascular pathways, i.e. the radial pathway for water uptake in young roots and in the cells between leaf veins and the evaporating surfaces of leaves. Non-vascular resistances can be 60 to 90 percent of the total plant resistance to water flow, even though the total path length may be less than 1 mm. Even though water flows though distances of 1 to 100 m in the vascular pathways from root vessel to stems to leaf vessels, the hydraulic resistance of this pathway rarely dominates.

The biggest resistance to water flow (liquid and vapour) from the soil to the atmosphere resides in the vapour transport phase. The rate of flow from the leaf to the atmosphere is determined primarily by the stomata, as discussed above.

THE COHESION–TENSION (C–T) THEORY AND XYLEM DYSFUNCTION

The C–T theory was proposed over 100 years ago by Dixon and Joly (1894), and some aspects of the C–T theory were put on a quantitative basis by van den Honert (1948) with the introduction of the Ohm's law analogue of sap flow in the soil–plant–atmosphere continuum (see page 24).

According to the C–T theory, water ascends plants in a metastable state under tension, i.e. with xylem pressure (P_x) more negative than that of a perfect vacuum. The driving force is generated by surface tension at the evaporating surfaces of the leaf, and the tension is transmitted through a continuous water column from the leaves to the root apices and throughout all parts of the apoplast in every organ of the plant. Evaporation occurs predominantly from the cell walls of the substomatal chambers due to the much lower water potential of the water vapour in air. The evaporation creates a curvature in the water menisci of apoplastic water within the cellulosic microfibril pores of cell walls. Surface tension forces consequently lower P_x in the liquid directly behind the menisci (the air–water interfaces). This creates a lower water potential, ψ, in adjacent regions, including adjoining cell walls and cell protoplasts.

The energy for the evaporation process comes ultimately from the sun, which provides the energy to overcome the latent heat of evaporation of the water molecules, i.e. the energy to break hydrogen bonds at the menisci.

Water in xylem conduits is said to be in a metastable condition when P_x is below the pressure of a perfect vacuum, because the continuity of the water column, once broken, will not rejoin until P_x rises to values above that of a vacuum. Metastable conditions are maintained by the cohesion of water to water and by adhesion of water to walls of xylem conduits. Both cohesion and adhesion of water are manifestations of hydrogen bonding. Even though air–water interfaces can exist anywhere along the path of water movement, the small diameter of pores in cell walls and the capillary forces produced by surface tension within such pores prevent the passage of air into conduits under normal circumstances. However, when P_x becomes sufficiently negative, air bubbles can be sucked into xylem conduits through porous walls.

This tension (negative P_x) is ultimately transferred to the roots, where it lowers ψ of the roots below the ψ of the soil water. This causes water uptake from the soil to the roots and from the roots to the leaves to replace water evaporated at the surface of the leaves. The Scholander–Hammel pressure bomb (Scholander et al., 1965) is one of the most frequently used tools for estimating P_x. Typically, P_x can range down to –2 MPa (in crop plants) or to –4 MPa (in arid zone species) and in some cases –10 MPa (in California chaparral species). Readers interested in learning more about the cohesion–tension theory of sap ascent may refer to Tyree (in press).

In some plants, water movement can occur at night or during rain events, when P_x is positive due to root pressure, i.e. osmotically driven flow from roots. Water flow under positive P_x is often accompanied by gutation, i.e. the formation of water droplets at leaf margins. But water is normally under negative pressure (tension) as it moves through the xylem towards the leaves during sunny days. The water is thus in a metastable condition and vulnerable to cavitation due to air entry into the water columns. Cavitation results in embolism (air blockage), thus disrupting the flow of water (Tyree and Sperry, 1989). Cavitation in plants can result from water stress, and each species has a characteristic 'vulnerability curve', which is a plot of the percentage loss of K_h in stems versus the xylem pressure potential, P_x, required to induce the loss. Vulnerability curves are typically measured by dehydrating large excised branches to known P_x. Stem segments are then cut under water from the dehydrated branches and, for the most part, the air bubbles remain inside the conduits. An initial conductivity measurement is made and compared with the maximum K_h after air bubbles have been dissolved. The vulnerability curves of plants, in concert with their hydraulic architecture, can give considerable insight into drought tolerance and water relations 'strategies'.

The vulnerability curves for a number of species are reproduced in Figure 2.8 (Tyree et al. 1994a). These species represent the range of vulnerabilities observed so far in over sixty species; 50 percent loss of K_h occurs at P_x values ranging from –0.7 to –11 MPa. Many plants growing in areas with seasonal rainfall patterns (wet and dry periods) appear to be drought evaders, while others are drought toleraters. The drought evaders avoid drought (low P_x and high percentage loss K_h) by having deep roots and a highly conductive hydraulic system. Alternatively, they evade drought by being deciduous. The other species frequently reach very negative P_x for part or all of the year and are shallow-rooted or grow in saline environments. These species survive by having a vascular system highly resistant to cavitations.

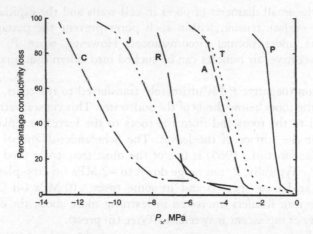

Figure 2.8 Vulnerability curves for various species. *y*-axis is percentage loss of hydraulic conductivity induced by the xylem pressure potential, P_x, shown on the *x*-axis. C = *Ceanothus megacarpus*, J = *Juniperus virginiana*, R = *Rhizophora mangle*, A = *Acer saccharum*, T = *Thuja occidentalis*, P = *Populus deltoides*.

Embolisms may be dissolved in plants if P_x in the xylem becomes positive or close to positive for adequate time periods (Tyree and Yang, 1992; Lewis *et al.*, 1994). Embolisms disappear by dissolution of air into the water surrounding the air bubble. The solubility of air in water is proportional to the pressure of air adjacent to the water (Henry's law). Water in plants tends to be saturated with air at a concentration determined by the average atmospheric pressure of gas surrounding plants. Thus, for air to dissolve from a bubble into water, the air in the bubble has to be at a pressure in excess of atmospheric pressure. If the pressure of water (P_x) surrounding a bubble is equal to atmospheric pressure (P_a), bubbles will naturally dissolve, because surface tension (τ) of water raises the pressure of air in the bubble (P_b) above P_a. In general, $P_b = 2\tau/r + P_x$, where r is the radius of the bubble. According to the cohesion theory of sap ascent, P_x is drawn below P_a during transpiration. Since $2\tau/r$ of a dissolving bubble in a vessel is usually <0.03 MPa, and since P_x is in the range of −0.1 to −10 MPa during transpiration, P_b is usually $\leq P_a$, and hence bubbles, once formed in vessels, rarely completely dissolve. Repair (i.e. dissolution) occurs only when P_x becomes large via root pressure. One notable exception to this generality has recently been found in *Laurus nobilis* shrubs, which appear to be able to refill embolised vessels even while P_x is at −1 MPa (Salleo *et al.*, 1996). The mechanism involved has evaded explanation.

FACTORS CONTROLLING THE RATE OF WATER UPTAKE AND MOVEMENT

Environmental conditions control the rate of water movement in plants. The dominant environmental factor is net solar radiation, as mentioned previously. Water movement can be explained by the solar energy budget of plants. We will now take a more quantitative look at the solar energy budget of plants at both the leaf and stand level.

Leaf-level energy budgets

Objects deep in space far away from solar radiation tend to be rather cold (about 5 K), whereas objects on Earth tend to be rather warm (270–310 K). The reason for the elevated temperature on Earth is the warming effect of solar radiation. The net radiation absorbed by a leaf (R_{NL}, W m^{-2}) can be arbitrarily divided into short-wave and long-wave radiation, with the demarcation wavelength being at 1 μm. Most short-wave radiation comes from the Sun, and most long-wave radiation comes from the Earth. As a leaf absorbs R_{NL}, the leaf temperature (T_L) rises, which increases the loss of energy from the leaf by three different mechanisms – black body radiation (B), sensible heat flux (H), and latent heat of vaporisation of water (λE, where E is the evaporative flux density of water due to transpiration (mol of water s^{-1} m^{-2}) and λ is the heat required to evaporate a mole of water (J mol^{-1})). R_{NL} can also be converted to chemical energy by the process of photosynthesis and to energy storage, but these two factors are generally small for leaves and are ignored in the equation below:

$$R_{NL} = B + H + \lambda E \qquad (2.13)$$

The B in equation (2.13) is the 'black body' radiation. All objects emit radiation. Very hot objects like the Sun emit mostly short-wave radiation, whereas cooler objects on Earth emit mostly long-wave radiation. The amount of radiation emitted increases with the kelvin temperature according to the Stefan–Boltzmann law: i.e. $B = e\sigma_{sb} T_L^4$, where e is the emissivity (≥ 0.95 for leaves), and σ_{sb} is the Stefan–Boltzmann constant = 5.67×10^{-8} W m^{-2} K^{-4}.

H in equation (2.13) is sensible heat flux density, i.e. the heat transfer by heat diffusion between objects at different temperatures. For leaves, the heat transfer is between the leaf and the surrounding air. The rate of heat transfer is proportional to the difference in temperature between the leaf and the air, so $H = k(T_L - T_a)$, where T_a is the air temperature and k is the heat transfer coefficient. Heat transfer is more rapid and hence k is larger under windy conditions than in still air. But the value of k is also determined by leaf size, shape and orientation with respect to the wind direction. Readers interested in more detail should consult Slatyer (1967) or Nobel (1991).

The factors determining E have already been given in equation (2.8), so if we combine equation (2.8) with the other equations for B and H we get:

$$R_{NL} = e\sigma_{sb}T_L^4 + k(T_L - T_a) + \lambda g_L(X_iT_L - X_o) \qquad (2.14)$$

Every term on the right side of equation (2.14) is a function of leaf temperature. This gives a clue about how the 'balance' is achieved in the solar energy balance equation. At any given R_{NL}, the value of T_L will rise or fall until the sum of B, H and λE equals R_{NL}. As R_{NL} increases or decreases, the value of T_L increases or decreases to re-establish equality. In practice, this equality is achieved with a leaf temperate near T_a; T_L is rarely less then 5 K below T_a or more than 15 K above T_a. Some examples of leaf energy budgets are shown in Figure 2.9. Once we know the value of T_L and solve the equation, we can calculate the value of E. If we sum the values of E for every leaf in a stand of plants then we can calculate stand-level water flow.

Equation (2.14) provides a good qualitative understanding of the dynamics of energy balance for a community of plants (= the stand level), because the evaporation rate from a stand is the sum of the evaporation from all the leaves in the stand. But equation (2.14) is of little practical value, because it is not possible to obtain values of R_{NL}, T_L, k and g_L for every leaf in a stand in order to compute the required sum!

Stand-level energy budgets

Another approach to energy budgets is to measure energy and matter flux in a region of air above an entire stand of plants (Figure 2.10). A net radiometer can be used to measure net radiation (R_N) above the stand. R_N is the total radiation balance (incoming long- and short-wave radiation minus outgoing

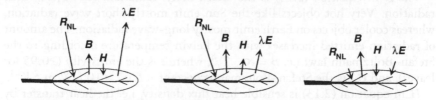

Figure 2.9 Solar energy budgets of leaves under different conditions. R_{NL} = net absorbed short- and long-wave radiation, B = black body radiation from leaves, H = sensible heat flux, λE = latent heat flux. Direction of arrow indicates direction of flux and length of arrow indicates relative magnitude. Specific conditions: Left: high transpiration rate, so leaf temperature is below air temperature because $B + \lambda E > R_{NL}$. Middle: intermediate transpiration rates, so leaf temperature is above air temperature. Right: during dew-fall $B > R_{NL}$ at night, so leaf temperature is below the dew-point temperature of the air.

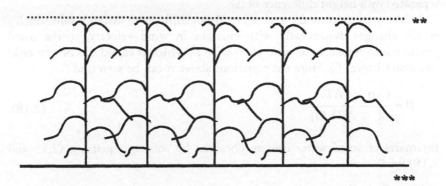

* T_{air}, RH%$_{air}$, and R_N measured at this level
** T_{crop} and RH%$_{crop}$ measured at this level
*** G measured at this level

Figure 2.10 Solar energy budget of a uniform stand of plants. Net radiation (R_N) is measured at location marked '*', air temperature and relative humidity is measured at locations marked '*' and '**'. Rate of heat storage in soil is measured at '***'.

long- and short-wave radiation, so at the leaf level $R_N = R_{NL} - B$). The energy balance equation for this situation is:

$$R_N = G + H + \lambda E \qquad (2.15)$$

The rate of heat storage in the soil, G (W m^{-2}), can be significant, because soil temperature can change by a few K on a daily basis in the upper few centimetres of soil. Heat storage rate is usually measured with a soil heat flux sensor. The H and λE terms are similar to those in equation (2.14), but the equations define the flux densities between the heights marked '*' and '**' in Figure 2.10. The flux of heat and water vapour across the distance ΔZ is controlled by vertical air convection (eddy), and the defining equations are:

$$H = -c_p \rho_a K_e \Delta T / \Delta Z \qquad (2.16)$$

and

$$\lambda E = \lambda K_e \Delta[H_2O] / \Delta Z \qquad (2.17)$$

where K_e is the eddy transfer coefficient, C_p is the heat capacity of air, ρ_a is the air density, ΔT is the difference in air temperature measured at the two levels in Figure 2.10 separated by a height difference of ΔZ, and $\Delta[H_2O]$ is the difference in water vapour concentration at the two levels in Figure 2.10 separated by a height difference of ΔZ.

It is easy to measure ΔT and $\Delta[H_2O]$ but difficult to assign a value for K_e, which changes dynamically with changes in wind velocity, so the usual practice is to measure the ratio of $H/\lambda E = \beta$, which is called the Bowen ratio (see also Chapter 6). From the equations above it can be seen that

$$\beta = \frac{C_p \rho_a}{\lambda} \frac{\Delta T}{\Delta[H_2O]}$$

(2.18)

Estimates of stand water use are obtained by solving equations (2.15 and 2.18) for E:

$$E = \frac{R_N - G}{\lambda(\beta + 1)}$$

(2.19)

An example of the total energy budget measured over a pasture is reproduced in Figure 2.11, the data in which show again that the main factor determining E is the amount of solar radiation, R_N, as in Figure 2.4. However, the leaves in a stand of plants do have some control over the value of E. As soils dry, and leaf water potential falls, stomatal conductance falls. This causes a reduction in E, an increase in leaf temperature and thus an increase in H.

Very intensive monitoring of climatic data is needed to obtain data for a solution of equation (2.19). Temperature, R_N and relative humidity have to be measured every second at several locations. Eco-hydrologists prefer to estimate E with a less complete data set. Fortunately, the Penman–Monteith formula (Monteith, 1964) permits a relatively accurate estimate of E under some restricted circumstances. The Penman–Monteith formula is derived from energy budget equations together with the assumption that some non-linear functions can be approximated as linear relations. After many obtuse steps in the derivation (Campbell, 1981), a formula of the following form results:

$$E = \frac{e'(R_N - G) + \rho_a C_p V_{pd} g_a}{\lambda\left(e' + \gamma\left[1 + \frac{g_a}{g_c}\right]\right)}$$

(2.20)

where e' is the rate of change of saturation vapour pressure with temperature at the current air temperature, V_{pd} is the difference between the vapour

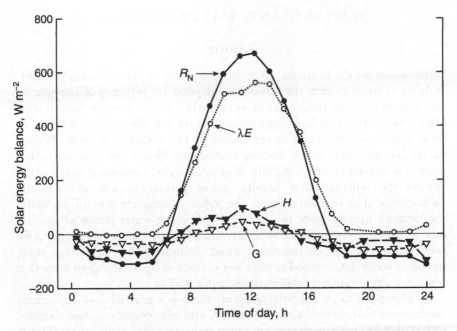

Figure 2.11 Solar energy budget values measured in a meadow. R_N = net solar radiation of the meadow, λE = latent heat flux, H = sensible heat flux and G = rate of heat storage in soil. Adapted from Slatyer (1967).

pressure of air at saturation and the current vapour pressure, γ is the psychrometric constant, g_a is the aerodynamic conductance and g_c is the canopy conductance of the stand. The problem with equation (2.20) is that g_a and g_c are both difficult to estimate and are not constant. The value of g_a depends on wind speed and roughness parameters that describe the unevenness at the boundary between the canopy and the bulk air. Surface roughness affects air turbulence and hence the rate of energy transfer at any given wind speed. Surface roughness changes as stands grow and is difficult to estimate in terrain with hills or mountains. The value of g_c is under biological control and difficult to estimate from climatic data. In one study on an oak forest in France, E was estimated independently by Granier sap flow sensors in individual oak trees. This permitted the calculation of g_c against season, morphological state of the forest and climate (Granier and Bréda, 1996). The value of g_c was found to be a function of global radiation, V_{pd}, leaf area per unit ground area, which changes with season, and relative extractable water, which is a measure of soil dryness. Once all these factors were taken into account, equation (2.20) did provide a reasonable estimate of half-hourly estimates of E over the entire summer.

WILTING AND WATERLOGGING

Wilting

Wilting denotes the limp, flaccid or drooping state of plants during drought. Wilting is most evident in leaves that depend on cell turgor pressure to maintain their shape and, therefore, occurs when turgor pressure falls to zero. Many plants maintain leaf shape through rigid leaf fibre cells. 'Wilting' in these species is considered to commence at the turgor loss point. Wilted plants generally have low E because stomata are closed and g_s is very small when leaf water potential, ψ_L, falls during drought. Continued dehydration beyond the wilting point usually causes permanent loss of hydraulic conductance due to cavitation in the xylem. Complete loss of hydraulic conductance usually leads to plant death, but the water potential causing loss of hydraulic conductance varies greatly between species (see Figure 2.8). Some plants in arid environments avoid drought either by having short reproductive cycles confined to brief wet periods or by having deep roots that can access deep sources of soil water (Kramer, 1983).

As plants approach the wilting point, there is a gradual loss in stomatal conductance and, hence, a reduction in E and photosynthetic rate (Schulze and Hall, 1982). Some species are much more sensitive than others (Figure 2.12B). The short-term effects of decreased ψ_L on transpiration are less dramatic than are long-term effects (Figure 2.12A). Long-term effects of drought are mediated by hormone signals from roots, which cause a medium-term decline in g_s and by changes in root morphology, e.g. loss of fine roots, suberisation of root surfaces, and formation of corky layers (Ginter-Whitehouse *et al.*, 1983). The morphological changes to roots cause an decrease in whole plant hydraulic conductance (K_p), so ψ_L becomes more negative at lower values of E because $\psi_L = \psi_{soil} - E/K_p$. Very severe drought can further lower K_p due to cavitation of xylem vessels.

Waterlogging

Waterlogging denotes an environmental condition of soil water saturation or ponding of water, which can last for a just a few hours or for many months. Plants absent from flood-prone sites are damaged easily by waterlogging. On the other hand, plants that inhabit flood-prone sites include species that can grow actively in flooded soils or species that survive flooding in a quiescent or dormant state (see Chapter 5). Paradoxically, the commonest sign of tobacco roots having an excess of water is the development of a water deficit in the leaves (Kramer, 1983; Kramer and Jackson, 1954); but flood-tolerant species are not so easily affected.

Flooding often affects root morphology and physiology. The wilting and defoliation that is found on flooding can be traced to an increased resistance

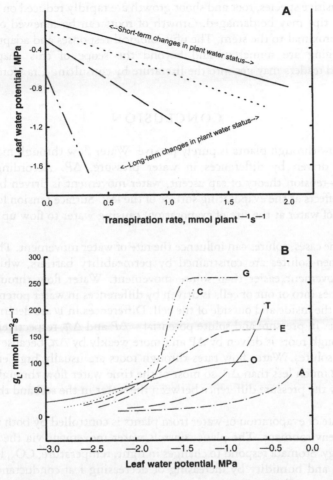

Figure 2.12 Effects of drought on transpiration, leaf water potential and stomatal conductance. A: Short-term and long-term effects of drought on leaf water potential and transpiration. The short-term effects are dynamic changes in leaf water potential that might occur in the course of one day. The long-term effects are associated with slow drying of soil over many days. B: Short-term effects of leaf water potential on stomatal conductance. Species represented: A = *Acer saccharum*, C = *Corylus avellana*, E = *Eucalyptus socialis*, G = *Glycine max*, T = *Triticum aestivum*. Adapted from Schulze and Hall (1982).

to water flux in the roots (Mees and Weatherley, 1957). In most flood-tolerant species, flooding induces morphological changes in the roots. These modifications usually involve root thickening, with an increase in porosity. The increase in porosity increases the rate of oxygen diffusion to root tips and thus permits continued aerobic metabolism in the inundated roots. In flood-sensitive species, root and shoot growth are rapidly reduced on flooding, and root tips may be damaged. Growth of roots can be renewed only from regions proximal to the stem. The physiological responses and adaptations to waterlogging are numerous and beyond the scope of this chapter, but interested readers may get into the literature by consulting Crawford (1982).

CONCLUSION

Water flow through plants is purely passive. Water flow through most of the plant is driven by differences in water pressure, ΔP. According to the cohesion–tension theory of sap ascent, water movement is driven by surface tension effects at the evaporating surface of the leaf. Surface tension lowers the pressure of water at the site of evaporation, causing water to flow up from the roots.

In some cases, solutes can influence the rate of water movement. These cases occur when solutes are constrained by permeability barriers, which make water movement easier than solute movement. Water flow through plant membranes into or out of cells is driven by differences in water potential, $\Delta \psi$, between the inside and outside of the cell. Differences in ψ are determined by differences in pressure and solute potential – ΔP and $\Delta \pi$, respectively. Water flow through roots is driven by ΔP and more weakly by $\Delta \pi$, because roots are leaky to solutes. Water flow rates through roots are usually high enough to make $\Delta \pi$ much less than ΔP, so most of the time water flow through roots is driven by the pressure difference between the water in the soil and the base of the plant.

The rate of evaporation of water from plants is controlled by both the plant and the environment. The plant controls water movement via the stomatal physiology. Stomata respond to changes in light, temperature, CO_2, leaf water potential and humidity by increasing or decreasing leaf conductance (g_L) to water vapour diffusion. The most important environmental factor controlling evaporation is then net radiation absorbed by plants.

REFERENCES

Bidwell, R.G.S. (1979) *Plant Physiology*, New York: Macmillan.
Campbell, G.S. (1981) Fundamentals of radiation and temperature relations, *Encyclopedia of Plant Physiology*, New Series Vol. 12A: 11–40, New York: Springer-Verlag.
Chang, J.-H. (1968) *Climate and Agriculture*, Chicago: Aldine Publishing Co.
Crawford, R.M.M. (1982) Physiological responses to flooding. *Encyclopedia of Plant Physiology*, New Series Vol. 12B: 453–477, New York: Springer-Verlag.
Davies, W.J. and Zhang, J. (1991) Root signals and the regulation of growth and

development of plants in drying soil, *Annual Review of Plant Physiology and Plant Molecular Biology* 42: 55–76.

Dawson, T.E. (1993) Hydraulic lift and water parasitism by plants: implications for water balance, performance, and plant–plant interactions, *Oecologia* 95: 565–574.

Dixon, H.H. and Joly, J. (1894) On the ascent of sap, *Philosophical Transactions of the Royal Society*, London, Series B, 186: 563–576.

Ginter-Whitehouse, D.L., Hinckley, T.M. and Pallardy, S.G. (1983) Spatial and temporal aspects of water relations of three tree species with different vascular anatomy, *Forest Science* 29: 317–329.

Granier, A. and Bréda, N. (1996) Modelling canopy conductance and stand transpiration of an oak forest from sap flow measurements, *Annales des Sciences Forestières* 53: 537–546.

Kramer, P.J. (1983) *Water Relations of Plants*, New York: Academic Press.

Kramer, P.J. and Jackson, W.T. (1954) Causes of injury to flooded tobacco plants, *Plant Physiology* 29: 241–245.

Lewis, A.M., Harnden, V.D. and Tyree, M.T. (1994) Collapse of water-stress emboli in the tracheids of *Thuja occidentalis L.*, *Plant Physiology* 106: 1639–1646.

Mees, G.C. and Weatherley, P.E. (1957) The mechanism of water absorption by roots. II. The role of hydrostatic pressure gradients. *Proceedings of the Royal Society*, London, Series B, 147: 381–391.

Monteith, J.L. (1964) Evaporation and environment, *Symposium of the Society for Experimental Biology* 19: 205–234.

Nobel, P.S. (1991) *Physiochemical and Environmental Plant Physiology*, New York: Academic Press.

Richards, J.R. and Caldwell, M.M. (1987) Hydraulic lift: substantial nocturnal water transport between soil layers by *Artemisia tridentata* roots, *Oecologia* 74: 486–489.

Salleo, S., LoGullo, M.A., De Paoli, D. and Zippo, M. (1996) Xylem recovery from cavitation-induced embolism in young plants of *Laurus nobilis*: a possible mechanism, *New Phytologist* 132: 47–56.

Scholander, P.F., Hammel, H.T., Bradstreet, E.D. and Hemmingsen, E.A. (1965) Sap pressure in vascular plants, *Science* 148: 339–346.

Slatyer, R.O. (1967) *Plant Water Relationships*, New York: Academic Press.

Schulze, E.D. and Hall, A.E. (1982) Stomatal responses, water loss and CO_2 assimilation rates of plants in contrasting environments. *Encyclopedia of Plant Physiology*, New Series Vol. 12B: 181–230.

Steudle, E. (1992) The biophysics of plant water: Compartmentation, coupling with metabolic processes, and flow of water in plant roots, in Somero, G.N. Osmond, C.B. and Bolis, C.L. (eds) *Water and Life: Comparative Analysis of Water Relationships at the Organismic, Cellular, and Molecular Levels*, Heidelberg: Springer-Verlag, pp. 173–204.

Steward, F.C. (1964) *Plants at Work*, Reading, Mass.: Addison-Wesley.

Tyree, M.T. (in press) The cohesion–tension theory of sap ascent: current controversies, *Journal of Experimental Botany*.

Tyree, M.T. and Ewers, F.W. (1991) The hydraulic architecture of trees and other woody plants, *New Phytologist* 119: 345–360.

Tyree, M.T. and Sperry, J.S. (1989) The vulnerability of xylem to cavitation and embolism, *Annual Review of Plant Physiology and Molecular Biology* 40: 19–38.

Tyree, M.T. and Yang, S. (1992) Hydraulic conductivity recovery versus water pressure in xylem of *Acer saccharum*, *Plant Physiology* 100: 669–676.

Tyree, M.T., Davis, S.D. and Cochard, H. (1994a) Biophysical perspectives of xylem evolution: is there a trade off of hydraulic efficiency for vulnerability to dysfunction? *Journal of the International Association of Wood Anatomists* 15: 335–360.

Tyree, M.T., Cruiziat, P. and Sinclair, B. (1994b) Novel methods of measuring hydraulic conductivity of tree root systems and interpretation using AMAIZED: A maize-root dynamic model for water and solute transport, *Plant Physiology* 104: 189–199.

van den Honert, T.H. (1948) Water transport in plants as a catenary process, *Discussions of the Faraday Society* 3: 146–153.

Yang, S., Liu, X. and Tyree, M.T. (in press) A model of stomatal conductance in sugar maple (*Acer saccharum* Marsh), *Journal of Theoretical Biology*.

Zotz, G., Tyree, M.T., Patino, S. and Carlton, M.R. (1998) Hydraulic architecture and water use of selected species from a lower montane forest in Panama. *Trees* 12: 302–309.

3

SCALES OF INTERACTION IN ECO-HYDROLOGICAL RELATIONS

Robert L. Wilby and David S. Schimel

INTRODUCTION

> parameters and processes important at one scale are frequently
> not important or predictive at another scale, and information
> is often lost as spatial [or temporal] data are considered at
> coarser scales of resolution.
>
> Turner, M.G. (1990)

This observation holds for all the geophysical and biological sciences, including eco-hydrology. Indeed, 'scale issues' have become the legitimate focus of a growing body of research in the environmental sciences and were even identified as a research priority by the US Committee on Opportunities in the Hydrological Sciences (James, 1995). According to Bloschl and Sivapalan (1995), the task of *linking* and *integrating* hydrological 'laws' at different scales has not yet been fully addressed, and doing so remains an outstanding challenge in the field of surficial processes. Several factors have contributed to the current awareness of scale issues in plant–water relations.

First, recent population growth, technological developments and economic activities have broadened the scale of human interference in hydrosystems. Historically, the hydrological discipline evolved from the technical demands of supplying societies with potable water for domestic and agricultural consumption, disposing of liquid wastes, and protecting land from inundations (Klemes, 1988; Wilby, 1997a). Such concerns were initially small-scale but have now attained global dimensions with respect to climate change impacts, regional land-use changes, and the potential for transboundary pollution incidents.

Second, most meteorological, hydrological and ecological processes are strongly heterogeneous in both space and time. However, our ability to observe and/or sample these processes is severely constrained by the available

39

technology, human and financial resources, and a general lack of standardisation in measurement techniques (Rodda, 1995). At the same time, modelling often demands long-term, spatially consistent descriptions of heterogeneous parameters such as albedo or soil hydraulic characteristics. As will be shown below, the advent of remote sensing and new technologies for visualising and handling spatial data have gone some way to addressing such needs.

Third, eco-hydrological theories and models are often highly scale-specific. Models constructed at one scale are not always transferable to a higher or lower temporal/spatial resolution, because the dominant processes may be different between the scales or because there may be strong non-linearities in the system behaviour. To date, the vast majority of eco-hydrological studies have focused on point (or plot) to catchment (or sub-regional) scale processes. Again, this is a reflection of the fact that small-scale studies are logistically easier to conduct than large-scale experiments, and that the perceived level of the anthropic impact has traditionally been hill-slope or catchment scale (Wilby, 1997b).

The following sections will elaborate upon these issues. Although the emphasis of this chapter will be upon *spatial* scales of interaction in plant–water relations, temporal scaling is implicit to all the following discussions given that small length-scale processes tend to operate at small-scale time intervals, and large length-scale processes at long time intervals. However, the primary aim of the chapter is to evaluate different methods of representing spatial heterogeneity in plant–water relations, and the extent to which these relations are transferable between plant and patch, and from patch to regional scales. Accordingly, key concepts such as characteristic process scales, heterogeneity, spatial organisation and scale interactions are introduced. These terms are discussed within the context of emerging 'spatial' technologies such as remote sensing, geographical information systems (GIS) and digital terrain models (DTMs). Three broad approaches to handling heterogeneity and linking between scales are then reviewed. First, methods of extrapolating plant-process models to patch and regional scales are considered with reference to leaf-area models and holistic scaling approaches such as fractals. Second, techniques for obtaining sub-grid-scale heterogeneity through empirical and/or statistical downscaling methods are described. Third, procedures for distributing sub-grid-scale parameters using higher-resolution, covariant properties such as topography are examined with reference to distributed eco-hydrological modelling. The concluding section briefly considers the significance of scale issues in research on regional deforestation, and studies of the potential terrestrial impacts of greenhouse-gas forcing (global warming).

CHARACTERISTIC OBSERVATION SCALES

According to Dooge (1988), hydrology as a scientific discipline could theoretically span fifteen orders of magnitude (Table 3.1), ranging from the scale of a cluster of water molecules (10^{-8} m) to the planetary scale of the global hydrological cycle (10^7 m). In practice, hydrological studies have traditionally favoured the catchment scale, or what Dooge refers to as the mesoscale or the lower end of the macroscale. Similarly, there has been a bias among ecologists towards experimental studies of biotic interactions within 'tennis-court-sized' field plots (Root and Schneider, 1995). Both hydrological and ecological scales of interest have, therefore, been in stark contrast to those of climatologists, who typically employ grid squares of the order 500×500 km in their models. For example, hydrological models are frequently concerned with small, sub-catchment (even hill-slope) scale processes, occurring on spatial scales much smaller than those resolved in general circulation models (GCMs). Conversely, GCMs deal most proficiently with fluid dynamics at the continental scale, yet incorporate regional and smaller-scale processes (Figure 3.1). As Hostetler (1994) has observed, the greatest errors in the parameterisation of *both* GCMs and hydrological models occur on the scale(s) at which climate and terrestrial-impact models interface. These scale-related sensitivities and mismatch problems are further exacerbated because they usually involve the most uncertain components of climate models, namely water vapour and cloud feedback effects (Rind *et al.*, 1992). Furthermore, mismatch problems have important implications for the credibility of impact studies driven by the output of models of climate change, especially as research into potential human-induced modifications to hydrological and ecological cycles is assuming increasing significance (Ehleringer and Field, 1993).

Therefore, in order to bridge the scales between climate, hydrological and ecological models, methods for both observing and representing sub-

Table 3.1 Spatial scales in hydrology.

Class	System	Typical length (m)
Macro	Planetary	10^7
	Continental	10^6
	Large catchment	10^5
Meso	Small catchment	10^4
	Sub-catchment	10^3
	Catchment module	10^2
Micro	Elementary volume	10^{-2}
	Continuum point	10^{-5}
	Molecular clusters	10^{-8}

Source: Dooge (1988).

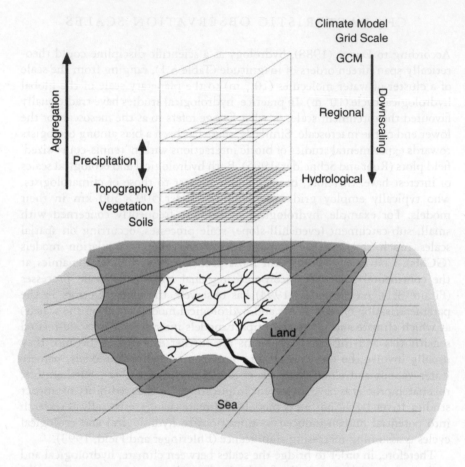

Figure 3.1 Conceptualisation of down-scaling and aggregation between atmospheric and hydrological models (Hostetler, 1994). Reproduced with permission of Kluwer Academic Publishing.

grid-scale heterogeneity, as well as linking parameter/state variables across disparate scales, are required. However, as Beven (1995: p. 268) observes: 'Hydrological science is constrained by the measurement techniques that are available at the present time. Hydrological theory reflects the scale at which measurements are relatively easy to make.'

In other words, a distinction should be made between the characteristic 'observation scale' and the 'process scale'. Prior to the 1970s, most environmental monitoring was focused on local-scale point processes and measurements (e.g. rainfall, pan evaporation, infiltration, groundwater levels, river flows). This favoured observation scale reflects the applied tradition of hydrology, the relative ease with which such data can be collected and the available technology for data storage and interrogation (see Boucher,

1997, for an overview). With a growing appreciation of the global dimension to environmental change, there have been moves to rationalise existing monitoring networks (Rodda, 1995) and to undertake coordinated, international experiments such as the Global Energy and Water Cycle Experiment, GEWEX (Anon., 1993). At the same time, the compilation of proxy and/or palaeoenvironmental data sets has enabled the reconstruction of hydrometeorological and palaeoecological time series, and has involved a greater range of scientific disciplines in eco-hydrological research (Barker and Higgitt, 1997). Similarly, archival and documentary evidence is increasingly employed to augment instrumental records and to construct homogeneous data sets for key hydrological variables such as rainfall (Jones and Conway, 1997) and temperature (Parker et al., 1992).

Despite these contributions, there is still a paucity of near-surface data for the oceans and many remote or mountainous regions of the world. However, remotely sensed (satellite and radar) information has provided hydrologists and ecologists with new opportunities to observe non-point processes at resolutions of a few metres to the global scale. Remote sensing now routinely provides data for hydrological model parameter estimation, computations of soil moisture and flow regimes, and for real-time flood forecasting (Schultz, 1988). Indeed, remote sensing is capable of quantifying all the key hydrological fluxes and storages of the water balance equation (i.e. precipitation, evapotranspiration, snow cover, runoff and soil moisture) to resolutions of 10–30 m (Engman and Gurney, 1991). This capability has facilitated what Shuttleworth (1988) terms 'macrohydrology': the study of global atmosphere–hydrosphere–biosphere relations and feedbacks. For example, Oki et al. (1995) used atmospheric vapour flux convergence in conjunction with atmospheric water balance calculations to estimate monthly discharge for nearly seventy large rivers and to produce global freshwater runoff volumes.

By calculating ratios of the upwelling land surface reflectance in the red and infrared portions of the electromagnetic spectrum, satellite sensors such as the Advanced Very High Resolution Radiometer (AVHRR) can also be used to derive areal averages of photosynthetically active biomass, albedo, canopy resistance, leaf area index and fractional vegetation cover (Xinmei et al., 1995). For example, the normalised difference vegetation index (NDVI) (Tucker and Sellers, 1986) has been used to study vegetation phenology (Justice et al., 1986) and to monitor vegetation changes over whole continents in relation to shifts in precipitation (Tucker et al., 1991) or El Niño-related droughts (Anyamba and Eastman, 1996). NDVI data are also used to monitor crops, and as a major biophysical indicator in drought and famine early warning systems such as the Global Early Warning System (GEWS) of the Food and Agricultural Organisation (FAO) (Henricksen, 1986; Hutchinson, 1991).

Two technological developments have afforded visualisation and quantitative modelling of eco-hydrological problems, namely geographical information systems (GIS) and digital terrain models (DTM). GIS systems allow the organisation and display of digitised terrain attributes such as soil type, vegetation or raw elevation data to represent the three-dimensional properties of a landscape. Such powerful data interrogation and visualisation techniques have been enabled by the increasing computational speed and data storage capabilities of modern PCs and work stations. Together, GIS and DTM have facilitated many new hydrological, geomorphological and biological applications (see Moore *et al.*, 1991). Although Grayson *et al.* (1993) asserted that GIS are 'hydrologically neutral' they have nonetheless allowed hydrologists to theorise and work in ways that previously would not have been practicable. McDonnell (1996) noted that valuable hydrological research has already been undertaken using GIS in the realms of parameter estimation, loosely coupled hydrological models, integrated modelling and the assembly of hydrological inventories. For example, recent applications of GIS and DTM include the estimation of hydrologically relevant geomorphological parameters from topographic information (Gyasi-Agyei *et al.*, 1995); physically based modelling of spatial variations in soil depth as a function of topographic position (Dietrich *et al.*, 1995); and the regionalisation (or classification) of catchment behaviour based on 'hydrological response units' (Flügel, 1995). According to Sivapalan and Kalma (1995), the availability of both GIS and DTM has revolutionised hydrology, but their use is still undergoing rapid development (often uncritically), and a number of scale problems associated with their use are often ignored. At the same time, GIS and DTM have also presented new opportunities for investigating the effect of scale and heterogeneity on eco-hydrological processes (see below).

PROCESS SCALES, HETEROGENEITY AND VARIABILITY

Earlier, a clear distinction was made between characteristic observation and process scales: the former is traditionally a point (or more recently a 'grid-box average'); the latter often exhibits variations across both space and time. As the following discussion will show, because the observation and process scales seldom correspond, inferences must be made concerning the nature of the heterogeneity, variability and process scaling. By convention, the term 'heterogeneity' is typically used to describe spatial variations in media properties (such as soil hydraulic conductivity), whereas 'variability' refers to fluxes that vary in space and/or time (such as rainfall).

For example, at the local scale (1 m), the dominant flow pathway may be via soil macropores; at the hill-slope scale (100 m), preferential flow may occur through high-conductivity soil horizons or pipes; at the catchment scale (10 km), the drainage network may reflect underlying differences in soil types; and at the regional scale (1000 km), the stream network density may correspond to variations in geology and/or climate. Similarly, variability in time is also present at a range of scales: at the event scale (hours or days), the shape of the flood hydrograph is governed by the characteristics of the precipitation event and by the receiving catchment; at the seasonal scale (months to years), the flow regime may be strongly influenced by physioclimatic controls on rainfall, snowmelt and evaporation; and at the long-term scale (decades or centuries), runoff may exhibit variability due to land-use, climate or anthropic changes. By way of an example, Figure 3.2 shows spatial *and* temporal scales of rainfall in relation to a range of hydrological problems. Once again, the mismatch between observation and process scales becomes evident when it is noted that most rainfall measurements are taken for daily or monthly totals at individual locations for

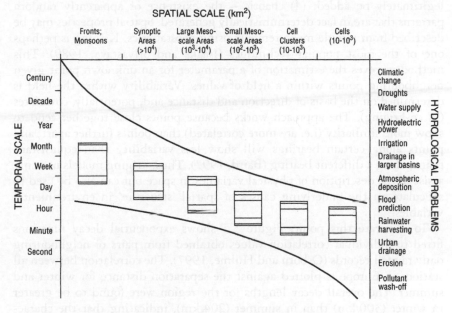

Figure 3.2 Spatial and temporal scales of rainfall in relation to a range of hydrological problems (Berndtsson and Niemczynowicz, 1988). The figure indicates that different precipitation mechanisms have different characteristic spatial and temporal scales. For example, synoptic-scale precipitation systems typically affect areas exceeding 10,000 km^2 and persist for periods of several days to a week. The precipitation generated by such atmospheric circulations can cause 'hydrological problems' such as modifications to the chemical mass balance of acidified catchments or flood generation in intermediate-sized drainage basins. The boxes indicate the confidence intervals about the *probable* mean temporal scales of the precipitation mechanism. The solid line provides the envelope of *possible* temporal scales. Reproduced with permission of Elsevier Science.

which homogeneous records seldom exceed thirty years. Although individual weather radar measurements provide superior spatial coverage (typically 100 km in the mid-latitudes), the networks are currently limited to the industrialised nations with records extending over even shorter periods (see Collier, 1991). Similarly, satellite data may potentially yield global coverage but are constrained by factors such as cost, standardisation, ground-truthing and interpretation.

According to Bloschl and Sivapalan (1995), heterogeneity and variability may be described using three attributes: (1) discontinuity – the existence of discrete zones (e.g. geological or biotic) in which the properties are uniform and predictable, yet between which there is marked disparity; (2) periodicity – the existence of a predictable cycle (e.g. annual runoff regime); (3) randomness – which is predictable only in terms of statistical properties such as the probability density function. To this list a fourth category might legitimately be added: (4) chaotic – the existence of apparently random patterns that are in fact deterministically generated. Spatial properties may be described from sample measurements using geostatistics. Kriging is perhaps one of the most popular techniques (Isaaks and Srivastava, 1989). This method involves the estimation of a parameter for an unknown point given neighbouring points within a field of values. Variability within the field is determined on the basis of direction and distance and, potentially, covariates (in co-kriging). The approach works because points close together tend to show more similarity (i.e. are more correlated) than points further apart, and points along certain bearings will show less variability than equidistant points along a different bearing (Baird, 1997). Thus, kriging models not only assist in the description of physical variables in space but can also be used to speculate on the underlying causes of spatial 'structure' in environmental variables.

To illustrate this point, Figure 3.3 shows exponential decay functions fitted to individual correlation values obtained from pairs of neighbouring daily rainfall records (Osborn and Hulme, 1997). The correlation between all stations in Europe is plotted against the separation distance for winter and summer. The overall decay lengths for the region were found to be greater in winter (300 km) than in summer (204 km), indicating that the characteristic scale of precipitation-causing weather is greater in winter than summer. This is due to the fact that a higher proportion of winter rainfall is of frontal origin, whereas the dominant mechanism in summer is convective rainfall. Given the dependence of the decay lengths on the relative importance of frontal versus convective rainfall, the correlation distances were also found to vary spatially as well as seasonally. Such techniques have important applications in the validation of GCM grid-box area-average precipitation frequencies and amounts, as well as for scaling of contrasting precipitation mechanisms.

Thus, the distance at which the spatial structuring or correlation in a

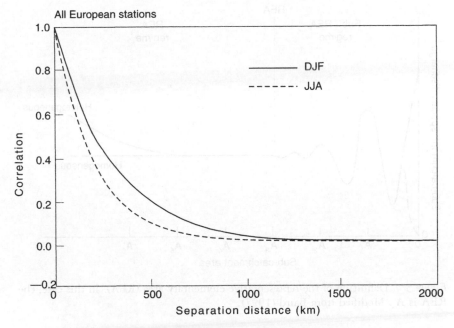

Figure 3.3 Correlation decay curves obtained for winter (DJF) and summer (JJA) daily rainfall between neighbouring stations across Europe. Adapted from Osborn and Hulme (1997).

variable breaks down can indicate that a new set of physical laws is required to describe the phenomenon at this length scale. This notion underlies the concepts of the representative elementary volume (REV) (Hubbert, 1940; Bear, 1972) (see Chapter 9) and, more recently, that of the representative elementary area (REA) (Wood *et al.*, 1988; 1990; Beven *et al.*, 1988), namely, that there exists a soil volume or spatial scale at which simple descriptions of governing processes may suffice (see also Chapter 9). Since measurements of streamflow are inherently spatial averages (Woods *et al.*, 1995), increasing the length scale (area or volume) effectively increases the sampling of hydrological forcings (rainfall and evaporation), hill slopes, soil properties and vegetation, which leads to a decrease in the difference between equivalent sub-catchment area responses (Figure 3.4). The REA is the domain scale at which the variance between hydrological responses for catchments of the same scale attains a minimum, and thus a fundamental building block for distributed hydrological modelling. At the scale of the REA, it is possible to neglect differences in spatial *patterns* of catchment properties (such as soil hydraulic conductivity), but it is still necessary to take spatial *variability* into account in terms of distribution functions (Beven *et al.*, 1988). However, as Wood (1995: p. 92) points out:

47

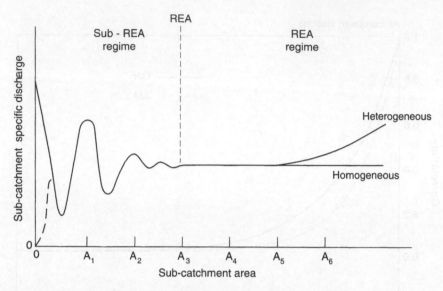

Figure 3.4 Definition of the representative elementary area (REA). In this case, the REA is A_3. Modified from Baird (1997).

this transition in scale, where pattern no longer becomes important but only its statistical representation, may occur at different spatial scales for different processes, may vary temporally for the same parameter (except during wet or dry periods) and may vary with the scale of the 'macroscale' model. Much of the research to date may indicate that the REA for catchment modelling is . . . [in the region of] 0.1 to 25 km².

Theoretically, an REA may be derived for any spatially averaged environmental parameter, but its precise value will be process-, region- and time-specific. For example, Blöschl *et al.* (1995) showed that the REA size is strongly governed by the correlation length of precipitation fields and by the storm duration.

The characterisation of spatial heterogeneities in land surface–atmosphere interactions at all length scales also remains a great challenge to hydro-meteorologists, because the characteristics of the atmosphere above dry and wet (vegetated) surfaces are significantly different due to the contrasting processes of energy redistribution at the ground surface. Avissar (1995) identified five land-surface characteristics that need to be specified accurately in atmospheric models, namely, stomatal conductance, soil surface wetness, surface roughness, leaf area index and albedo. Under stable weather conditions, spatial-variability of any of these characteristics of the order of 100 km was found to induce mesoscale circulations in models of the transport of heat and moisture in the planetary boundary layer. Such circulations were observable for length scales as low as 10 km and were sufficient to promote non-linear feedbacks arising from cloud formation and precipitation. In a similar analysis, Shuttleworth (1988) made a distinction between disorganised and organised variability, also for length scales of 10 km (Figure 3.5). Under conditions of disorganised variability at length scales of 10 km or less, fluid flow and mixing processes naturally integrate the land–atmosphere feedbacks, but under conditions of organised variability at length scales greater than 10 km, there may be an organised (convective) response in the atmosphere that alters the effective value of surface properties at GCM sub-grid scales.

Both studies suggest that spatial variability in land-surface parameters over length scales of the order 10 km is sufficient to induce non-linear feedbacks,

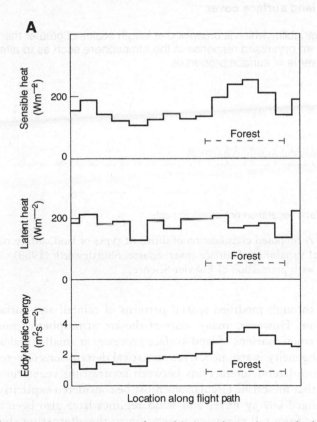

*Figure 3.5*A: Aircraft measurements of sensible heat, latent heat and eddy kinetic energy made at a height of 100 m, these being the median values for three flights between 11:45 and 15:00 GMT on 16 June 1986. The flight path was over a disorganised mixture of agricultural crops apart from the (organised) change to pine forest for the portion indicated by the broken line.

B

Type 'A' land surface cover

Exhibits disorganised variability at length scales of 10 km or less:
Gives no apparent organised response in atmospheric boundary layer

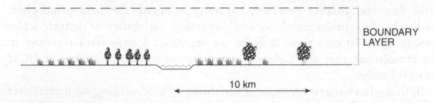

BOUNDARY
LAYER

10 km

Type 'B' land surface cover

Exhibits variability which is organised at length scales of greater than 10 km:
may give an organised response in the atmosphere such as to alter the
effective value of surface properties

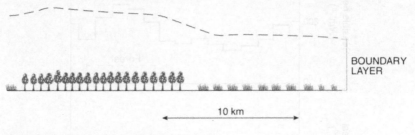

BOUNDARY
LAYER

10 km

Note: Surface vegetation not drawn to scale

*Figure 3.5*B: A proposed classification of different types of land surface based on
organisational variability of surface cover. Source: Shuttleworth (1988).
Reproduced with permission of Elsevier Science.

principally through modified spatial patterns of rainfall and surface energy
redistribution. However, many state-of-the-art atmospheric models have
adopted parameterisations of land-surface processes at smaller scales at which
spatial homogeneity is assumed. Even statistical descriptions of heterogeneity
fail to capture lateral interactions between contrasting vegetation surfaces,
suggesting that mesoscale circulations must be considered explicitly (Avissar,
1995). Changed energy fluxes and wind regimes have also been associated
with changes from tall evergreen vegetation to the alternating short annual
vegetation and bare soils associated with agriculture. For example, Hobbs
(1994) has described profound changes in nutrient, energy and water fluxes
as a result of the fragmentation of native perennial vegetation in Western
Australia by the introduction of predominantly annual crops and pastures.

Similarly, Xinmei *et al.* (1995) demonstrated that the clearance of native vegetation modified the albedo, surface roughness and canopy resistance, leading to weaker vertical transport of heat and water vapour, and thence convective rainfall over agricultural regions. Furthermore, as Veen *et al.* (1996) observed, forest edges are special, high-flux environments; therefore, heterogeneous terrain with a high frequency of forest edges will have a significant impact on the atmosphere at the landscape scale.

The preceding comments and theoretical considerations highlight the complexity of the task of characterising heterogeneity and of relating characteristic processes across different spatial scales. Indeed, it may be argued that, because each catchment has a unique physiography and history of human intervention, the search for universal scaling 'laws' for discharge and evaporation may be a cross-disciplinary problem (Beven, 1995). Add to this the deficiencies of the observational data sets for certain regions, time and space scales, and the task appears even more insurmountable. However, the pressing concerns of global climate and environmental change dictate that solutions, albeit pragmatic ones, be found to such 'scale problems'. Accordingly, the following sections examine the development of three contrasting techniques for linking plant–water relations across scales.

EXTRAPOLATION AND HOLISTIC SCALING TECHNIQUES

Extrapolation and 'scaling-up' (or 'bottom-up') techniques make the fundamental assumption that meso and sub-mesoscale laws and equations can be used to describe processes at larger scales. In other words, empirical relationships at the plant scale are applicable at the canopy, patch and biome scales. The macroscale parameters used in the meso and sub-mesoscale equations are often termed 'effective parameters'. These parameters must characterise all points within a model domain, such that a model based on the uniform parameter field yields the same response as a model based on a heterogeneous parameter field (Blöschl and Sivapalan, 1995). However, as Dawson and Chapin (1993: p. 318) point out, there is always the danger that '[too much] information about detailed mechanisms may be inefficient, incorporating excessive detail and ignoring other aspects that are critical to understanding processes at the higher levels.'

Thus, processes underlying larger-scale responses may be obscured by noisy or unrelated local variations. For example, Figure 3.6 shows spatial variations in 'greenness' obtained from a composite image (visible red, near infrared and near middle infrared) of the LANDSAT Thematic Mapper in which each pixel represents an area of 30 m^2. Figure 3.6A shows the entire image covering the Derbyshire Peak District, Nottinghamshire and parts of north Leicestershire, UK. At this scale, the principal factors governing the vegetation distribution

A

Figure 3.6 LANDSAT (1984) Thematic Mapper composite image showing variations in land cover at A: the regional scale, Derbyshire Peak District and north Nottinghamshire, UK; B: the mesoscale, River Trent floodplain, east of Uttoxeter, Staffordshire, UK; and C: the local scale, a field in the River Trent floodplain, east of Uttoxeter, Staffordshire, UK. Since the original image was a colour composite, the darkness of the pixels is not directly related to 'greenness'. Rather, the image is used to demonstrate that coherent spatial patterns and variations in 'greenness' or 'vigour' of the surface vegetation occur at different scales.

are geological and climatological. A clear distinction between the vegetation overlying the gritstone and limestone is evident in the northwest corner of the image. To the northeast, the geology is sandstone, while the south and southwest comprise Keuper marl overlying Coal Measures. At the mesoscale (catchment), variations in vegetation reflect land management practices and proximity to the river floodplain (Figure 3.6B). Finally, at the scale of individual fields (Figure 3.6C) spatial variations in the vegetation primarily reflect soil moisture, which in turn is governed by local topography, field drainage and soil properties. Hence, it is improbable that the significance of

B

C

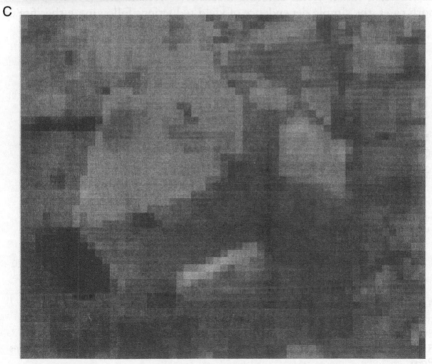

geological or climatological controls to spatial variations in the vegetation cover evident at the regional scale (Figure 3.6A) could be inferred from the observable information at the field scale (Figure 3.6C).

Nonetheless, the most straightforward up-scaling approach is to assume that the parameters at the meso and sub-mesoscale are the same as those at larger scales. The scaling problem then becomes a matter of multiplying by the new length scale or catchment area. For example, Kuczera (1985) modelled observed reductions in river catchment runoff following bushfires in the eucalyptus forests of the Melbourne water supply district, Australia, as a function of stand age:

$$Y_t = L_{max}K(t - 2)e^{1-K(t-2)} \qquad \text{if } t > 2 \text{ else } Y_t = 0 \qquad (3.1)$$

where Y_t is the average yield reduction (mm) relative to mature mountain ash forests t years after the bushfire, L_{max} is the maximum yield reduction (mm) and $1/K$ is the time to maximum yield reduction (years). Regional relationships were then established between the two parameters L_{max} and K using forest descriptors such as the percentage of the catchment area covered by regrowth eucalyptus forest. The model was subsequently adapted by Wilby and Gell (1994) to simulate the effect of piecemeal forest harvesting on downstream runoff yields entering a sensitive wetland site (Figure 3.7). In both cases, the simple dependence of yield reductions following either fire or

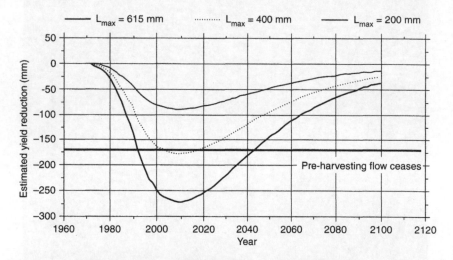

Figure 3.7 Simulated yield reduction curves for selected L_{max} values following twenty years of eucalypt forest harvesting upstream of Tea-tree Swamp, Errinundra Plateau, Victoria, Australia. Source: Wilby and Gell (1994).
Reproduced with permission of International Association of Hydrological Sciences Press.

harvesting (equation (3.1)) reflects empirical variations in sapwood area and throughfall losses with the age of forest regrowth. Field measurements of transpiration flows indicate that stand water use is strongly related to the amount of sapwood area per unit area of forest. Similarly, canopy conductance is proportional to the product of the average leaf conductance and the leaf area index (LAI, a ratio of leaf area to ground area, which is in turn a function of the stand basal area) (Rogers and Hinckley, 1979). Thus, Haydon *et al.* (1996) found that the sapwood area of eucalpyt overstorey in the central highlands of Victoria, Australia, reached a peak of 10.5 m^2 ha^{-1} at age 15 years and declined gradually to 2.5 m^2 ha^{-1} at age 200 years, whereas the peak interception loss of 25 percent occurred at age 30 years, declining to 17 percent at age 200 years.

Hatton and Wu (1995) present an alternative scaling theory, which predicts the nature of the water vapour flux/leaf area relationship by combining elements of ecological field theory, the hydrological equilibrium theory and a standard treatment of the soil–plant–atmosphere continuum in the form:

$$Q = aIA + b\psi A^f \qquad (3.2)$$

where Q is the vapour flux (m^3), I is radiation intercepted by the canopy (MJ m^{-2}), ψ is the soil water potential (MPa) (see Chapter 9), A is the leaf area (m^2), f is a scaling exponent, and a and b are lumped parameter coefficients. However, the use of equation (3.2) to extrapolate individual tree water consumption to stand water consumption entails several assumptions: (1) a strong relationship is assumed to exist between the leaf area of individual trees/stands and the site water balance; (2) each tree tends towards an equilibrium between its size (leaf area) and its local environment such that at no time do resources in local abundance remain untapped; (3) the site is assumed to be homogeneous with regard to soils and climate; (4) the site is fully occupied and tending towards hydrological equilibrium between available moisture and available energy; (5) individual plants react in parallel with all others in the stand; (6) the simple demand function based on irradiance ignores advective contributions to evaporative demand. The principal advantage of the technique is that it does not assume a single fixed relationship between leaf area and water flux; instead, tree water use extrapolated on the basis of leaf area is affected by temporally variable, non-linear stress responses (Figure 3.8). The sampling variability of parameters a and b was found to depend on plot homogeneity, although there is a degree of generality of a between species. Therefore, making use of the equation to scale across heterogeneous landscapes requires spatial distributions of soil water potential and irradiance, and local leaf area distributions. As will be shown below, spatial patterns of soil and energy environments may be obtained from topographic indices.

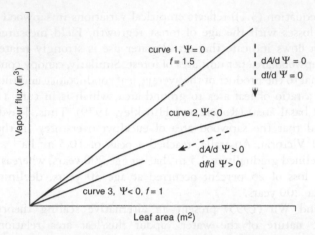

Figure 3.8 Forms of the theoretical scaling relationship between tree leaf area and water flux. With unlimited soil water, the relationship is linear and the slope is limited only by irradiance (curve 1). As the soil dries, the retention of leaves creates a quasi-equilibrium in which the leaf efficiency of the larger trees declines (curve 2). If a drought is of sufficient extent and duration, the trees will drop leaves such that the relationship is in a new, lower hydrological equilibrium, and the curve may once again be linear (curve 3). Source: Hatton and Wu (1995). Reproduced with permission of John Wiley & Sons Ltd.

Despite their limitations, small-scale (*c.* 0.1 hectares) field studies of plant–water relations remain appealing because this is the scale at which species interactions are most readily observable and at which data have been traditionally collected by ecologists (and hydrologists too!). Forest 'gap' models seek to integrate dynamic plant responses to environmental constraints defined at the level of the individual plant (Shugart and Smith, 1996). Key processes in gap models include (1) species-specific growth rates, which depend upon the net photosynthetic rate of the tree per unit area of leaves; (2) spatial positioning and competition for light between individual trees; a range of environmental constraints such as soil moisture and fertility; (3) disturbances due to fire or flooding, and light availability; (4) mortality, which is often age- or stress-related and modelled stochastically; and (5) establishment, which again is represented stochastically and may be a function of environmental constraints. Such forest patch models have been used extensively in studies of global climate change (see, for example, Lauenroth, 1996; Post and Pastor, 1996; Bugmann *et al.*, 1996).

Scaling up patch models via statistical sampling procedures assumes that a generalised patch model is able to simulate the dynamics of entire ecosystems as well as individual species. However, there is a discontinuity between patch- and ecosystem-scale models (Shugart and Smith, 1996). At the patch scale, individual plant recruitment, growth and mortality are in a state of quasi-cyclical disequilibrium, whereas at the scale of the ecosystem the same processes are represented using mass balance approaches, which assume an equilibrium status for whole plant communities. Furthermore, at the patch scale there are feedbacks between the canopy composition and recruitment. Representation of processes such as dispersal, fire and insect outbreaks requires a spatial framework for *linking* patches. Such considerations suggest that for continental-scale applications it is desirable to replace the individual species in patch models with plant functional types that are based on life forms, physiology and rooting distributions with depth (Bugmann and Fischlin, 1996; Coffin and Lauenroth, 1996).

Jarvis and McNaughton (1986) and Schimel *et al.* (1991) have suggested that extrapolation techniques are also limited by the fact that plants do not respond uniformly to external environmental stresses, that there are feedback processes external to the leaf scale that operate at the ecosystem scale, and that there are major difficulties in forecasting the behaviour of (non-ideal) complex systems. For example, increasing concentrations of atmospheric CO_2 are assumed to have the dual effect of enhancing photosynthetic activity and improving the efficiency of plants' water use (Sellers *et al.*, 1996). This may be valid at the scale of individual plants, but CO_2 fertilisation confers advantages selectively within ecosystems, depending on species. At the forest scale, spatial and temporal variations in the microclimate (in particular the relative humidity) of the canopy will have a negative feedback on evapotranspiration rates of individual plants. This selectivity will also modify internal nutrient and water fluxes.

An alternative extrapolation philosophy is to derive scale-independent 'laws of nature' by employing dimensional analysis or self-similarity techniques (Blöschl and Sivapalan, 1995). Fractals provide a mathematical framework for the treatment of apparently complex shapes that display similar patterns over a range of space/time scales. According to Mandelbrot (1983) many attributes of nature exhibit a property known as statistical self-similarity, whereby subcomponents of an object, pattern or process are statistically indistinguishable from the whole. Fractals have found widespread application in hydrology, geomorphology and climatology (see, for example, the *Journal of Hydrology*, Volume 187, Issues 1–2, Gao and Xia, 1996; or Goodchild and Mark, 1987). It has even been suggested that fractals are indicative of underlying processes of organisation, as in the case of the geometry of drainage networks, which reflect minimum energy expenditure (Rinaldo *et al.*, 1992), and the hierarchical structuring of mid-latitude mesoscale atmospheric convective systems (Perica and Foufoula-Georgiou, 1996). However, simple fractal properties are seldom applicable across all temporal and/or spatial scales and it is unclear as to how multifractals should be interpreted relative to observable hydrological processes (Wilby, 1996).

STATISTICAL DOWNSCALING
TECHNIQUES

The International Geosphere–Biosphere Programme (IGBP) and the GEWEX Continental Scale International Project (GCIP) were established with the specific mandate to investigate the complex interactions between the physical and biological components of the planet and their responses to anthropic change. A major focus of the BAHC (Biological Aspects of the Hydrologic Cycle) component of the IGBP has been the development of tools for generating the high-resolution meteorological inputs required for subsequent use in eco-hydrological or forest-gap models (Bass, 1996). Wilby and Wigley (1997) suggest that such 'downscaling' techniques may be subdivided into four methodological categories, namely, regression methods, weather-pattern-based approaches, stochastic weather generators, and limited-area modelling.

Regression methods were among the earliest down-scaling approaches (e.g. Kim et al., 1984; Wigley et al., 1990). These approaches generally involve establishing linear or non-linear relationships between sub-grid-scale or single-site parameters such as rainfall and coarser-resolution (GCM grid scale) predictor variables such as geostrophic vorticity, a measure of the strength of atmospheric cyclonicity (e.g. Conway et al., 1996). Among regression methods it is also reasonable to include artificial neural network (ANN) approaches, since the internal weights of an ANN model emulate non-linear regression coefficients (Hewitson and Crane, 1996). Having derived a regression equation or trained an ANN to relate the observed local and regional climates, the equations may then be 'forced' using regional-scale climate data obtained from a GCM operating in either a 'control' or 'perturbed' state. For example, von Storch et al. (1993) used a canonical correlation technique to relate sea level pressure patterns in the North Atlantic to winter rainfall in the Iberian Peninsula. The predicted rainfall for multiple stations was then validated using data derived from the closest GCM grid points.

Atmospheric circulation (or atmospheric pressure pattern) downscaling methods typically involve relating observed station or area-average meteorological data statistically to a given, objectively or subjectively defined, weather classification scheme (see Yarnal, 1993). The downscaling relationship is obtained by fitting generalised probability distribution functions to observed data such as the probability of a wet day following a wet day, or the mean wet day amount associated with a given atmospheric circulation pattern (see Bardossy and Plate, 1992; Hughes and Guttorp, 1994). For example, Figure 3.9 indicates that the cyclonic mean wet day amount size distribution for a station at Kempsford, UK, is well represented by an exponential function. Daily precipitation series may be further dis-aggregated by month or season, or by the dominant precipitation mechanism

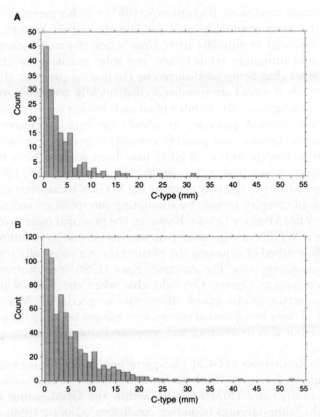

Figure 3.9 Frequency diagrams of wet day rainfall amounts (mm) at Kempsford in the Cotswolds, UK, 1970–1990 corresponding to the cyclonic (C-type) Lamb weather type on days A: without fronts, and B: with at least one front. Note the increased frequency of precipitation events under frontal weather conditions, particularly for days with rainfall exceeding 10 mm.

(Wilby *et al.*, 1995). In either case, a new precipitation (or meteorological) time series can be generated stochastically by applying input sequences of daily weather types to the observed conditional probability distribution functions. These 'forcing' weather pattern series are typically generated using either Monte Carlo techniques (Wilby, 1994) or the pressure fields of GCMs (Matyasovszky *et al.*, 1994). Although the majority of such studies have focused on daily precipitation, series of daily circulation patterns may be used to downscale other variables such as temperature, evaporation and ultraviolet radiation, or multivariate processes such as floods, droughts, acid deposition, smog, ozone and atmospheric particulates (Bass, 1996).

Stochastic weather generators share many attributes of conventional circulation-based downscaling models but differ in their means of application to future climate conditions. Richardson's (1981) weather generator (WGEN) model is the most commonly used for climate impact studies: this was originally designed to simulate daily time series of precipitation amount, maximum and minimum temperature, and solar radiation for the present climate. Rather than being conditioned by circulation patterns, all variables in the Richardson model are simulated conditionally according to previous precipitation occurrence. At the heart of all such models are first- or multiple-order Markov renewal processes in which, for each successive day, the precipitation occurrence (and possibly amount) is governed by outcomes on previous days. Models such as WGEN have been adapted for a number of climate change impact studies (e.g. Wilks, 1992; Mearns *et al.*, 1996). There is also the possibility of spatially distributing WGEN parameters across landscapes, even in complex terrain, by combining interpolation techniques and DTMs (see VEMAP study below). However, the principal issue involving the application of WGEN or other stochastic weather generators to future climates has been the method of adjusting the parameters in a physically realistic and internally consistent way. For example, Katz (1996) demonstrated, using daily observations at Denver, Colorado, that when the WGEN parameters are varied, certain unanticipated effects can be produced. Modifying the probability of daily precipitation occurrence changed not only the mean daily temperature but also its variance and autocorrelation in possibly unrealistic ways.

Given the limitations of GCM grid-point predictions for regional climate change impact studies, the final downscaling option is to embed a higher-resolution limited-area climate model within the GCM, using the GCM to define the (time-varying) boundary conditions (Giorgi, 1990; Mearns *et al.*, 1995). Although limited area models (LAMs) can produce climates for 20–50 km horizontal grid spacing and 100–1000 m vertical resolution there are several acknowledged limitations of the approach. LAMs still require considerable computing resources and are as expensive to run as a global GCM. Furthermore, they are somewhat inflexible in the sense that the computational demands apply each time that the model is transferred to a different region. Above all, the LAM is completely dependent upon the veracity of the GCM grid-point data that are used to drive the boundary conditions of the region – a problem that also applies to circulation-driven downscaling methods.

As Table 3.2 indicates, the confident application of statistical downscaling techniques to the study of plant–water relations is constrained by several as yet unresolved issues. Perhaps the most serious limitation is the way in which downscaling compels the use of 'passive' atmosphere–vegetation models; current methodologies do not incorporate feedbacks between the land surface and climate, or permit scale interactions. Furthermore, the majority of downscaling tools relate point or areal averages to regional scale phenomena according to the level of data available. However, what is often required for

Table 3.2 Outstanding challenges to the confident application of statistical
downscaling techniques.

- The whole ideology of downscaling presupposes that, as a result of human-
induced global warming, there will be significant (and predictable) changes in
the downscaling predictor variables (such as the frequencies of daily weather
patterns).

- To date, most downscaling studies have been conducted for daily or monthly
precipitation in temperate, mid-latitude regions of the Northern Hemisphere;
relatively few have tackled semi-arid or tropical locations. There has also been
a topographic bias towards low-altitude sites.

- There is a need for more general weather classification systems used in downscaling. Most weather classification schemes are inherently parochial because of
the important controlling influences of regional- and local-scale factors such as
topography or ocean/land distributions.

- The majority of downscaling approaches generate time series of data for only a
few hydrologically relevant parameters, most commonly precipitation. There is
a need for techniques that downscale internally consistent multivariate data
which preserve covariance among parameters and auto-correlation within time
series.

- Current downscaling approaches seldom capture climate variability at all
temporal or spatial scales. For example, even within a single circulation pattern,
the precipitation statistics may vary considerably from year to year.

- This type of non-stationarity can be accounted for by developing more complex
statistical models (within the constraints of model reliability imposed by
data availability). This, however, puts greater pressure on the driving GCMs to
provide reliable predictions for a greater range of variables.

- Downscaling is a uni-directional modelling technique. Local- and regional-
scale hydrological or ecological responses are forced by mesoscale predictor
variables. Current techniques have no means of incorporating terrestrial feedbacks into the driving climate models.

Source: Adapted from Wilby and Wigley (1997).

impact assessment is a means of interpolating between data-rich regions to
data-poor sites. The remaining two sections will address these important scale
issues.

DISTRIBUTING TECHNIQUES

Some commentators have suggested that a general theory of hydrological
scaling is unlikely to exist and that sub-grid-scale heterogeneity should be
treated statistically (Beven, 1995: p. 278):

the hydrological system is greatly influenced by external historical and geological forcings. This greatly limits the potential for scale-invariant behaviour and it is suggested that it may be better to recognise explicitly that scale-dependent models are required in which the characterisation of the heterogeneity of responses is posed as a problem of sub-grid scale parameterisation.

As was noted earlier, hydrological and ecological measurements tend to have a coarser spatial than temporal resolution, prompting the widespread use of interpolation techniques for the space domain. Perhaps the most thoroughly discussed interpolation problem in hydrology is the spatial estimation of rainfall from point rain-gauge measurements (see, for example, Shaw, 1994; Reed and Stewart, 1994). The problem arises because the measurements on which the interpolation techniques are based are too widely spaced relative to the natural variability of rainfall. For example, the average spacing of hourly rain-gauges in the relatively dense monitoring network of the British Isles is between 15 and 30 km (Faulkner and Reynard, personal communication), whereas the length scales of individual storm cells are of the order of 10 km (see Figure 3.2). Although the UK Meteorological Office's MORECS predictions of potential and actual evapotranspiration have a grid spacing of 40 km, rainfall radar typically has a resolution of 2 km, and remotely sensed AVHRR satellite imagery pixels of 1 km, the widespread use of these data sources is constrained by their high cost and short record lengths relative to surface networks (e.g. AVHRR provides about fifteen years of data).

Given the constraints of both remote and surface monitoring systems, and the natural heterogeneity of surface processes, one solution is to correlate the variable of interest to an auxillary variable whose spatial distribution is more readily obtained. The spatial distribution of the dependent variable is then inferred from the spatial distribution of the covariate (Blöschl and Sivapalan, 1995). To date, one of the most widely used covariates in eco-hydrology has been topography, for which data with a resolution of 50 m are often available in digital form. The availability of digital elevations and other environmental variables such as soils and geology, coupled with GIS and DTM software for manipulating data, has favoured the development of predictive mapping techniques in the last twenty years. As Table 3.3 indicates, topography can be used to infer the spatial distribution of a wide variety of secondary variables. For example, the precipitation–elevation regressions on independent slopes model (PRISM) (Daly et al., 1994) simulates precipitation at 10 km grid scales by, first, dividing the terrain into topographical facets of similar aspect; second, regressing point precipitation data against elevation for each facet by region; and third, employing these regression equations spatially to extrapolate station precipitation to cells that have similar facets.

Beven's (1987) solution to the problem of sub-grid-scale hydrological parameterisation was the use of spatial distribution functions to represent

Table 3.3 Examples of topographically derived hydrometeorological variables.

- *Elevation*: daily and seasonal precipitation totals; snow characteristics including snow water equivalent; wind speeds; air temperature; soil depth; air and soil water chemistry.

- *Slope*: incident solar radiation; soil properties and preferential flow pathways; lateral subsurface flow rate; soil moisture; transpiration rates; soil water chemistry.

- *Aspect*: incident solar radiation; precipitation totals; snow accumulation and melting; wind speeds; air temperature; soil moisture; transpiration rates; atmospheric wet and dry deposition.

- *Upslope catchment area*: precipitation and runoff volume; soil moisture; soil surface properties; stream network order and main channel length.

- *Slope curvature*: snow accumulation; subsurface water flow; soil moisture; soil litter accumulation; soil erosion/deposition rates; rill initiation; soil depth and texture; water-holding capacity; nutrient availability.

local-scale processes governing runoff production. For example, TOP-MODEL (Beven and Kirkby, 1979) simulates the scale-dependent dynamics of the storm hydrograph using the distribution function of an index of hydrological similarity, $\ln(\alpha T_0 \tan\beta)$, where α is the area draining through a point, $\tan\beta$ is the local slope angle at that point and T_0 is the local downslope transmissivity at soil saturation. These properties, as well as the spatial distribution functions for incoming rainfall intensities, vegetation processes such as throughfall, soil properties and downslope water velocities, may all be inferred from DTM of the catchment topography. Quinn *et al.* (1995) and Beven (1995) have extended TOPMODEL to represent interactions between downslope subsurface flows and water availability for transpiration. Figure 3.10 shows the key elements of their simple patch model, which represents areas of the landscape with broadly similar responses in terms of evapotranspiration. The model differs from conventional one-dimensional surface vegetation–atmosphere models in one respect: subsurface water fluxes are able to enter and leave the patch via upslope and downslope drainage, respectively. The variability in runoff production, soil moisture and evapotranspiration is then calculated from a linear combination of patches of varying topographical, meteorological, soil, geological and vegetation characteristics. Although this allows for more realistic modelling of the patch in the context of the landscape, the spatial distribution of many of the vegetation and soil properties listed in Figure 3.10 must still be specified from the results of micrometeorological experiments and available field data. Furthermore, it must be recognised that there may be many permutations of patches that give similar hydrological responses in terms of observed evapotranspiration or runoff. This equifinality in model behaviour (see

63

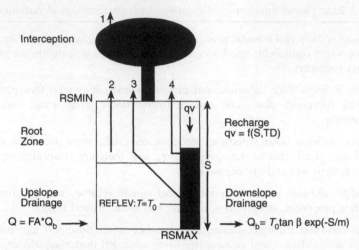

Processes in the patch model

1. Evaporation from the interception store (canopy resistance = 0)
2. Evapotranspiration from the root zone store (RSMIN<canopy resistance<RSMAX)
3. Evapotranspiration from the water table when in root zone
4. Evapotranspiration supplied by capillary rise from the water table

Parameters of the patch model

RA (s/m)	Aerodynamic resistance
RSMIN (s/m)	Minimum dry canopy resistance
RSMAX (s/m)	Maximum dry canopy resistance at wilting point
MAXINT (m)	Interception storage capacity
FA	Fractional upslope area
REFLEV (m)	Reference level for soil transmissivity
$T_0\tan\beta$ (m-2)	Product of saturated transmissivity at reference level and effective downslope hydraulic gradient
m (m)	Transmissivity profile (and recession curve) parameter
TD (h/m)	Effective wave speed per unit of deficit for recharge
SOIL	Soil type for capillary rise calculations

Figure 3.10 Schematic representation of a simple patch model including definitions of all model parameters. Source: Beven (1995). Reproduced with permission of John Wiley & Sons Ltd. (Please note: the model equations are not explained here and interested readers should consult the original paper.)

Chapter 9) is further exacerbated by the number and weighting of patches selected.

These parameterisation issues might be resolved by calibrating semi-distributed hydrological models such as TOPMODEL using the products of predictive vegetation mapping. Indeed, predictive vegetation mapping already embraces many of the principles inherent in topographic hydrological models (e.g. Palmer and Van Staden, 1992). Founded in ecological niche theory and gradient analysis, the approach predicts vegetation composition across a landscape given interpolated environmental variables that are related to physiological tolerances and are derived from digital soil or elevation data

(Table 3.3). For example, Saunders and Bailey (1994) assessed the significance of topography-induced heterogeneity in solar radiation budgets (versus the diffusing effects of orographic cloud) to spatial variability in the vegetation coverage of alpine tundra. Similarly, Haines-Young and Chopping (1996) reviewed the use of landscape indices as a means of predictive mapping for biodiversity and conservation planning.

However, as Figure 3.11 implies, a clear distinction must be made between *potential* (or 'climax') natural vegetation and *actual* (observed) vegetation distributions. The latter reflects variables related to the disturbance history (such as fire, grazing or other anthropic influences) as well as the underlying environmental gradients (such as soil moisture, light and nutrients). Furthermore, as Franklin (1995) notes, predictive vegetation mapping assumes static, equilibrium, whole-mosaic models of the spatial distribution of biota. Such models can be useful for understanding or predicting biogeographic distributions of species' realised niches at the landscape scale but cannot simulate transient ecosystem dynamics, because biotic relationships between species are liable to change through time. As has been shown previously, vegetation–atmosphere and vegetation–soil–moisture feedbacks operating across several scales of interaction may also modify the sub-mesoscale environment occupied by the biotic assemblage. In this respect, vegetation patterns shape and are shaped by environmental gradients. Perhaps the most significant potential contributions of predictive vegetation models will be in the construction of internally consistent baseline data sets or sub-grid-scale parameterisation schemes for use in hydrological models. Rather than predicting vegetation assemblages, it might be more useful, from a hydrological perspective, to map vegetation structure or physiognomy (as in Mackey, 1993; Kleidon and Heimann, in press).

Finally, improved representations of topography have also been incorporated within the latest generations of LAMs. These models now have the capability of simulating smaller-scale atmospheric features such as orographic precipitation (e.g. Segal *et al.*, 1994) and may ultimately provide atmospheric data for impact assessments that reflect the natural heterogeneity of the climate at regional scales (Hostetler, 1994). For example, Zeng and Pielke (1995) used extensive numerical simulations with the Regional Atmospheric Modeling System developed at Colorado State University to demonstrate that topography can induce mesoscale sensible heat, moisture and momentum fluxes that can be larger than, and have a different vertical structure from, the turbulent fluxes of a typical GCM grid box. Meanwhile, Mearns *et al.* (1995) concluded that errors in the frequency and intensity fields of daily precipitation produced by the NCAR Community Climate Model (CCM) and NCAR/Pennsylvania State Mesoscale Model (MM4) were due to inadequate representation of topography, even with a horizontal resolution of 60 km.

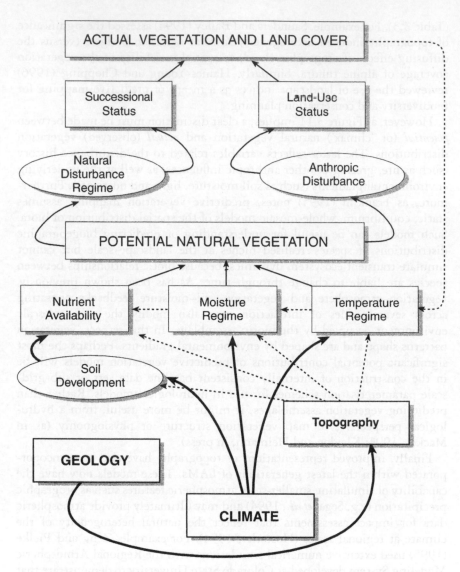

Figure 3.11 Conceptual model showing the relationship between direct gradients (nutrients, moisture, temperature), their environmental determinants (climate, geology, topography) and potential natural vegetation, and the processes that mediate between the potential and actual vegetation cover (the latter being observed by a remote-sensing device). The thickness of the lines is indicative of the strength of the relationships, while circles represent suites of processes. For example, the natural disturbance regime might include fire, flooding, disease, windfall, etc. Modified from Franklin (1995). Reproduced with permission of Edward Arnold Publishers Ltd.

SCALES AND SCALE 'MISMATCH' BETWEEN ECOLOGICAL AND HYDROLOGICAL PROCESSES

A major theme of research for the past decade has been to understand the consequences of climate variability and potential climate change on eco-hydrological systems. In any assessment of climate impacts on ecosystems, processes mediated through the hydrological cycle must come to the fore: droughts, floods and changes to moisture stress on vegetation. However, most of our knowledge of the atmosphere's spatial characteristics is of low spatial resolution (Giorgi and Marinucci, 1996). This is a classic issue in linking atmospheric GCMs to the land surface, where typical GCM grid cells represent many thousands of square kilometres, and most ecological and hydrological processes are best understood on length scales of metres to small river catchments (a few tens of square kilometres). Resolution is also an issue in analysing observations. A recent project, the Vegetation and Ecosystem Modelling and Analysis Project (VEMAP, 1995), produced a set of down-scaled climatologies for the USA as input into eco-hydrological models. The analysis was conducted on an 0.5° latitude × longitude grid, quite coarse for ecosystem studies, yet resulting in over 3000 grid cells. Even in the USA, where observations are relatively dense, meteorological station densities average no more than two per grid cell. While there exist a few areas in the USA and Europe where patterns of mean climate and interannual variability may be observed directly with a resolution of a few kilometres, in most regions precipitation and surface temperatures may only be estimated from observations with length resolutions of 50–100 km. Thus, large-scale studies (e.g. biome or drainage basin scale) must rely on downscaling to account for catchment-scale influences of topography on temperature, precipitation, radiation and other terrain-dependent aspects of climate (see above). Down-scaling procedures intrinsically produce a reasonable portrayal of mean conditions, which are strongly constrained by effects of elevation and orography on precipitation processes and temperature. They do a much worse job with precipitation rates, which depend upon the dynamics of atmospheric processes that intrinsically occur on scales smaller than the resolution of typical observing systems: processes on the scale of convective and precipitating cloud systems. Thus, capturing even the statistics of storm intensity at a regional scale can be challenging.

There are thus two 'scale mismatches' between climate and hydrology: first, between large-scale atmospheric features and small-scale land surface attributes, in effect, between the resolution of climate models and observations, and the scale of topographic features. This means that terrain-dependent features of climate – adiabatic effects, inversions and precipitation – must be inferred indirectly from models and observations. Second, there is a mismatch between small-scale atmospheric features (convective and precipitating clouds) and regional land systems, which makes capturing the occurrence and statistics of extreme events over large areas difficult.

In considering eco-hydrological systems, there are also key scale 'matches'. The river catchment or landscape has long been a scale at which both ecologists and hydrologists could work. Hill slopes provide ordered sequences of soil variability, moisture availability and vegetation that have been exploited by hydrologists, ecologists and soil scientists (Schimel *et al.*, 1985; Burke *et al.*, 1989). River catchments are a key unit in understanding hydrological processes such as evapotranspiration, runoff generation and baseflow: they are also an integrated unit for ecosystem studies, contributing to the understanding of nutrient budgets and transformations, water use, primary production and disturbance effects (Borman and Likens, 1979). Catchment studies provide both a scale at which variability within landscapes in ecological and hydrological processes may be studied (Moldan and Cerny, 1994), and through hydrological and hydrochemical gauging, an integrated scale where the aggregate interactions of climate, hydrology and ecology may be observed and modelled. For example, regional to global nitrogen studies have just begun, and it is evident that gains and losses of N are highly coupled to water budgets (via leaching) and to soil moisture variations, the key control over trace gas losses (Mosier *et al.*, 1997; Parton *et al.*, 1996; Figure 3.12). Given the crucial importance of carbon and nitrogen budgets in contemporary ecology and limnology (Schimel *et al.*, in press; Hornberger *et al.*, 1994), the catchment scale must be a central focus of research as the scale where phenomena such as leaching and soil moisture storage may be linked to the biogeochemistry of C and N.

While much of the material reported in this chapter has dealt with interpreting atmospheric information in terms of scales of hydrological and ecological processes, land-surface processes also affect climate. There is now an extensive literature documenting the effects of large-scale modification by the land surface of climate (e.g. Melillo *et al.*, 1996). Studies have indicated that deforestation may affect climate via changes to evapotranspiration and the Bowen ratio, via albedo and roughness changes, and via changes to land-surface patch structure, causing mesoscale effects (Dickinson, 1992; Pielke *et al.*, 1991; Henderson-Sellers and McGuffie, 1995). Other studies have looked at global-scale effects of the changing land surface on climate via sensitivity studies examining an artificial perturbation to surface conductance (Thompson and Pollard, 1995), simulated effects of doubled atmospheric CO_2 concentrations (Sellers *et al.*, 1996), and by a comparison of global 'potential natural' to actual vegetation (Chase *et al.*, 1996). Clearly, feedbacks from land-surface hydrology, especially via evapotranspiration, can have substantial effects on climate. Studies of global hydroclimatological feedbacks using models require that the surface processes be correctly forced, meaning that changes to the fluxes at the land–atmosphere interface be correct with regard

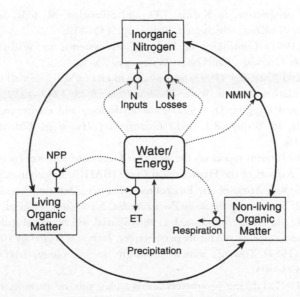

Figure 3.12 Schematic illustration of the coupling of water, nitrogen cycling and carbon in ecosystems. Principal features of these coupled controls are that (a) water controls the inputs and outputs of nitrogen (N), (b) increasing net primary productivity (NPP) allows more of the N flux through the system to be captured into organic matter, (c) increasing organic N stocks allows for more N mineralisation, supporting more NPP, and (d) increasing precipitation allows for both more NPP and more N cycling, thus water and nutrient limitation of NPP tend to become correlated. Note that the circles indicate key water/energy-driven processes involved in the nitrogen cycle. Source: Schimel *et al.* (in press).

to the natural scale of the land surface. They also require that land-surface fluxes be correctly aggregated back to the scale of the atmospheric processes simulated, taking into account any processes that arise at sub-grid scales (Zeng and Pielke, 1995). Simulating the coupled effects of land surface and climate change will be a 'grand challenge' for interdisciplinary earth science in the twenty-first century.

REFERENCES

Anon. (1993) The Global Energy and Water Cycle Experiment (GEWEX), *WMO Bulletin* 42: 20–27.

Anyamba, A. and Eastman, J.R. (1996) Interannual variability of NDVI over Africa and its relation to El Niño/Southern Oscillation, *International Journal of Remote Sensing* 17: 2533–2548.

Avissar, R. (1995) Scaling of land–atmosphere interactions: an atmospheric modelling perspective, in Kalma, J.D. and Sivapalan, M. (eds) *Scale Issues in Hydrological Modelling*, Chichester: Wiley, pp. 435–452.

Baird, A.J. (1997) Continuity in hydrological systems, in Wilby, R.L. (ed.) *Contemporary Hydrology*, Chichester: Wiley, pp. 25–58.

Bardossy, A. and Plate, F.J. (1992) Space–time model for daily rainfall using atmospheric circulation patterns, *Water Resources Research* 28: 1247–1259.

Barker, P.A. and Higgitt, D. (1997) Palaeohydrology and environmental change in drylands, in Wilby, R.L. (ed.) *Contemporary Hydrology*, Chichester: Wiley pp. 273–316.

Bass, B. (1996) Interim report on the Weather Generator Project, Focus 4 of IGBP Biospheric Aspects of the Hydrological Cycle (BAHC). Environmental Adaption Research Group, Atmospheric Environment Service, Ontario, Canada.

Bear, J. (1972) *Dynamics of Fluids in Porous Media*, New York: Elsevier, 764 pp.

Berndtsson, R. and Niemczynowicz, J. (1988) Spatial and temporal scales in rainfall analysis – some aspects and future perspectives, *Journal of Hydrology* 100: 293–313.

Beven, K.J. (1987) Towards a new paradigm in hydrology, *IAHS Publication* No. 164, 393–403.

Beven, K.J. (1995) Linking parameters across scales: subgrid parameterizations and scale dependent hydrological models, in Kalma, J.D. and Sivapalan, M. (eds) *Scale Issues in Hydrological Modelling*, Chichester: Wiley, pp. 263–282.

Beven, K.J. and Kirkby, M.J. (1979) A physically-based variable contributing area model of basin hydrology, *Hydrological Sciences Bulletin* 24: 43–69.

Beven, K.J., Wood, E.F. and Sivapalan, M. (1988) On hydrological heterogeneity – catchment morphology and catchment response, *Journal of Hydrology* 100: 353–375.

Blöschl, G. and Sivapalan, M. (1995) Scale issues in hydrological modelling: a review, in Kalma, J.D. and Sivapalan, M. (eds) *Scale Issues in Hydrological Modelling*, Chichester: Wiley, pp. 9–48.

Blöschl, G., Grayson, R.B. and Sivapalan, M. (1995) On the representative elementary area (REA) concept and its utility for distributed rainfall-runoff modelling, in Kalma, J.D. and Sivapalan, M. (eds) *Scale Issues in Hydrological Modelling*, Chichester: Wiley, pp. 71–88.

Bormann, F.H., and Likens, G.E. (1979) *Pattern and Process in a Forested Ecosystem: Disturbance, Development and the Steady State Based on the Hubbard Brook Ecosystem Study*, New York: Springer-Verlag, 253 pp.

Boucher, K. (1997) Hydrological monitoring and measurement methods, in Wilby, R.L. (ed.) *Contemporary Hydrology*, Chichester: Wiley, pp. 107–149.

Bugmann, H.K.M. and Fischlin, A. (1996) Simulating forest dynamics in a complex topography using gridded climatic data, *Climatic Change* 34: 201–211.

Bugmann, H.K.M., Yan, X., Sykes, M.T., Martin, P., Lindner, M., Desanker, P.V. and Cumming, S.G. (1996) A comparison of forest gap models: model structure and behaviour, *Climatic Change* 34: 289–313.

Burke, I.C., Reiners, W.A., and Schimel, D.S. (1989) Organic matter turnover in a sagebrush steppe landscape, *Biogeochemistry* 7: 11–31.

Chase, T.N., Pielke, R.A., Kittel, T.G.F., Nemani, R. and Running, S.W. (1996) Sensitivity of a general circulation model to change in leaf area index, *Journal of Geophysical Research* 101 (D3): 7393–7408.

Coffin, D.P. and Lauenroth, W.K. (1996) Transient responses of North American grassland to changes in climate, *Climatic Change* 34: 269–278.

Collier, C.G. (1991) Weather radar networking in Europe – The Commission of the European Communities' COST-73 project, *WMO Bulletin* 40: 303–307.

Conway, D., Wilby, R.L. and Jones, P.D. (1996) Precipitation and air flow indices over the British Isles, special issue of *Climate Research* 7: 169–183.

Daly, C., Neilson, R.P. and Phillips, D.L. (1994) A statistical-topographic model for mapping climatological precipitation over mountainous terrain, *Journal of Applied Meteorology* 33: 140–158.

Dawson, T.E. and Chapin, F.S. (1993), in Ehleringer, J.R. and Field, C.B. (eds) *Scaling Physiological Processes: Leaf to Globe*, New York: Academic Press.

Dickinson, R.E. (ed.) (1992) *The Geophysiology of Amazonia: Vegetation and Climate Interactions*, New York: Wiley.

Dietrich, W.E., Reiss, R., Hsu, M.L. and Montgomery, D.R. (1995) A process-based model for colluvial soil depth and shallow landsliding using digital elevation data, in Kalma, J.D. and Sivapalan, M. (eds) *Scale Issues in Hydrological Modelling*, Chichester: Wiley, pp. 141–158.

Dooge, J.C.I. (1988) Hydrology in perspective, *Hydrological Sciences Journal* 33: 61–85.

Ehleringer, J.R. and Field, C.B. (eds) (1993) *Scaling Physiological Processes: Leaf to Globe*, New York: Academic Press.

Engman, E.T. and Gurney, R.J. (1991) Recent advances and future implications of remote sensing for hydrologic modelling, in Bowles, D.S. and O'Connell, P.E. (eds) *Recent Advances in the Modelling of Hydrologic Systems*, Netherlands: Kluwer Academic Publishers, pp. 471–495.

Flügel, W. (1995) Delineating hydrological response units by Geographical Information System analyses for regional hydrological modelling using PRMS/MMS in the drainage basin of the River Bröl, Germany, in Kalma, J.D. and Sivapalan, M. (eds) *Scale Issues in Hydrological Modelling*, Chichester: Wiley, pp. 181–194.

Franklin, J. (1995) Predictive vegetation mapping: geographic modelling of bio-spatial patterns in relation to environmental gradients, *Progress in Physical Geography* 19: 474–499.

Gao, J. and Xia, Z. (1996) Fractals in physical geography, *Progress in Physical Geography* 20: 178–191.

Giorgi, F. (1990) Simulation of regional climate using a Limited Area Model nested in a General Circulation Model, *Journal of Climate* 3: 941–963.

Giorgi, F. and Marinucci, M.R. (1996) Improvements in the simulation of surface climatology over the European region with a nested modeling system, *Geophysical Research Letters* 23 (3): 273–276.

Goodchild, M.F. and Mark, D.M. (1987) The fractal nature of geographic phenomena, *Annals of the Association of American Geographers* 77: 265–278.

Grayson, R.B., Blöschl, G., Barling, R.D. and Moore, I.D. (1993) Process, scale and constraints to hydrological modelling in GIS, in Kovar, K. and Nachtnebel, H.P. (eds) *Application of Geographic Information Systems in Hydrology and Water Resources Management*, Report 211, Wallingford: IAHS.

Gyasi-Agyci, Y., Willgoose, G. and De Troch, F.P. (1995) Effects of vertical resolution and map scale of digital elevation models on geomorphological parameters used in hydrology, in Kalma, J.D. and Sivapalan, M. (eds) *Scale Issues in Hydrological Modelling*, Chichester: Wiley, pp. 121–140.

Haines-Young, R. and Chopping, M. (1996) Quantifying landscape structure: a review of landscape indices and their application to forested landscapes, *Progress in Physical Geography* 20: 418–445.

Hatton, T. and Wu, H. (1995) Scaling theory to extrapolate individual tree water use to stand water use, in Kalma, J.D. and Sivapalan, M. (eds) *Scale Issues in Hydrological Modelling*, Wiley: Chichester, pp. 283–296.

Haydon, S.R., Benyon, R.G. and Lewis, R. (1996) Variation in sapwood area and throughfall with forest age in mountain ash (*Eucalyptus regnans* F. Muell), *Journal of Hydrology* 187: 351–366.

Henderson-Sellers, A. and McGuffie, K. (1995) Global climate models and 'dynamic' vegetation changes, *Global Change Biology* 1: 63–75.

Henricksen, B.L. (1986) Reflections on drought: Ethiopia 1983–1984, *International Journal of Remote Sensing* 7: 1447–1451.

Hewitson, B.C. and Crane, R.G. (1996) Climate downscaling: techniques and application, *Climate Research* 7: 85–95.

Hobbs, R.J. (1994) Fragmentation in the wheatbelt of Western Australia: landscape-scale problems and solutions, in Dover, J.W. (ed.) *Fragmentation in Agricultural Landscapes*, Proceedings of the third annual IALE (UK) conference, held at Myersough College, Preston, 13–14 Sept. 1994.

Hornberger, G.M., Bencala, K.E. and McKnight, D.M. (1994) Hydrological controls on dissolved organic carbon during snowmelt in the Snake River near Montezuma, Colorado, *Biogeochemistry* 25: 147–165.

Hostetler, S.W. (1994) Hydrologic and atmospheric models: the (continuing) problem of discordant scales, *Climatic Change* 27: 345–350.

Hubbert, M.K. (1940) The theory of groundwater motion, *Journal of Geology* 48: 785–944.

Hughes, J.P. and Guttorp, P. (1994) A class of stochastic models for relating synoptic atmospheric patterns to regional hydrologic phenomena, *Water Resources Research* 30: 1535–1546.

Hutchinson, F.C. (1991) Use of satellite data for famine early warning in sub-Saharan Africa, *Journal of Climate* 1: 240–255.

Isaaks, E.H. and Srivastava, R.M. (1989) *An Introduction to Applied Geostatistics*, New York: Oxford University Press.

James, L.D. (1995) NSF research in hydrological sciences, *Journal of Hydrology* 172: 3–14.

Jarvis, P.G. and McNaughton, K.G. (1986) Stomatal control of transpiration: scaling up from the leaf to the region, *Advances in Ecological Research* 15: 1–49.

Jones, P.D. and Conway, D. (1997) Precipitation in the British Isles: an analysis of area-average data updated to 1995, *International Journal of Climatology* 17: 427–438.

Justice, C., Holben, B.N. and Gwynne, M.D. (1986) Monitoring East African vegetation using AVHRR data, *International Journal of Remote Sensing* 7: 1453–1474.

Katz, R.W. (1996) Use of conditional stochastic models to generate climate change scenarios, *Climatic Change* 32: 237–255.

Kim, J.W., Chang, J.T., Baker, N.L., Wilks, D.S. and Gates, W.L. (1984) The statistical problem of climate inversion: determination of the relationship between local and large-scale climate, *Monthly Weather Review* 112: 2069–2077.

Kleidon, A. and Heimann, M. (submitted) A method of determining rooting depth from a terrestrial biosphere model and its impacts on the global water and carbon cycle, *Global Change Biology*.

Klemes, V. (1988) A hydrological perspective, *Journal of Hydrology* 100: 3–28.

Kuczera, G. (1985) *Prediction of Water Yield Reductions following Bushfire in Ash-mixed Species Eucalypt Forest*, Melbourne and Metropolitan Board of Works Report no. MMBW-W-0014, Melbourne, Australia.

Lauenroth, W.K. (1996) Application of patch models to examine regional sensitivity to climate change, *Climatic Change* 34: 155–160.

Mackey, B.G. (1993) Predicting the potential distribution of rain-forest structural characteristics, *Journal of Vegetation Science* 4: 43–54.

Mandelbrot, B.B. (1983) *The Fractal Geometry of Nature*, New York: W.H. Freeman.

Matyasovszky, I., Bogardi, I. and Duckstein, L. (1994) Comparison of two GCMs to downscale local precipitation and temperature, *Water Resources Research* 30: 3437–3448.

McDonnell, R.A. (1996) Including the spatial dimension: using geographical information systems in hydrology, *Progress in Physical Geography* 20: 159–177.

Mearns, L.O., Giorgi, F., McDaniel, L. and Shields, C. (1995) Analysis of daily variability of precipitation in a nested regional climate model: comparison with observations and doubled CO_2 results, *Global and Planetary Change* 10: 55–78.

Mearns, L.O., Rosenzweig, C. and Goldberg, R. (1996) The effect of changes in daily and interannual climatic variability on ceres-wheat: a sensitivity study, *Climatic Change* 32: 257–292.

Melillo, J.M., Prentice, I.C., Farquhar, G.D., Schulze, E.-D. and Sala, O.E. (1996) Terrestrial biotic responses to environmental change and feedbacks to climate, in Houghton, J.T., Meira Filho, L.G., Callander, B.A., Harris, N., Kattenberg, A. and Maskell, K. (eds) *Climate Change 1995: The Science of Climate Change* (Contribution of Working Group I to the Second Assessment Report of the Intergovernmental Panel on Climate Change), Cambridge: Cambridge University Press, pp. 445–482.

Moldan, B. and Cerny, J. (1994) *Biogeochemistry of Small Catchments. A Tool for Environmental Research*, SCOPE 51, New York: Wiley, 419 pp.

Moore, I.D., Grayson, R.B. and Ladson, A.R. (1991) Digital terrain modelling: a review of hydrological, geomorphological and biological applications, *Hydrological Processes* 5: 3–30.

Mosier, A.R., Parton, W.J., Valentine, D.W., Ojima, D.S., Schimel, D.S. and Heinemeyer, O. (1997) CH_4 and N_2O fluxes in the Colorado shortgrass steppe 2. Long-term impact of land use change, *Global Biogeochemical Cycles* 11: 29–42.

Oki, T., Musiake, K., Matsuyama, H. and Masuda, K. (1995) Global atmospheric water balance and runoff from large river basins, in Kalma, J.D. and Sivapalan, M. (eds) *Scale Issues in Hydrological Modelling*, Chichester: Wiley, pp. 411–434.

Osborn, T.J. and Hulme, M. (1997) Development of a relationship between station and grid-box rainday frequencies for climate model validation, *International Journal of Climatology*, in press.

Palmer, A.R. and Van Staden, J.M. (1992) Predicting the distribution of plant communities using annual rainfall and elevation: an example from southern Africa, *Journal of Vegetation Science* 3: 261–266.

Parker, D.E., Legg, T.P. and Folland, C.K. (1992) A new daily central England temperature series, 1772–1991, *International Journal of Climatology* 12: 317–342.

Parton, W.J., Mosier, A.R., Ojima, D.S., Valentine, D.W., Schimel, D.S., Weier, K. and Kulmala, A.E. (1996) Generalized model for N_2 and N_2O production from nitrification and denitrification, *Global Biogeochemical Cycles* 10: 401–412.

Perica, S. and Foufoula-Georgiou, E. (1996) Linkage of scaling and thermodynamic parameters of rainfall: results from midlatitude mesoscale convective systems, *Journal of Geophysical Research* 101 (D3): 7431–7448.

Pielke, R.A., Dalu, G.A., Snook, J.S., Lee, T.J. and Kittel, T.G.E. (1991) Nonlinear influence of mesoscale land use on weather and climate, *Journal of Climate* 4: 1053–1069.

Post, W.M. and Pastor, J. (1996) LINKAGES – an individual-based forest ecosystem model, *Climatic Change* 34: 253–261.

Quinn, P.F., Beven, K.J. and Culf, A. (1995) The introduction of macroscale hydrological complexity into land surface–atmosphere transfer function models and the effect on planetary boundary layer development, *Journal of Hydrology* 166: 421–445.

Reed, D.W. and Stewart, E.J. (1994) Inter-site and inter-duration dependence of rainfall extremes, in Barnett, V. and Turkman, K.F. (eds) *Statistics for the Environment 2*, Chichester: Wiley, pp. 125–146.

Richardson, C.W. (1981) Stochastic simulation of daily precipitation, temperature and solar radiation, *Water Resources Research* 17: 182–190.

Rinaldo, A., Rodriguez-Iturbe, I., Rigon, R., Bras, R.L., Ijjasz-Vasquez, E. and Marani, A. (1992) Minimum energy and fractal structures of drainage networks, *Water Resources Research* 28: 2183–2195.

Rind, D., Rosenzweig, C. and Goldberg, R. (1992) Modelling the hydrological cycle in assessments of climate change, *Nature* 358: 119–122.

Rodda, J.C. (1995) Guessing or assessing the world's water resources, *Journal of the Institute of Water and Environmental Managers* 9: 360–368.

Rogers, R. and Hinckley, T.M. (1979) Foliar mass and area related to current sapwood area in oak, *Forest Science* 25: 298.

Root, T.L. and Schneider, S.H. (1995) Ecology and climate: research strategies and implications, *Science* 269: 334–341.

Saunders, I.R. and Bailey, W.G. (1994) Radiation and energy budgets of alpine tundra environments of North America, *Progress in Physical Geography* 18: 517–538.

Schimel, D.S., Braswell, B.H. and Parton, W.J. (in press) Equilibration of the terrestrial water, nitrogen and carbon cycles, *Proceedings of the National Academy of Science USA*.

Schimel, D.S., Kittel, T.G.F. and Parton, W.J. (1991) Terrestrial biogeochemical cycles: global interactions with atmosphere and hydrology, *Tellus* 43AB: 188–203.

Schimel, D.S., Stillwell, M.A. and Woodmansee, R.G. (1985) Biogeochemistry of C, N, and P in a soil catena of the shortgrass steppe, *Ecology* 66: 276–282.

Schultz, G.A. (1988) Remote sensing in hydrology, *Journal of Hydrology* 100: 239–265.

Segal, M., Alpert, P., Stein, U., Mandel, M. and Mitchell, M.J. (1994) Some assessments of $2 \times CO_2$ climatic effects on water balance components of the eastern Mediterranean, *Climatic Change* 27: 351–371.

Sellers, P.J., Bounoua, L., Collatz, G.J., Randall, D.A., Dazlich, D.A., Los, S.O., Berry, J.A., Fung, I., Tucker, C.J., Field, C.B. and Jensen, T.G. (1996) Comparison of radiative and physiological effects of doubled atmospheric CO_2 on climate, *Science* 271: 1402–1406.

Shaw, E.M. (1994) *Hydrology in Practice* (third edition), London: Chapman & Hall.

Shugart, H.H. and Smith, T.M. (1996) A review of forest patch models and their application to global change research, *Climatic Change* 34: 131–153.

Shuttleworth, W.J. (1988) Macrohydrology – the new challenge for process hydrology, *Journal of Hydrology* 100: 31–56.

Sivapalan, M. and Kalma, J.D. (1995) Scale problems in hydrology: contributions of the Robertson Workshop, in Kalma, J.D. and Sivapalan, M. (eds) *Scale Issues in Hydrological Modelling*, Chichester: Wiley, pp. 1–8.

Thompson, S.L. and Pollard, D. (1995) A global climate model (GENESIS) with a land-surface transfer scheme (LSX). Part 2: CO_2 sensitivity, *Journal of Climate* 8: 1104–1121.

Tucker, C.J. and Sellers, P.J. (1986) Satellite remote sensing of primary production, *International Journal of Remote Sensing* 7: 1133–1135.

Tucker, C.J., Dregne, H.W. and Newcomb, W.W. (1991) Expansion and contraction of the Sahara desert from 1980 to 1990, *Science* 253: 299–301.

Turner, M.G. (1990) Spatial and temporal analysis of landscape patterns, *Landscape Ecology* 4: 21–30.

Veen, A.W.L., Klaassen, W., Kruijt, B. and Hutjes, R.W.A. (1996) Forest edges and the soil–vegetation–atmosphere interaction at the landscape scale: the state of affairs, *Progress in Physical Geography* 20: 292–310.

VEMAP members (1995) Vegetation/ecosystem modeling and analysis project: comparing biogeography and biogeochemistry models in a continental-scale study of terrestrial ecosystem responses to climate change and CO_2 doubling, *Global Biogeochemical Cycles* 9: 407–437.

von Storch, H., Zorita, E. and Cubasch, U. (1993) Downscaling of global climate change estimates to regional scales: an application to Iberian rainfall in winter-time, *Journal of Climate* 6: 1161–1171.

Wigley, T.M.L., Jones, P.D., Briffa, K.R. and Smith, G. (1990) Obtaining sub-grid scale information from coarse resolution general circulation model output, *Journal of Geophysical Research* 95: 1943–1953.

Wilby, R.L. (1994) Stochastic weather type simulation for regional climate change impact assessment, *Water Resources Research* 30: 3395–3403.

Wilby, R.L. (1996) The fractal nature of river channel networks: applications to flood forecasting in the River Soar, Leicestershire, *East Midlands Geographer* 19: 59–72.

Wilby, R.L. (1997a) The changing roles of hydrology, in Wilby, R.L. (ed.) *Contemporary Hydrology*, Chichester: Wiley, pp. 1–24.

Wilby, R.L. (1997b) Beyond the river catchment, in Wilby, R.L. (ed.) *Contemporary Hydrology*, Chichester: Wiley, pp. 317–346.

Wilby, R.L. and Gell, P.A. (1994) The impact of forest harvesting on water yield: modelling hydrological changes detected by pollen analysis, *Hydrological Sciences Journal* 39: 471–486.

Wilby, R.L. and Wigley, T.M.L. (1997) Downscaling General Circulation Model output: a review of methods and limitations, *Progress in Physical Geography*, in press.

Wilby, R.L., Barnsley, N. and O'Hare, G. (1995) Rainfall variability associated with Lamb weather types: the case for incorporating weather fronts, *International Journal of Climatology* 15: 1241–1252.

Wilks, D.S. (1992) Adapting stochastic weather generation algorithms for climate change studies, *Climate Change* 22: 67–84.

Wood, E.F. (1995) Scaling behaviour of hydrological fluxes and variables: empirical studies using a hydrological model and remote sensing data, in Kalma, J.D. and Sivapalan, M. (eds) *Scale Issues in Hydrological Modelling*, Chichester: Wiley, pp. 89–104.

Wood, E.F., Sivapalan, M. and Beven, K.J. (1990) Similarity and scale in catchment storm response, *Reviews of Geophysics and Space Science* 28: 1–18.

Wood, E.F., Sivapalan, M., Beven, K.J. and Band, L. (1988) Effects of spatial variability and scale with implications to hydrologic modelling, *Journal of Hydrology* 102: 29–47.

Woods, R.A., Sivapalan, M. and Duncan, M. (1995) Investigating the representative elementary area concept: an approach based on field data, in Kalma, J.D. and

Sivapalan, M. (eds) *Scale Issues in Hydrological Modelling*, Chichester: Wiley, pp. 49–70.

Xinmei, H., Lyons, T.J. and Smith, R.C.G. (1995) Meteorological impact of replacing native perennial vegetation with annual agricultural species, in Kalma, J.D. and Sivapalan, M. (eds) *Scale Issues in Hydrological Modelling*, Chichester: Wiley, pp. 401–410.

Yarnal, B. (1993) *Synoptic Climatology in Environmental Analysis: A Primer*, London: Belhaven Press.

Zeng, X. and Pielke, R.A. (1995) Landscape-induced atmospheric flow and its parameterization in large-scale numerical models, *Journal of Climate* 8: 1156–1177.

4

PLANTS AND WATER IN DRYLANDS

John Wainwright, Mark Mulligan and John Thornes

INTRODUCTION

It could be argued that the understanding of eco-hydrological processes is critical in drylands for two reasons. First, in most cases the aridity of the drylands means that the supply of water is the dominant control on the growth and maintenance of plants. Second, most drylands are characterised by extreme variability in water availability. Plants must adapt to use this variable source. In addition, large amounts of soil erosion during storm events can remove the uppermost, relatively fertile and moisture-retaining parts of the soil profile and thus enhance the importance of moisture availability as the dominant control on dryland vegetation. In this chapter, we will explore these interrelationships and feedback mechanisms at a number of spatial and temporal scales. In doing so, we have divided the material into three major sections.

First, we consider a number of basic eco-hydrological processes in drylands. At the largest spatial scale, we consider soil–vegetation–atmosphere transfers in drylands. This is followed by a more detailed consideration of the relationships between plants and soil water in relation to climatic controls and the processes that control the movement of water into and out of the soil. The structures and patterns of dryland vegetation that develop from these interactions are then discussed. Finally, we look at the plant controls on runoff and sediment production, which reinforce many of the patterns and processes considered earlier.

Second, the means of modelling eco-hydrology in drylands are considered. The use of models to develop and test our understanding of processes and their interactions is considered by us a fundamental part of our methodology. We consider general progress in modelling such interactions, and note some important factors that must be accounted for in developing models of dryland eco-hydrology.

Third, we consider variations in dryland eco-hydrology through space and time. The origins of several of the present-day drylands are considered and their evolution through time assessed in order to understand their dynamics over much longer timescales, particularly in relation to Quaternary climatic change. Through the Holocene, human activity becomes a further – if not the most – significant control on dryland eco-hydrology, and thus the general lines of such activity are considered. The understanding of processes through simulation models, as discussed in the previous sections, is then used to present an analysis of present and future change in drylands.

This chapter has a strong bias towards the Mediterranean and the American southwest. We have not attempted to cover every dryland area in detail – for which a whole volume would be required – but have drawn from examples with which we are most familiar, and made linkages to or comparisons with other drylands where possible.

ECO-HYDROLOGICAL PROCESSES IN DRYLANDS

Dryland climates

Drylands are characterised by low rainfall, ranging from zero or near zero for the South American coastal hot deserts and the Libyan deserts (Evenari, 1985a) to more than 700 mm per annum for large parts of the European Mediterranean. In the driest lands, fog and dew precipitation can contribute significantly to the water balance at night. This is particularly true for coastal areas of the Peruvian–Chilean deserts and the Namib, which may be termed humid deserts in spite of their near zero rainfall. Clearly, precipitation is not the only factor that distinguishes drylands from other climates; rather it is the combination of low rainfall and high potential evaporation (due to high solar radiation receipt, clear skies and low atmospheric humidity) that characterises drylands. Rainfall is controlled primarily on a latitudinal basis, with the most extensive hot deserts occurring in low-latitude zones of atmospheric subsidence. In these areas the global circulation prevents large-scale convective activity, since the dominant atmospheric movement is downwards. Where convection is low, cloud formation will be rare and so radiation receipt will be high. Since little water is available for evaporation, most of this radiation will yield sensible rather than latent heat and thus temperatures will be high, while rainfall is very low. Where these conditions exist over very large areas, the possibility of advected rainfall is also low, giving rise to hot, dry deserts such as those of the Sahara–Sahel, the Middle East and central Australia.

Many smaller areas with desert-like conditions exist and occur in more complex climatic situations such as the European Mediterranean, the climate of which is primarily controlled by the 'struggle' between the Azores high-pressure cell and westerly low-pressure systems. Desert-like conditions can also occur as a result of orographic effects (southwest South America) and land–sea interactions controlled by ocean currents (southwest USA, southwest South Africa and southwest and south Australia). In these zones, drylands may be characterised by infrequent or highly variable rainfall rather than very low overall rainfall. Because they span such a large range of conditions and because spatial and temporal variability is high, it is difficult to list the detailed characteristics of dryland climates in general (Table 4.1).

Large-scale eco-hydrological processes in drylands

As a result of their low but variable annual rainfall and high potential evaporation, drylands are characterised by a low biological potential (UNEP, 1977; 1991; UNEP/ISRIC, 1992). Plant cover is often less than 30 percent, and above-ground biomass often approaches 1 g m^{-2} per mm annual rainfall. For most drylands, this means above-ground biomass ranges between 300 and 700 g m^{-2}, though this value is subject to strong seasonal, inter-annual and long-term variability in response to climatic forcing.

The land surface represents the boundary between the relatively uniform and highly dynamic fluid that is the atmosphere and the much more

Table 4.1 General characteristics of dryland climates.

Rainfall – near zero in the hot deserts to >700 mm per annum for large parts of the northern Mediterranean.

Rainfall seasonality – a range of responses from erratic to highly seasonal.

Rainfall variability – sometimes erratic but often showing clear inter-annual and decadal variability (hydrological drought).

Precipitation – in the driest lands dew and fog can contribute 40–300 mm per annum to the precipitation, especially in coastal zones (Evenari, 1985a).

Temperature – Hot deserts have absolute maxima between 45 and 47 °C, with mean annual values between 20 and 25 °C (but 16–19 °C for coastal deserts). Hot deserts are characterised by a low seasonal variability of temperature. Mediterranean climates show much more seasonality of temperature, with the hottest months often in excess of 30 °C and sometimes in excess of 40 °C, while the coldest months can be well below zero.

Radiation receipt – the combination of mid- to low-latitude locations and lack of cloud cover leads to a high radiation receipt.

Potential evaporation – high as a result of the high radiation receipt, although low water availability and low surface cover means that actual evapotranspiration will be low.

---

static and spatially complex lithosphere. It is at the land surface that the differences between these two systems are resolved. The characteristics of the land surface determine the energy (Dirmeyer and Shukla, 1994), moisture (Walker and Rowntree, 1987; Shukla and Mintz, 1982) and momentum exchange (Sud et al., 1985; Smith et al., 1994) between the two systems. These exchanges can affect, and in turn are affected by, climatic and eco-hydrological processes. Since the surface is the interface between climatic forcing and eco-hydrological response, the characteristics of the surface also determine the nature and intensity of surface and subsurface hydrological processes as well as the dynamics of the vegetation cover.

The low cover of vegetation means that soil–vegetation–atmosphere transfers (SVATs) in drylands are particularly complex (Shuttleworth and Wallace, 1985; Noilhan et al., 1997; Pelgrum and Bastiaanssen, 1996). SVATs in drylands are complicated in two main ways. First, the low and discontinuous vegetation cover presents the atmosphere with a very complex surface when compared with a continuous forest cover or grassland character-istic of more humid environments. Second, the boundary layer behaviour of dryland soils and dryland vegetation is quite distinct, and our understanding of hydrological and micrometeorological feedback between the two remains elementary (see Sene, 1996).

A number of crude general circulation model (GCM) sensitivity experiments in various ecosystems have suggested that the characteristics of the surface have an important influence on the medium- and long-term dynamics of the climate system (Wilson et al., 1987; Meehl and Washington, 1988; Koster and Suarez, 1994; Marengo et al., 1994; Eltahir and Bras, 1994; Polcher and Laval, 1994; Stockdale et al., 1994). Most of these experiments have been carried out within the context of large-scale field experiments in the same ecosystem (Bolle et al., 1993; Goutorbe et al., 1994). Most of the early experiments of this type dealt with simple surfaces such as continuous forest cover (Blyth et al., 1994; Shuttleworth et al., 1991) and grassland (Betts et al., 1992; Hall and Sellers, 1995). Some of the later experiments considered more complex surfaces such as North African crops and tiger-bush (HAPEX-Sahel: Goutorbe et al., 1994), Mediterranean shrublands and rain-fed and irrigated agriculture (Echival Field Experiment in a Desertification-Threatened Area – EFEDA I and II: Bolle et al., 1993). The early experiments looked at the impact of large-scale changes in surface cover associated with deforestation and afforestation on boundary layer structure and SVAT fluxes of water, energy and momentum. They were mainly *deforestation* experiments (O'Brien, 1996; Sud et al., 1996; Lean and Rowntree, 1997). It was clear, numerically and empirically, that these gross changes had a significant impact on local, regional and, potentially, global climates.

The later experiments looked at much more subtle surface changes associated with the progressive reduction of the biological potential of the land in response to climatic aridification and agricultural intensification.

81

These *desertification* experiments (Gouturbe *et al.*, 1994; Bolle *et al.*, 1993; Xue and Shukla, 1996) produced evidence to support the hypothesised feedback between aridification and degradation of the land surface. Reduced vegetation cover, higher albedo, lower latent heat loss and higher sensible heat loss result from aridification and lead to less turbulence and cloud formation, in turn causing further aridification of the climate (Figure 4.1). The most significant characteristics of the vegetation cover in this process are plant albedo (vegetation type, cover, leaf area index), the degree of atmospheric coupling of the stomata (i.e. the degree to which stomatal opening is controlled by atmospheric variables as opposed to soil moisture availability; see Chapter 2) and the seasonal phenology of the vegetation cover. Anthropic land-use change in drylands has accelerated in recent decades and can thus have a direct impact on climate because of this climate − land-surface properties − climate feedback loop (Henderson Sellers, 1994; Dale, 1997).

Plant−soil−water relationships in drylands

The water budget of any part of a dryland hill slope can be expressed as the difference between water inputs at a point (precipitation, run-on and seepage) and water outputs at the same point (evapotranspiration, runoff and drainage). The most important components of the budget for most drylands are precipitation and evapotranspiration, and so the balance between the two, expressed as the ratio of precipitation to potential evapotranspiration, is a good measure of aridity.

Rainfall

Precipitation in drylands is low in volume, irregular and low in frequency, and often high in intensity. The irregularity of precipitation can be as important in controlling plant−soil−water relationships as the long-term mean total precipitation, since plants require a constant supply of water for the maintenance of biological activity, and most species have no way of storing water surpluses. The irregularity of precipitation in drylands can be expressed as the coefficient of variation of annual total precipitation, and this is commonly of the order of 30 percent for semi-arid Spain (Mulligan, 1996a), 25 percent for Mediterranean France, and around 35 percent in New Mexico (Wainwright, in press, a). In more arid areas, these values are commonly exceeded. Rumney (1968) gives values in excess of 40 percent for much of the Sudano−Sahelian belt, parts of Mongolia, the southwest coast of Africa and parts of Peru, Bolivia and Argentina.

The characteristics of precipitation in drylands determine to a large extent the hydrological fate of the same precipitation. These characteristics include (1) the timing of rainfall both seasonally and with respect to previous events; (2) the intensity of rainfall relative to the rate of surface infiltration of water;

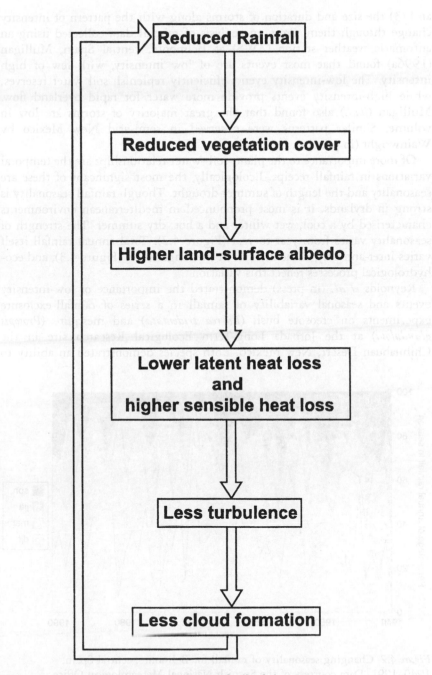

Figure 4.1 The aridification–desertification feedback loop.

and (3) the size and duration of storms along with the pattern of intensity change through them. From an analysis of rainfall data collected using an automatic weather station (AWS) at Belmonte, central Spain, Mulligan (1996a) found that most events are of low intensity, with few of high intensity. The low intensity events efficiently replenish soil water reserves, while high-intensity events provide more water for rapid overland flow. Mulligan (*ibid.*) also found that the great majority of storms are low in volume. Similar patterns were observed in semi-arid New Mexico by Wainwright (in press, a).

Of more importance to the plant–soil–water relationships are the temporal variations in rainfall receipt. Ecologically, the most significant of these are seasonality and the length of summer drought. Though rainfall seasonality is strong in drylands, it is most pronounced in mediterranean environments characterised by a cool, wet winter and a hot, dry summer. The strength of seasonality varies from year to year (Figure 4.2). Total annual rainfall itself varies inter-annually, decadally and in the longer term (Figure 4.3), and eco-hydrological processes reflect this variation.

Reynolds *et al.* (in press) demonstrated the importance of low-intensity events and seasonal variability of rainfall in a series of rainfall-exclosure experiments on creosote bush (*Larrea tridentata*) and mesquite (*Prosopis glandulosa*) at the Jornada Long-Term Ecological Research site in the Chihuahuan Desert, New Mexico. Both species demonstrated an ability to

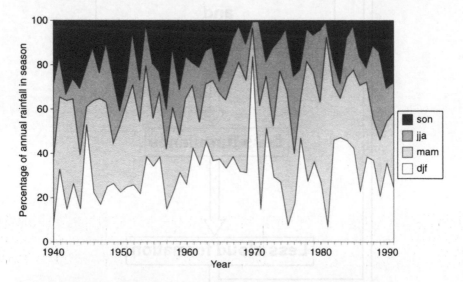

Figure 4.2 Changing seasonality of rainfall for Belmonte, central Spain, 1940–1991. Data courtesy of the Spanish National Meteorological Office. son = September, October, November; djf = December, January, February; mam = March, April, May; jja = June, July, August.

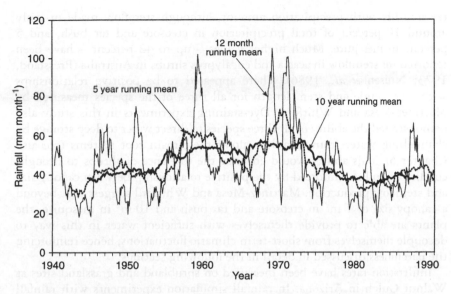

Figure 4.3 Rainfall variability for Belmonte, central Spain (1940–1991). Data courtesy of the Spanish National Meteorological Office.

compensate for summer drought by using moisture from (low-intensity) winter rainfall and for winter drought by exploiting summer rainfall. This may be one means by which shrub species are able to out-compete annuals in this environment.

Throughfall, infiltration and runoff

The relationship between rainfall, soil moisture and runoff production can be considered in relation to a series of experiments carried out on semi-arid shrubland and grassland in Arizona and New Mexico. Martinez-Mesa and Whitford (1996) looked at processes of throughfall, stemflow and root channelisation of water in three shrub communities at Jornada. Throughfall in creosote bush averaged 56 percent of rainfall throughout the year with a range of values from 22 to 83 percent. The average for tar bush (*Flourensia cernua*) was 53 percent in summer and 58 percent in winter, with a range from 9 to 73 percent. Mesquite had less seasonal variability, with 64 percent of rainfall occurring as throughfall in summer and 62 percent in winter, with values ranging between 36 and 89 percent. These values seem comparable with measurements on shrub vegetation from elsewhere in the world. Martinez-Mesa and Whitford were also able to demonstrate that the percentage throughfall in all of these species increases rapidly for low rainfall rates and reaches the mean value asymptotically beyond event totals of between 2.5 and 3 mm. Their stemflow measurements show a similar relationship with precipitation amount, although stemflow made up only around 10 percent of total precipitation in creosote and tar bush, and 5 percent in mesquite. Much higher values – up to 42 percent – have been reported for stemflow in acacia and eucalyptus shrubs in Australia (Pressland, 1973; Nulsen *et al.*, 1986). There appeared to be positive relationships with stem angle and stem length for all three of the species measured by Martinez-Mesa and Whitford. Dye-staining experiments in this study also demonstrated the ability of all three species to direct water to deep storage by channelling water from stemflow along their main root systems (see also Chapter 6). This ability would enhance the tolerance of shrubs to drought conditions and is reinforced by the positive relationship between canopy size and stemflow production. Martinez-Mesa and Whitford suggest that beyond a canopy size of 1 m² in creosote and tar bush and 10 m² in mesquite, the plants are able to provide themselves with sufficient water in this way to decouple themselves from short-term climatic fluctuations, hence reinforcing the patchiness discussed in detail in the following sections.

Infiltration rates have been measured on shrubland and grassland sites at Walnut Gulch in Arizona. In rainfall simulation experiments with rainfall rates of 72 mm h⁻¹, Parsons *et al.* (1996b) found mean final infiltration rates of 41.1 mm h⁻¹ (σ = 18.4 mm h⁻¹) on shrubland dominated by creosote in association with species such as *Acacia constricta*, *Dasylirion wheeleri*, *Rhus microphylla* and *Yucca baccata*. Grassland dominated by various species of grama (*Bouteloua* spp.) had mean final infiltration rates of 39.7 mm h⁻¹ (σ = 14.7 mm h⁻¹). Although these distributions are not significantly different, these gross rates hide important variability at the slope scale. While the grass cover is relatively continuous, the shrub cover is highly patchy (see below), so that the shrubland slopes are made up of areas that are either dominated by shrubs or bare and covered by desert pavement. Surface runoff in the pavement areas in the inter-shrub zones is significantly higher, with an inverse relationship between pavement cover (which is itself an inverse function of vegetation cover on the shrubland) and final infiltration rate. Because of the interconnectedness of these pavement areas, the shrubland slopes tend to produce more runoff than the grassland, despite the similarities in gross final infiltration rates (Abrahams *et al.*, 1995). More recent experiments at the Jornada Long-Term Ecological Research site demonstrate comparable results for the grasslands and shrublands there (Abrahams *et al.*, forthcoming; Schlesinger *et al.*, in press). Significantly lower infiltration rates than those quoted have been measured in other semi-arid contexts, particularly on sparsely vegetated marl badlands (Scoging and Thornes, 1979; Scoging, 1982; but see also Wainwright, 1996c).

Fires, which can be prevalent in mediterranean and other dryland eco-systems, can also affect the infiltration properties of dryland soils (Naveh, 1967, 1975). Using rainfall simulation experiments, Imeson *et al.* (1992) were able to demonstrate a reduction in infiltration rates of between 50 and

70 percent for burnt compared with unburnt Mediterranean forests in Catalonia. Apart from the change of soil structure caused by the fire, these authors point to the importance of leaf litter at the surface and in the upper part of the mineral soil. Although hydrophobic, the leaf litter is usually rather loose, allowing free passage of water. Upon burning, the litter layer is lost and hydrophobic resins from the burnt litter coat mineral particles in the surface soil. Once this burnt mineral soil is exposed, infiltration is dramatically reduced. Therefore, fire affects the ability of water to infiltrate into the soil and become available for plants to regrow.

Evapotranspiration

Evapotranspiration is driven primarily by the evaporative power of the atmosphere, which is a function of the net radiation available for latent heat flux, although temperature is often used as a surrogate variable (see also Chapter 2). The partitioning of net radiation between sensible and latent heat flux is markedly dependent upon the amount of available water and the water potential thus produced in the soil–plant–atmosphere continuum (SPAC). The ratio of actual to potential evapotranspiration (E_{ta}/E_{tp}) can provide a valuable index of the stomatal control exerted during daylight hours on the transpiration from the leaf canopy. Specht (1972) calculated annual mean E_{ta}/E_{tp} for evergreen woody vegetation in Australia and South Africa and found values ranging from 0.18 to 0.56, where values >0.2 indicate sufficient soil moisture to allow rapid growth of understorey vegetation.

Because E_{ta}/E_{tp} is a measure of stomatal activity it is also a measure of the rate of potential photosynthesis. Where E_{ta}/E_{tp} is high, stomata are open and photosynthetic activity and growth rates are high. Stomatal closure reduces CO_2 uptake and thus productivity. The reduced productivity reduces the leaf mass that can be maintained and thus causes a reduction in the leaf area index, causing reductions in evapotranspiration. In this way, carbon gain and water loss are locked in a tight feedback control loop. To model the dynamics of hydrology or plant growth in drylands usefully, one must take account of this feedback and develop coupled hydrology–growth models.

At the drier end of the climatic spectrum, plants are much more strongly coupled with their hydrological environment, since it is even more marginal in terms of their survival. On a continuum of aridity from semi-deserts (300 to 400 mm per year) through true deserts (less than 120 mm per year) to extreme deserts (less than 70 mm per year), one can observe distinct changes in the plant community. Tree-like species do not occur where rainfall is below 400 mm, while the 120 mm isohyet is the boundary between steppe (semi-desert) and desert vegetation (Shmida, 1985). The 70 mm isohyet corresponds roughly to the boundary between 'diffuse' and 'contracted' vegetation. Diffuse vegetation is that with a low cover (<10 percent) of sparse dwarf shrubs (chamaephytes) and, in the rainy season, a flush of annuals. Contracted vegetation is restricted to wadis, where it can take advantage of accumulations of water at depth. At the wetter end of the spectrum, steppe (semi-desert) merges into grassland and shrubland, savannah and woodland. In all of these communities, vegetation cover and plant size are positively correlated with rainfall (Shmida, 1972; Beatley, 1974, 1976). Species diversity is also related to rainfall, with maximum species richness occurring in the semi-desert zone between desert and forest ecosystems. However, edaphic variability and ecological stability are also important – ecologically stable areas are thought to have fewer species, though this is highly contentious (Tilman, 1994; Whitford, 1997).

Evenari (1985b) describes the adaptations of plants to the desert environment. He distinguishes the poikilohydrous plants (lichens and algae) and the homoiohydrous plants. Poikilohydrous plants have a number of physiological adaptations to extreme aridity, which are outlined in Table 4.2. Lichens can be a very significant component of dryland vegetation. However, most desert plants fall into the broad class of homoiohydrous but can be further classified as arido-passive perennials, arido-passive pluviotherophytes, arido-active perennials and biseasonal annuals. These types possess a number of adaptations to aridity, which are fully described elsewhere (*ibid.*).

Plant structures and patterns

Spatial structure

Perhaps the most striking characteristic of dryland vegetation is its low cover. As one moves along a continuum of aridity from $P/E_{tp} > 1$ down to $P/E_{tp} < 0.3$, the potential vegetation cover breaks up from total cover through a series of broken canopies to a situation in which vegetation cover is very patchy. The patches are characterised by a high density of stems in a small area and a consequent build-up of sediment – typically derived from rain splash or aeolian activity – and organic material beneath the vegetation (Parsons *et al.*, 1992; Sánchez and Puigdefábregas, 1994; Puigdefábregas and Sánchez, 1996;

Table 4.2 Features of poikilohydrous plants (after Evenari, 1985b).

1. The capacity to take up water from rain, dew or water vapour in the air (at >70% humidity).
2. The capacity to equilibrate with the hydration level of their environment and to survive extreme and prolonged desiccation without damage.
3. The capacity to enter an anabiotic state (with extreme cold and heat resistance) and become metabolically inactive after desiccation without injury.
4. Rapid switching on and off of metabolic activity according to water availability.
5. Relatively high rate of photosynthesis at low temperatures and low light intensities.

Wainwright *et al.*, 1995; in press). The resulting increase in surface elevation may prevent erosion either by diverting overland flow and/or by obstructing the flow and reducing the local slope, leading to an increase in the amounts of infiltration of water and deposition of sediment generated on bare patches upslope of the plants (Cerdà, 1997). This process enhances the soil moisture storage and the soil moisture storage *capacity* of the patches at the expense of upslope bare areas. This kind of environmental 'manipulation' has been identified in drylands from Israel (Valentine and Nagorcka, 1979) through the Mediterranean to North Africa (Lefever and Lejeune, 1997) and sub-Saharan Africa (MacFadyen, 1950; Grove, 1957; Worral, 1960), as well as in the American southwest (Schlesinger *et al.*, 1990), Patagonia (Rostagno and del Valle, 1988) and Australia (Mabbutt and Fanning, 1987) and can lead to the development of hill-slope-scale, and even regional, patterned structures on sloping terrain. Dunkerley (1997) has suggested on the basis of model simulations and evidence from eastern Australia that such patterns are relatively stable under, and may even be enhanced by, drought and/or grazing pressure. Davis and Burrows (1994) have also proposed that the repeated occurrence of fires can be an important factor in the development of vegetation patchiness in drylands.

Patch development

While a number of plant-scale advantages of clumping can be recognised, the ecological mechanism for the phenomenon remains poorly understood, although recent theoretical (van der Maarel, 1988) and modelling (Sánchez and Puigdefábregas, 1994; Puigdefábregas and Sánchez, 1996; Jeltsch *et al.*, 1996) studies have helped in understanding the processes involved. In addition to increasing the available soil moisture and soil moisture capacity through overland flow trapping, clumping can have a number of other eco-hydrological impacts, some of which are presented in Table 4.3.

Plant controls on runoff and sediment movement

General considerations

There is a complex and often non-linear interaction between vegetation, hill-slope hydrological processes and erosion. Despite this, many forest engineers in drylands still promote the long-held, and somewhat simplistic, idea that degradation and soil erosion can be reduced and hydrological processes adjusted by encouraging the growth of forests.

The core of the forest protection approach is based on more than 100 years of paired catchment experiments in the USA and the UK and the repeated empirical observation that sediment yields are less from forested catchments and well-vegetated plots. A critical review of the confused signals coming

Table 4.3 The eco-hydrological consequences of vegetation clumping.

1. Reduction of soil evaporation loss through shading of the root zone soil.
2. Enhancement of root zone moisture supply through the concentration of stem-flow.
3. Increase of the infiltration of direct rainfall and reduction of overland flow loss.
4. Provision of a low air temperature and high-humidity environment around the majority of leaves, which may reduce transpirational loss.
5. Concentration of nutrient reserves in the surface soil from (drought deciduous) leaf loss, thereby reducing nutrient losses.
6. Reduction of grazing pressure by excluding access to all but the edge of the patch.
7. Provision of a shaded environment for small animals, enhancing their contribution to the nutrient budget of the plant and also increasing the potential soil moisture storage by their burrowing.
8. Reduction of the expenditure on below-ground biomass by bringing resources to the roots rather than growing roots to the resources.

from this research reveals that the main causes of ambiguity have been the failure to obtain truly comparable paired catchments, the difficulty engendered by comparing different serial rainfall inputs to the catchments and the differences in both treatment and responses of the catchments. In this latter instance, Obando (1996) has shown in a modelling study of sites in southern Spain that changes in the location of matorral removal and re-establishment can have quite significant impacts on the pattern of water and sediment yield (Figure 4.4). The arguments for forest protection are well known. It is suggested that the forest canopy reduces the drop size and the energy of falling rain, both of which are correlated with erosion amounts and rates. However, Brandt (1986, 1989) was able to show, for both UK deciduous forests and Amazonian forest, that drop sizes and, in some cases, total kinetic energy at the surface can be substantially increased under forest. These results suggest that it is the protection afforded to the soil by the root matter and leaf litter layer rather than the canopy itself that reduces soil erosion. With lower-level canopies, however, Brandt demonstrated that the throughfall energy was significantly reduced when compared with rainfall energy, although certain events did still produce enhanced throughfall energies. Although throughfall in semi-arid shrubs in New Mexico was measured as being between 9 and 89 percent of the rainfall (Martinez-Mesa and Whitford, 1996: see above), the resulting energy produced was not measured. However, the results of Parsons *et al.* (1992) on differential splash rates between and under shrubs with a similar size and structure in Arizona suggest that there is a significant energy reduction in this setting. Thus the effectiveness of vegetation protection against erosion is a function of canopy size, height and spatial pattern, although more work is required to define the exact relationships.

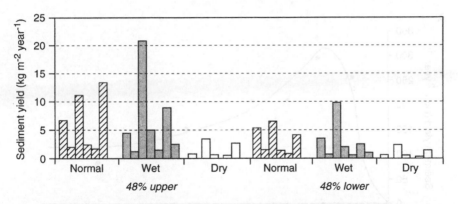

Figure 4.4 The effect of different spatial distributions of vegetation cover on predicted erosion rate in the study of Obando (1996). Both cases have a 48 percent vegetation cover of matorral, concentrated in either the upslope or downslope area.

It has been observed that lower plant life forms (algae and lichens) can have a dramatic effect on erosion rates by increasing the surface stability (Alexander and Calvo, 1990). As discussed above, such forms are an important component of many dryland ecosystems; thus this effect may be widespread. Surface stability can be quickly lost by grazing and negligent soil management. However, the debate about the impact of grazing has to be treated with caution, not least because animals are highly selective in their choice of food, and also because most of the enhanced runoff can be attributed to compaction by hooves. In addition, grazing may induce better fertilisation of soil, so that in the Sahel vegetation may be better established closer to water points than further away (Warren and Khogali, 1992). Moreover, there are some species for which grazing *improves* the cover characteristics with respect to erosion by increasing leaf density and encouraging outward growth of bushy species such as *Quercus coccifera* (Barbero *et al.*, 1990).

Regional scale

At regional scales it is argued that, with zero rainfall, there must be zero runoff and therefore no erosion, but that as the rainfall and runoff increase, the sediment yield must also increase to a peak at *c*. 300–500 mm of effective rainfall (i.e. semi-arid environments) (Langbein and Schumm, 1958; Figure 4.5A). Further rainfall increases the vegetation cover, causing a steep slump in sediment yield (Figure 4.5A). Douglas (1967) argued that total denudation might well follow the same curve, but that in tropical climates the high solution rates and throughput of soil water would produce much higher denudation, despite the increased vegetation cover. Walling and Webb (1983) also qualified the Langbein and Schumm (1958) curve on the basis of a much greater and more carefully filtered sample size (Figure 4.6).

91

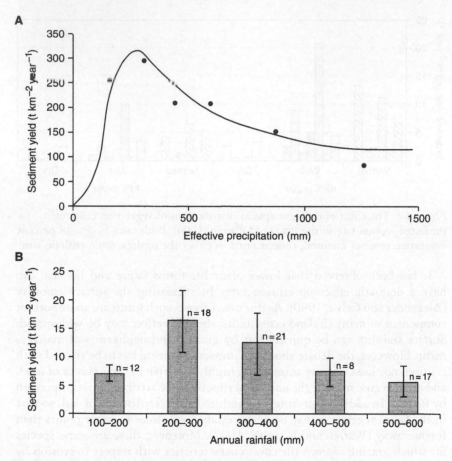

Figure 4.5 Relationships between precipitation and sediment yield. A: Relationship between effective annual precipitation and sediment yield proposed by Langbein and Schumm (1958). B: Relationship between annual precipitation and sediment yield as measured on 76 Mediterranean sites under shrubland vegetation by Kosmas *et al.* (1997).

A relationship similar to that of Langbein and Schumm has recently been confirmed empirically for Mediterranean-type climates in both the Old and the New Worlds (Inbar, 1992) and for the Mediterranean by the experimental sites in the MEDALUS project (Kosmas *et al.*, 1997) (Figure 4.5B). Inbar (1992) demonstrated that the peak of the curve occurred at around 300 mm of annual precipitation for catchments in coastal Israel and Spain, with corresponding erosion rates of 310 t km^{-2} year^{-1}. The subsequent decrease in sediment yield with increasing annual precipitation is of the order of 50 t km^{-2} year^{-1} for each 100 mm of additional precipitation. Peak rates in the mediterranean climates of California and Chile are of the order of 500 to

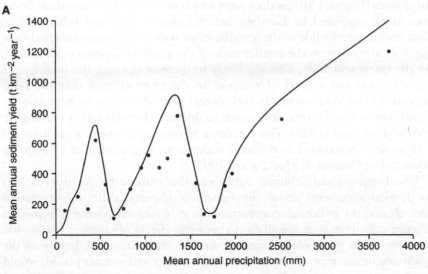

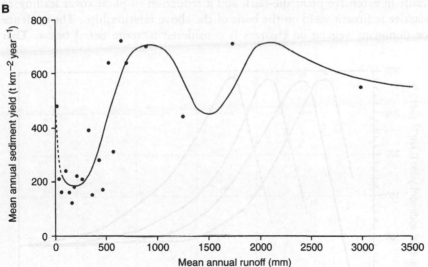

Figure 4.6 Generalised relationships of Walling and Webb (1983) showing the mean annual sediment yield as a function of A: mean annual precipitation and B: mean annual runoff.

600 t km^{-2} year^{-1}, with the peak occurring at 400 mm of annual precipitation, and a subsequent more rapid decrease of around 100 t km^{-2} year^{-1} for each 100 mm of additional precipitation. These differences were attributed to relief and human impact, demonstrating the need to look at local conditions, as discussed below. The results of Kosmas *et al.* (1997) for Mediterranean

shrublands (Figure 4.5B) produce rates which are an order of magnitude lower than those suggested by Langbein and Schumm (1958) and Inbar (1992). This may in part be due to the less disturbed nature of the experimental sites but may also relate to the smaller scale of the study of Kosmas *et al.* (1997). At the catchment scale, rates are likely to increase through the inclusion of gully erosion and erosion of material in the channel zone that has been deposited in previous events. In this context, the role of riparian vegetation is significant in the dynamics of erosion in dryland channels and is thoroughly reviewed by Graf (1988). The problems of moving between measurements and models constructed at different scales is an important area of ongoing research (see Chapter 3; Zhang *et al.*, 1997).

The Langbein and Schumm curve was also utilised by Schumm (1965) to develop arguments about the impact of climatic shifts on vegetation and thence on sediment concentrations in rivers and terrace formation (Figure 4.7). However, speculation about the effects of climatic change has been rife in the palaeoenvironmental reconstruction literature, largely on the questionable assumptions that water stress in drier and warmer periods would result in extensive plant die-back and a reduction in plant cover leading to massive sediment yield on the basis of the above relationships. The evidence for dominant vegetation changes is considered in more detail below. That

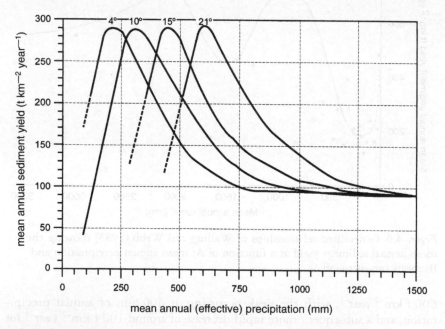

Figure 4.7 Relationships proposed by Schumm (1965) relating annual precipitation to sediment yield as a function of different temperatures. Temperatures given in °C.

reduction of vegetation produces increases in sediment yield, however, is well documented from a number of carefully carried out investigations of human clearance. Examples of clearances include those in mid-Holocene Europe (Starkel, 1983; 1987), when the Aulehm deposits were created in river valleys, and those in the Mediterranean (e.g. Laval *et al.*, 1991; van Andel *et al.*, 1986; see also below). One of the problems in providing unequivocal statements about past changes comes from the timescales of observation. As noted above, many major erosional events in a Mediterranean context occur as a result of plant die-back following drought on an annual to decadal cycle, or as a result of intense storms following a period of seasonal drought (Kirkby and Neale, 1987). However, the resolution available throughout most of the Holocene in terms of the palaeobotanical record for the region is generally of the order of centuries.

A further significant problem here is that, while the spatial character of growth under increased precipitation and its impact on vegetation can be successfully modelled (Thornes 1990), the spatial patterning of die-off is rather inadequately modelled because of the difficulty that death is often partial in plants (twigs, leaves), so that the spatial occurrence of bare patches is spotty rather than systematic in development. Thus in desertification studies, the concept of the 'rolling forward' of the Sahel as plant death occurs under stress or human management cannot be justified. More sophisticated modelling inevitably leads to more complex conclusions about the effects of climatic change on degradation. For example, Woodward (1994) has modelled the impacts of future scenarios of Mediterranean climate change on vegetation cover in the Iberian Peninsula using physically based ecophysio-logical models of photosynthesis, respiration, growth, allocation and death of plants on the basis of their functional type. The work produced a plant growth model that was coupled to hydrological and geomorphological models at the hill-slope scale (Kirkby *et al.*, 1996). The results indicate that the hypothesised doubling of CO_2 and 15 percent reduction in winter rainfall would act differentially on bushes and grasses, and that the most marginal climatic situations would experience the greatest effects (i.e. those areas having an average dry season of seven months without rain). The results indicate losses in biomass of up to 50 percent, which would be expected to produce a dramatic change in sediment yield.

Slope or plot scale

At smaller scales, variability in the controls of erosion such as vegetation leads to much greater spatial and temporal variability in rates. For this reason, local-scale estimates of erosion must be based on a thorough understanding of the processes. The erosion process can be divided into three interacting elements. Erosion by splash occurs due to the impact of raindrops on the surface, which can entrain particles in droplets that rebound from the surface.

Once overland flow is generated, it can transport sediment either in a rather diffuse manner in the undissected inter-rill zone or in a more concentrated way once incision leads to the formation of rills and, ultimately, gullies.

Erosion by splash

The splash mechanism is effective for soil particles less than about 12 mm in maximum dimension (Wainwright, 1992; Wainwright et al., 1995). Transport distances are typically less than a metre and decline exponentially in all directions from the source although with a net downslope movement (Savat and Poesen, 1981). Parsons et al. (1994) demonstrated using rainfall simulations that splash rates increased and then decreased during a first rainfall event on a semi-arid grassland plot at Walnut Gulch, with peak rates of 9.0×10^{-5} g m^{-2} s^{-1}. During a second event carried out three days later, the rates continued to decline. This pattern was interpreted as relating initially to the increased wetting of the soils, facilitating their movement by reducing cohesion, followed by a subsequent decrease in available soil for transport. Thus, the importance of splash might be expected to decrease, ceteris paribus, during the period of a rainy season at a particular site. As noted below, there is an important interaction between splash and sparse vegetation canopies in drylands. Soil and litter can commonly be splashed underneath a shrub canopy, but the splash process is considerably reduced beneath the canopy because of the relatively small proportion of throughfall (see above), and thus little material is splashed back out into the inter-shrub areas. Through time, this build-up of soil and litter contributes significantly to the formation of mounds (Parsons et al., 1992; Wainwright et al., 1995; in press). The rapidity with which this occurs is partly a function of splash rates in the inter-shrub areas, which can be an order of magnitude higher than those in the grassland areas (7.2×10^{-3} g m^{-2} s^{-1}; Parsons et al., 1991).

Erosion by inter-rill overland flow

Overland flow in the inter-rill zone is able to move larger particles – Abrahams et al. (1988) observed movement of particles as large as 53 mm on steep slopes – typically for much greater distances than the splash process. Tracer experiments on the Walnut Gulch grassland suggested that although a large proportion of material is redeposited within 25 cm of its origin, travel distances of almost 3 m were recorded in a single event, again with an exponential decline in the amount of material transported with distance (Parsons et al., 1993). These experiments suggest that the transport is intermittent and that it would take a number of events for soil particles to travel the length of the slope and enter the channel system. Abrahams et al. (1991) demonstrated that inter-rill erosion is determined by sediment detachment by raindrops, as the flows themselves have insufficient energy to detach soils. The capacity to transport such detached material is then defined by the flow depth, which is itself highly variable due to the significant surface irregularities on typical dryland slopes (Abrahams and Parsons, 1991; Parsons *et al.*, 1994; 1996a). Raindrop detachment is itself a negative exponential function of flow depth (Torri *et al.*, 1981), leading to the development of spatial patterns of erosion. On both the grassland and shrubland slopes at Walnut Gulch, the erosion first increases from the slope divide as flow transport capacity increases, and then decreases as the raindrop detachment decreases, because of the general downslope increase of flow depth (Abrahams *et al.*, 1991; Parsons *et al.*, 1996a). In both cases, there are feedbacks between the vegetation, flow dynamics and erosion. On the shrubland, the high surface runoff from the inter-shrub areas increases the erosion in these areas, which accentuates the relief of the shrub-topped mounds and equally reduces the likelihood of recolonisation of the inter-shrub areas, because of the removal of water and nutrient resources (Parsons *et al.*, 1996b; Schlesinger *et al.*, in press). The grassland is characterised more by a negative feedback wherein the more diffuse distribution of grass clumps causes overland flow to be slower and less concentrated. Sediment is thus deposited behind the clumps, often forming a series of microtopographic steps, which in turn enhance the diffusion of the overland flow (Parsons *et al.*, 1996a). Thus, despite the lower vegetation cover on the grassland (33 percent compared with 44 percent on the shrubland), erosion rates are lower, with peak rates of 0.23 g m^{-2} s^{-1} compared with 0.53 g m^{-2} s^{-1} on the shrubland. It can be seen that inter-rill overland flow is several orders of magnitude more effective at transporting sediment than splash (see also Wainwright, 1996b and c for Mediterranean examples).

Erosion by concentrated overland flow

Erosion in concentrated overland flow occurs when the flow gains sufficient energy to detach sediment itself, thus causing incision into the soil surface, forming rills and ultimately gullies. In this case, the detachment of soil by raindrops is no longer a limiting factor. Rilling occurs on the shrubland at Walnut Gulch and is apparently controlled by the occurrence of greater extremes of flow depth in the downslope direction (Parsons *et al.*, 1996a). This is probably a further consequence of the feedback described above, where flow is increasingly constrained into the inter-shrub areas as it travels downslope. Unpublished work at the Jornada site suggests that shrub spacing is a dominant control on this process (Charlton, personal communication).

The example discussed in detail above suggests that simple models that show a negative exponential relationship between vegetation cover and erosion – as, for example, demonstrated by Elwell and Stocking (1976) for simple crops – cannot be said to hold in all cases in drylands. Rogers and Schumm (1991) have also shown experimentally that the negative exponential relationship breaks down at vegetation covers of 15 percent or less. Morgan (1996) has discussed differences in the form of the relationship and its parameters in relation to canopy structure as part of the European Soil-Erosion Model (EUROSEM) project. It would appear that further work on this problem is required that takes account of the canopy structure of dryland plants and their spatial distribution on a slope. Given that many drylands have plant covers of less than 30 percent, as noted above, this point is of critical importance.

As with runoff production, the consequences of fire are an important consideration in relation to erosion, principally because of the increase in runoff described above, the decrease in protection of the soil surface by the vegetation canopy, and changes in the soil structure. Vega and Diaz Fiernos (1987) showed a difference of erosion rates, with only 1.5 t ha^{-1} year^{-1} produced on a moderately burnt *Pinus pinaster* woodland, compared with 22 t ha^{-1} year^{-1} on an intensely burnt area in northwest Spain. On *Quercus ilex* plots in Catalonia, Soler and Sala (1992) measured erosion rates of 0.017 t ha^{-1} year^{-1} on unburnt and 0.162 t ha^{-1} year^{-1} on burnt plots, compared with 0.024 t ha^{-1} year^{-1} where the plots had simply been cleared. Fire can also lead to a loss of nutrients from the system by a number of mechanisms (Christensen, 1994), which, combined with soil loss, has important consequences for vegetation regrowth (Soler and Sala, 1992; Kutiel and Inbar, 1993).

MODELLING ECO-HYDROLOGICAL PROCESSES IN DRYLANDS

Models for understanding dryland eco-hydrology

Numerical models (see Chapter 9) are increasingly being used as an aid to field experimentation and theoretical analysis for understanding the eco-hydrological dynamics of drylands. Models are applied at scales from the GCM grid cell, such as the land-surface sensitivity experiments discussed above and in Chapter 3, through to the individual plant scale. Models are applied for time periods from a few hours for process studies and event simulation through to hundreds of years for climate change experiments. A number of hydrological (for example, Braud *et al.*, 1995, Kirkby and Neale, 1987, Kirkby, 1990), ecological (for example, MARIOLA, Uso-Domenech *et al.*, 1995) and coupled eco-hydrological models (for example, PATTERN, Mulligan, 1996a and b; MEDALUS, Kirkby *et al.*, 1996; FOREST-BGC, Kremer *et al.*, 1996; Thornes, 1990) have been developed for dryland environments.

The particular characteristics of dryland climates, surfaces and plant communities require a particular set of modelling strategies appropriate to these conditions.

Model timesteps

Drylands are characterised by long periods of hydrological inactivity with short, intensive events in which a great deal of change occurs. In modelling terms, this requires the specification of a high-resolution timestep (perhaps a minute or less) in order to represent accurately rapid rate processes (particularly Hortonian overland flow) during an extreme event but also low-resolution timesteps for integration of fluxes outside these periods of activity. In this way, a variable timestep rather than a fixed one is most appropriate to dryland eco-hydrological modelling. The timestep should respond to the rate of the most rapid acting flux or process so that it is inversely proportional to this rate. A variable timestep is computationally efficient and hydrologically precise.

Where ecological processes are coupled with hydrological ones, and particularly where processes of the energy budget are included, an additional, concurrent, timestep is required. The timestep should reflect the slower biological processes and is commonly of the order of hours or days. Hourly timesteps have the advantage of increased precision but require hourly data, which are often not available. Daily timesteps, on the other hand, are easier to parameterise but can lead to significant error because they do not take into account the significant differences between daytime processes and night-time processes. These can be particularly important in drylands, where dew inputs can be high during the night and where water stress changes significantly between day and night. The analysis in Table 4.4 is the result of an application of the Penman–Monteith equation (Monteith, 1973) to the same data set, which is integrated first on an hourly basis, second on a daily basis, and third on a day–night basis with one timestep from sunrise to sunset and the next timestep from sunset to the following sunrise. Clearly, there are significant differences in the calculated evapotranspiration between these situations. If we take the hourly results as those that most closely represent reality, it is clear that a day–night timestep significantly improves the accuracy of estimation of evapotranspiration when compared to a daily timestep. The accuracy of using the day–night timestep also varies through the year, as indicated by the June figures compared with those for January.

Specific characteristics of drylands

A number of specific physical characteristics of drylands must also be accounted for within eco-hydrological models. First, drylands tend to have an open canopy with a complex pattern of vegetated and bare patches. This makes the use of standard evapotranspiration equations developed for closed canopies difficult, and more appropriate models are those developed for sparse crops (Shuttleworth and Wallace, 1985) and patchy vegetation (Wallace and Holwill, 1997; Wood, 1997).

99

Table 4.4 The effect of data resolution on evapotranspiration estimates. These model simulations of transpiration were generated from the same data set, with the model integrated at the different timesteps outlined.

Transpiration (mm/30 days)

	January	June
Hourly	9.6	35.0
Daily	20.9	54.9
Day/night	9.7	40.7

Second, drylands tend to exhibit extreme spatial variability in soil, surface and vegetation properties (Wood, 1997), and the spatial scale of modelling must take this into account – either the scale must be small enough to encompass the dominant patch-scale variability or significant resources need to be put into up-scaling the soil and vegetation properties (Zhang *et al.*, 1997) (see Chapter 3). Using a distributed model of runoff for the Walnut Gulch sites discussed above, Parsons *et al.* (1997) demonstrated that the use of cell sizes of greater than about 0.5 × 0.5 m seriously degrades the results produced. Such models also seem highly sensitive to the quality of information about the spatially distributed parameters. Again, this suggests that high-quality model results demand a large investment in terms of field parameterisation. In the case of Parsons *et al.*, empirical relationships were used based on measurable surface properties, but clearly this leads to the development of site-specific models that are difficult to generalise. Thus effective model parameterisation is another area that clearly needs attention.

Third, drylands are characterised by poor, often skeletal soils of low organic matter content. In Mediterranean drylands, these soils tend to be very high in stone content, and this significantly changes both the surface and subsurface hydrological fluxes. Within-profile stones have a relatively low porosity, *c.* 30 percent, compared with inter-stone soil porosities (*c.* 40–60 percent), and much of this 30 percent is unavailable for water storage. As a result, the presence of stones within a soil profile effectively reduces the porosity of the profile and thus soil moisture storage capacity (Childs and Flint, 1990, Mulligan, 1996a). Water and other resources tend to concentrate in the inter-stone spaces. In addition to this, the surface area of pore space at the soil–bedrock interface is reduced by the presence of stones, and this can reduce the areal rate of recharge to the bedrock. Finally, stones within the soil profile may support the soil structure and in these cases, or in cases where there is significant diurnal and seasonal temperature fluctuation, the soil may become removed from the stone surfaces, providing a preferential flow path through the stone 'skeleton'. These effects have been experimentally tested (Childs and Flint, 1990) and modelled (Mulligan, 1996a).

The stone content at the soil surface is a function of the profile stone content and the rate of removal of surface sediment. Surface stones in drylands have been widely studied in the field and laboratory (Poesen, 1986; de Lima and de Lima, 1990; Poesen et al., 1990; Poesen and Bunte, 1995) and their hydrological effects modelled (Mulligan, 1996a). They have a significant effect on hydrological fluxes, on the surface energy budget and on the development of the vegetation cover. The two basic hydrological impacts of stones concern infiltration and soil evaporation. The presence of surface stones effectively seals a proportion of the surface from infiltration since some of the soil pore space at the surface is covered by an impermeable stone. This means that during a rainfall event, water falling on the stone-covered part runs off as stone flow onto the bare part of the soil. This increases the rate at which water is delivered to the inter-stone soil surface by an amount that is proportional to the stone–soil area ratio. A 50 percent cover of stones (common for large areas of the semi-natural Mediterranean shrublands and very common in the desert-pavement-covered inter-shrub areas of the American southwest) effectively doubles this rate of water delivery during rainfall. This can lead to high overland flow generation, leading to enhanced erosion of the inter-stone spaces and removal of soil to expose further stone surfaces. The importance of this mechanism depends upon whether the stones are simply resting on the soil surface or whether they are partly buried within the soil. Stones resting on the surface will tend to leak rainfall into the underlying pore space and will reduce surface sealing and raindrop impact, thereby potentially increasing infiltration and reducing overland flow, whereas embedded stones tend to have the opposite effect (for a more detailed review, see Baird, 1997).

Both types of surface stone cover prevent energy from reaching the soil pore space and as such act as a mulch preventing soil evaporation. This can lead to the development of ecologically important surface wet patches under high stone covers. Where the vegetation cover is low, a high stone cover can lead to extremely low soil evaporation even from wet soils, but where vegetation cover is high this mechanism is ineffectual since an alternative route for evaporation is present through the plants.

DRYLAND ECO-HYDROLOGY IN SPACE AND TIME

Evolution of plant cover

The first development of plants currently associated with the Mediterranean region occurred during the Pliocene. In the eastern Mediterranean, there is evidence from Israel that the Pliocene vegetation was characterised by a cool, temperate woodland (Horowitz and Horowitz, 1985). In the later part of the Pliocene, tree cover decreased, being made up of *Pinus*, *Abies orientalis* and later *Quercus* with a dominance of steppe species in sites in the south of the country from about 3.5 Ma, suggesting the onset of environmental gradients similar to those of the present day (Levin and Horowitz, 1987). Before 3.2 Ma, the northwestern part of the Mediterranean had been characterised by tropical vegetation species such as *Taxodium*, *Myrica*, *Symplocos* and *Nyssa* in the lowlands, suggesting high moisture conditions, and lower moisture conditions in the uplands, as demonstrated by species such as *Engelhardia*, *Carya*, *Rhioptelea*, *Hammamelis* and *Embolanthera* (Suc, 1984). Similar associations are found from before the Messinian salinity crisis, suggesting a great stability in this pattern (Suc and Bessais, 1990). Mediterranean evergreen and pine species are present in the early Pliocene in southern Italy between 4.5 and 3.2 Ma, and few tropical species were found (Bertoldi *et al.*, 1989). However, the period from 2.6 to 2.2 Ma is marked by cyclic changes in vegetation groupings, with deciduous forest, tropical humid forest, coniferous forest and steppe vegetation all present (Combourieu-Nebout, 1993). Rare Taxodiaceae and subtropical species are still present in the Villafranchian type area of northern Italy between 2.9 and 2.2 Ma (Carraro *et al.*, 1996). The latter part of this sequence and those of southern France from 2.3 Ma show an increase of steppe species (Suc and Zagwijn, 1983). From 2.1 to 1.7 Ma, the first fully Mediterranean-type vegetation zonation is found in southern France. This is made up of *Olea* and *Ceratonia* with *Pistacia*, *Phillyrea* and *Myrtus* at the lowest altitudes; *Phillyrea* and *Quercus ilex* with *Olea*, *Carpinus orientalis* and *Rhamnus* at middle altitudes; deciduous oak and *Carpinus* with *Carya*, *Ulmus* and *Zelkova* at higher altitudes; and finally *Cedrus*, *Pinus*, *Abies*, *Picea* and *Tsuga* making up a mountain forest (Suc, 1984). In southern Italy and Sicily, the same period sees the final disappearance of the subtropical Taxodiaceae (Baggioni *et al.*, 1981; Bertoldi *et al.*, 1989), although they are still present in Calabria (Combourieu-Nebout, 1993) and as late as 1.3 Ma in central Italy (Rosa Attolini *et al.*, 1988), albeit most probably as relict populations in local refugia.

Thus, the major lines of the modern vegetation show a complex development over a period of two million years. The development seems to occur earlier in the east and south than in the west and north. On the general trend are superimposed a number of cycles of vegetation change. The period of time over which these changes occurred is marked by important climate changes within the region that also reflect broader global changes. Oxygen isotope data for the western Mediterranean suggest that temperatures were up to 3 °C higher than present until 3.2 Ma (Thunell, 1979) or 3.1 Ma (Vergnaud-Grazzini *et al.*, 1990), when there was a relatively sudden drop in winter sea surface temperatures of 1.2 °C (Keigwin and Thunell, 1979). This change represents the global onset of cooling and sea level fall as seen in the Pacific (Shackleton and Opdyke, 1977) and Atlantic (Clemens and Tiedemann, 1997). A second phase of cooling occurred around 2.5 Ma, with oscillations until around 2.1 or 2.0 Ma, after which there was a general downward trend in temperatures until the onset of the larger-scale glacial–interglacial cycles around 900 ka (Vergnaud-Grazzini *et al.*, 1990; Thunell, 1979; Thunell and Williams, 1983). Estimated sea surface temperatures for the period from 2 to 1 Ma range from values similar to those of the present day to 2 °C higher than at present (Thunell, 1979). Data for the reconstruction of the moisture regime independently of the vegetation are much more tenuous. Vergnaud-Grazzini *et al.* (1990) have suggested that the later Pliocene was marked in the Tyrrhenian basin by the gradual onset of a seasonally contrasted rainfall pattern. This was superimposed on conditions that were already relatively dry. Such dry conditions may be reflected by the vegetation associations in Italy and Israel recorded during this period. If this is the case, it may be that the development of the Mediterranean associations are related to the development of highly seasonal water regimes at the same time as the relatively slight temperature changes experienced. The adaptation of Mediterranean species such as *Quercus ilex* to seasonal drought has been noted by a number of authors (e.g. Quézel, 1978; Daget, 1980; Terradas and Savé, 1992; Acherar and Rambal, 1992). Furthermore, the ability of species such as *Quercus ilex* to control microclimates and therefore accompanying understorey plants (Barbero *et al.*, 1992) may have played an important part in the selection of the species that now make up Mediterranean associations.

The origins of Saharan vegetation are discussed by Maley (1980). The development of the Sahara itself took place from the early Pliocene, around 5 Ma. Later Pliocene samples from the Hoggar and from Egypt suggest that most of the elements of the modern vegetation were in place by this time, although the period saw the presence of a larger number of tree species than are now found even in the Hoggar uplands. Le Houérou (1997) suggests that during the various wetter periods of the Quaternary, the Sahara was an important refuge for Mediterranean-type plants. In East Africa, the initial development of dryland vegetation took place in the later Miocene as a consequence of uplift, which caused the development of large rain shadow areas (van Zinderen Bakker, 1983). The development of savannah seems to have taken place here between 2 and 1.8 Ma.

A relatively continuous vegetation sequence covering the last 3 Ma in northern California is recorded at Tulelake (Adam *et al.*, 1989). Vegetation cover is dominated here by forest with *Pinus*, Taxodiaceae, Cupressaceae, Taxaceae (TCT) and oak, again representing relatively wet conditions. Around 2.9 and especially 2.5 Ma, there are fluctuations in the cover with the expansion of species such as *Artemisia* and desert shrubs. From *c.* 1.65 Ma, there are a series of extensive dry phases with *Artemisia*, desert shrubs and grasses, alternating with more wooded spells. In this case, the development of the more arid vegetation is similar to that found in the Mediterranean, with a relatively long transitional phase with oscillating groupings, although there are obvious differences, notably that the TCT species remain within the region.

Thus, in many places, present dryland associations developed first in the later Pliocene. In some locations, the new associations seem to have stabilised by the time of the onset of Pleistocene cooling, whereas in others, notably the Mediterranean, such associations continued to develop against this backdrop of climatic change. Fluctuations in response to such longer-term changes are considered in the following section for the Mediterranean region.

Climate variations and plant response: a Mediterranean case study

As noted above, there were oscillations superimposed on the Pliocene climatic trend of general cooling over a period of at least one million years. High-resolution pollen data from the Semaforo section in southern Italy suggest the presence of five cycles of vegetation change in response to such oscillations between 2.6 and 2.2 Ma (Combourieu-Nebout, 1993). As well as the tropical species, which expand during the cooler (and therefore wetter) phases, there are also periods of deciduous and coniferous forest, as well as steppe vegetation indicative of much drier conditions. Similarly, four oscillations are found in northern Italy from 2.9 to 2.2 Ma (Carraro et al., 1996), and four cycles of wet to dry conditions in Israel between 2.6 and 2.0 Ma (Horowitz, 1989). Similar oscillations are noted in the early Pleistocene (Suc, 1984) with changes between steppe and Mediterranean forest vegetation.

The core from Tenaghi Philippon in northern Greece preserves a sequence that extends over the last 975 ka, providing a record of the response to the large climatic changes of the later Pleistocene (Wijmstra, 1969; Wijmstra and Smit, 1976; van der Wiel and Wijmstra, 1987a, b). The changes in the sequence have been studied by Mommersteeg et al. (1995) by grouping the pollen into distinct associations that occur repeatedly, and by looking at the representation of tree species and those species indicative of marsh and open water vegetation. Four associations are present through the entire sequence: (1) mixed deciduous forest, (2) mixed coniferous–deciduous and mixed evergreen–deciduous forest, (3) open coniferous–deciduous forest, and (4) steppe species. By analogy with modern groupings, these probably represent (1) wet coastal or montane Mediterranean, (2) seasonally dry Mediterranean, (3) cooler and relatively wet, and (4) cold dry conditions, respectively. The modern climate at the site is of the seasonally dry Mediterranean type (2). Comparison with marine oxygen isotope series suggests some important changes through time. Before c. 530 ka, the warmer climatic periods tend to be associated with wetter vegetation types (1) or (3). The association of seasonally dry conditions (2) in the warmer periods takes place only after this time, which may reflect climate changes due to the shift in the Earth's orbital activity after c. 600 ka. Even so, it is still relatively infrequent, making up only one-fifth of the total period. Thus, although we have seen that the species making up this association in the Mediterranean were in place since the Pliocene, the conditions that have led to their full development are relatively recent. Dry conditions have more commonly been associated with open, steppe vegetation (often with *Artemisia*, as currently found in the higher Mediterranean mountains and the arid zones of North Africa: Le Houérou, 1986). From the total sequence of 975 ka, this association is present for 400 ka. Mommersteeg *et al.* (1995) demonstrated the presence of orbital frequencies relating to precession and obliquity effects, which they interpreted as being due to movements in the inter-tropical convergence zone (ITCZ) causing rainfall changes within the Mediterranean basin. Horowitz (1989) also postulated the existence of monsoon-related changes in the Pleistocene oscillations between Mediterranean woodland and steppe in the long sequences from Israel.

Further interesting observations about the longer-term response to climate change can be obtained by analysis of the changes between the different vegetation associations outlined by Mommersteeg *et al.* A Markov chain analysis of these transitions shows a number of patterns of change through time (Figure 4.8). The mixed Mediterranean forest indicative of seasonally dry conditions (2) occurs only after the presence of the deciduous forest (1) or open forest (3). This suggests that wetter conditions are characteristic of transitional phases, and that such conditions are necessary to produce a substrate upon which the mixed forest can be competitive once drier conditions are re-established. The steppe vegetation can follow the mixed Mediterranean forest, showing that the reverse process is not necessarily true, although it is slightly more common to have a transitional open forest phase before steppe conditions fully develop.

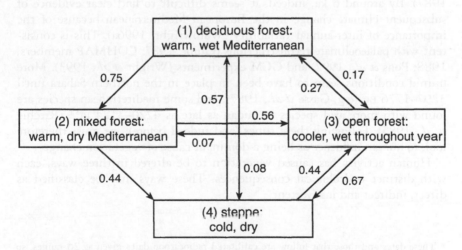

Figure 4.8 Markov transition probabilities of transitions of vegetation types observed in the Tenaghi Philippon core. Vegetation types and their timing defined after Mommersteeg *et al.* (1995).

Another long sequence from Ioannina in northern Greece shows similar stages of development at a higher resolution (Tzedakis, 1994). Over the last 423 ka, there have been nine forested phases, which show a similar pattern of development. *Quercus* and *Ulmus/Zelkova* develop early after warming, followed by *Carpinus betulus, C. orientalis* and *Ostrya carpinifolia* and finally *Abies*, which is often accompanied by *Fagus*. Species such as *Olea*, *Buxus* and *Juniperus* tend to be common throughout, albeit in small quantities. The intervening cold periods are marked by an initial steppe with Graminae, Chenopodiaceae and *Artemisia*, a desert-steppe with dominant *Artemisia* and a variety of shrub species, and a steppe–forest transition with rapidly increasing *Pinus* and *Quercus*. Similar patterns have also been observed in long sequences from Italy (Follieri *et al.*, 1988; Watts *et al.*, 1996) and Spain (Pons and Reille, 1988).

The data from Tenaghi Philippon and Ioannina suggest that the Mediterranean mixed forest vegetation as we see it today is a relatively recent phenomenon. Although a general succession is demonstrated at Ioannina, there may be an inherent instability due to the lateness and infrequency of occurrence of such associations. Their vulnerability to disturbance is discussed below.

Anthropic effects on dryland plants

Following the development of the deciduous then mixed Mediterranean forest in the late Pleistocene and early Holocene, it becomes increasingly difficult to derive climate-induced change in the region (e.g. Vernet and Thiébault, 1987). By around 6 ka, indeed, it seems difficult to find clear evidence of subsequent climate change in the European Mediterranean because of the importance of inter-annual variability (Wainwright, 1996a). This is consistent with palaeoclimate reconstructions (Guiot, 1987; COHMAP members, 1988; Pons *et al.*, 1992) and GCM experiments (Wright *et al.*, 1993). More humid conditions seem to have been in place in the northern Sahara until 3297–1776 BC (e.g. Gasse *et al.*, 1987), and some mediterranean species are found with more arid species in Sudan as late as 4220–3124 BC* (Ritchie *et al.*, 1985). Following these times (and indeed before), we must therefore look to human influence as being a dominant cause of vegetation change.

Human activity has caused vegetation to be altered in three ways, each with distinct hydrological consequences. These ways may be classified as direct, indirect and inadvertent.

---

* These dates and those that follow are calibrated radiocarbon dates given as 2σ ranges, so that the true date has a 95 percent probability of falling within the given interval. Calibration is according to the program CALIB 3.0 of Stuiver and Reimer (1993), using the data of Stuiver and Pearson (1993) and Pearson and Stuiver (1993).

Direct modification of species assemblage

Direct modifications occur either by clearance and deforestation or by deliberate planting or introduction. Widespread clearance is suggested from pollen evidence, for example in Attica, by 4072–3541 BC (Turner and Greig, 1975) and possibly even slightly earlier in northern Greece (Willis, 1994). The impact of forest clearance is described above. Deforested landscapes in the northwestern Mediterranean led to a number of erosional phases in the second millennium BC (Wainwright, in press, b). Similar phases are recorded in the Argolid in southern Greece in the third millennium BC (Pope and van Andel, 1984) and as early as the mid-sixth millennium BC in Macedonia (van Andel *et al.*, 1990). Such episodes are indicative of greater runoff and possibly also greater recharge into alluvial aquifers, which may also have affected the development of riparian vegetation. The runoff and erosion may have been accentuated by the use of fire as a means of vegetation clearance, which has a long history in the region (Naveh, 1975).

Deliberate planting includes not only cereals, with consequences for runoff production, especially during the period after harvest when the ground surface is relatively bare and usually crusted, but also vines, olives and other tree crops. Peaks in vine and olive cultivation occur in the mid-second millennium BC, during the first millennium BC, and from the early Mediaeval period (Wainwright, 1997, after Andrieu *et al.*, 1995). Vine cultivation can lead to increased runoff generation due to the presence of bare areas between the crops (Wainwright, 1996b), although this may have been reduced in the past with more traditional *cultura promiscua* techniques, in which tree crops were interplanted with wheat or legumes, providing a more extensive ground cover. On the other hand, olive cultivation, particularly with the development of terraces from the third millennium BC in the Levant and slightly later in southern Greece (van Andel and Runnels, 1987), may have been important in reversing the processes of runoff and sediment production, and the development of irrigation would have had important effects on the hill-slope hydrology. Indeed, Kosmas *et al.* (1997) have presented figures for average soil losses from experimental plots of 0.008 t ha^{-1} year^{-1} under olive cultivation, compared with 1.43 t ha^{-1} year^{-1} under vines, 0.18 t ha^{-1} year^{-1} under wheat and 0.07 t ha^{-1} year^{-1} for shrubland. Other tree crops such as chestnut and walnut, which spread in the later first millennium BC (Andrieu *et al.*, 1995), may have had a similar effect even without terrace cultivation. Tree crops such as almond, which spread into Spain during the Roman or Moorish period, are unfortunately undetectable by pollen analysis (*ibid.*) but may cause increased runoff if planted in wide rows, as recent experience has shown, although the same proviso as with vine rows may also apply.

Indirect modification of the species present

Indirect modifications may arise as a result of the development of new habitats relating to agro-pastoral land use. Most commonly this takes the form of grasses and shrubs making up matorral associations, or steppe-like groupings, which Sturdy *et al.* (1994) have pointed out bear a resemblance to the associations that characterise the cold, dry conditions that dominate the Mediterranean throughout most of the Pleistocene. This may relate to a feedback where less cover and more runoff, coupled with reduced soil moisture storage as a result of erosion and soil compaction, leads to the development of more arid soils and a progressive inability to support trees. Blumler (1993) has also pointed out the importance of grazing pressure leading to similar outcomes (see also Thornes, 1988, for a conceptual approach).

Inadvertent modifications

Inadvertent modifications may be said to occur where certain species are given a competitive advantage as a result of environmental modifications. The widespread clearances recorded in the pollen sequences of the Mediterranean are often preceded by an expansion of a *Quercus ilex* forest or maquis (see Wainwright and Thornes (forthcoming) for a summary), which is usually synchronous with population expansion and small-scale agriculture, as shown by the archaeological record. Barbero *et al.* (1992) note that disturbance cycles must take place at least every hundred years or so to stop such an association developing into the full mixed oak forest. While the mechanism for this is not fully known, it may be due to the fact that *Quercus ilex* is able to outcompete other species in periods of water stress, which may again be the result of small-scale clearances. The water stress may have been accentuated by the use of fire for clearance, and this would have reinforced the competitive advantage of *Quercus ilex*, which re-sprouts rapidly following burning (Naveh, 1975).

MODELLING THE ECO-HYDROLOGICAL PRESENT AND FUTURE OF DRYLANDS: A MEDITERRANEAN CASE STUDY

Eco-hydrological impact of climatic variability and climate change in drylands

Climatic variability is very pronounced in many drylands. In addition, many drylands are expected to undergo significant climate change in the next few centuries. For the Mediterranean there is confidence in a general warming of the same order as the expected global warming (conservatively, 0.5–1.4 °C) and decreases in precipitation in the southern Mediterranean, although there are great uncertainties around the timing, pattern and magnitude of these changes (Wigley, 1992).

A number of workers have attempted to model the impact of climatic variability and change on semi-arid systems. Uso-Domenech *et al.* (1995) used a multiple regression-based model to look at the impact of climatic deterioration on a Mediterranean community. They concluded that, subjected to increasingly harsh conditions, the first plants to perish are those with low woody biomass (the *r*-strategists). Kremer *et al.* (1996) simulated the response of semi-arid systems in the northwestern USA to a 2 °C temperature increase coupled with a 10 percent increase and decrease of precipitation. While the invasive grass community tolerated the climatic change, the native sagebrush suffered decreased stomatal conductance and thus decreased photosynthesis, leading to a decrease in net primary productivity, which in turn led to increases in soil moisture. However, when natural climate variability was incorporated, the response was quite different, with both communities surviving the climate change by adjusting levels of biomass production.

This result is an important one. Since dryland systems exist in highly variable climatic conditions it is not surprising that they have developed mechanisms to cope with this kind of change. Not least of these mechanisms are the important feedbacks between plant biomass and soil water discussed earlier, but characteristics of the environment, including patchiness and the abundance of stones, are also important.

Figure 4.9 shows the hydrological impact of the climatic variability expressed in the rainfall time series (see Figure 4.3) for a station in Castilla la Mancha, central Spain, using the PATTERN hydrology and vegetation growth model (Mulligan, 1996a, b). Clearly the highly variable rainfall has an impact on infiltration and thus recharge (drainage) and soil moisture, all of which show clear inter-annual and decadal variability. Evapotranspiration tends to be less variable, responding to longer-term changes rather than the extreme events. In addition to hydrological change, climatic variability also forces large changes in plant competition, expressed in Figure 4.10 as the leaf area index for the three functional types present. Clearly there is high temporal variability on a number of scales, and during the wettest period there is a switch in competitive ability between the dwarf shrubs and the shrubs. These large changes in leaf area index and thus plant cover will lead to significant growth and decline of patches, with implications for a number of (non-simulated) ecological and surface processes.

At the plant level, Figure 4.11 shows the structural adjustment of the shrub functional type over the period of simulation. The feedback between water resources and growth leads to changes in structure to minimise water loss and maximise growth. It is clear from these last three figures, and the recent work by Kremer's group (1996), that the response to climate variability in dryland systems is likely to be important.

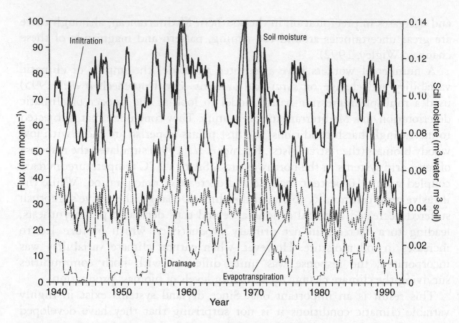

Figure 4.9 Hydrological output of PATTERN model for matorral site with Belmonte rainfall data, 1940–1991.

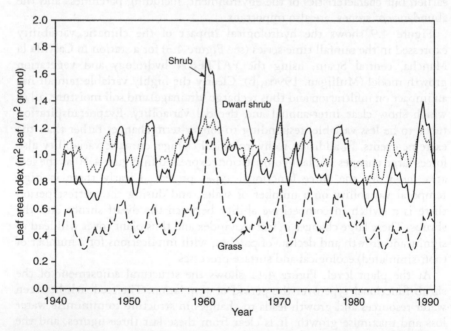

Figure 4.10 Leaf area index for main matorral functional types from PATTERN model using Belmonte rainfall data, 1940–1991.

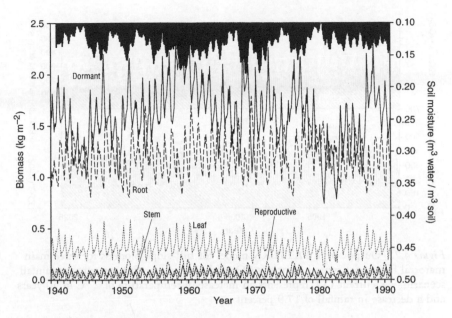

Figure 4.11 Plant structural properties for shrub functional type from PATTERN model using Belmonte rainfall data, 1940–1991.

The inter-annual and decadal variability in drylands can be much higher than the rather slower long-term changes expected from climatic change. This is illustrated in Figure 4.12, which shows the impact of GCM-derived rainfall scenarios for the Mediterranean on the hill-slope biomass of the dominant maquis functional types – grass, dwarf shrub and shrub. There is a high degree of seasonal, inter-annual and long-term variability in the biomass for the three functional types, and these patterns of change are as strong as the pattern expressed by the climatic change from the $1 \times CO_2$ series to the $2 \times CO_2$ series (an increase of temperature of 2 °C and a decrease in rainfall of 17.9 percent in fifty years). This vegetation response will have significant impacts on the spatial structure of the plant community and on hydrological and geomorphological processes. The small magnitude of the change compared with the variability may mean that the impacts of climate change are less rather than more significant than the equivalent response in a much less climatically dynamic and less marginal system such as temperate forests.

Change in magnitude and frequency

In addition to changes in the annual, monthly or daily total or mean of climatic variables, climate change is also expected to lead to changes in the frequency and magnitude of climatic events, particularly droughts and storms. These changes are often not incorporated in analyses and models of

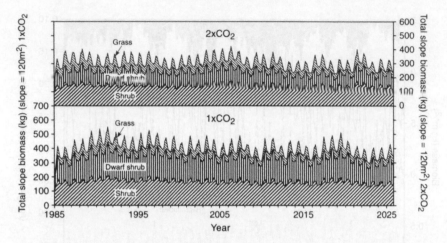

Figure 4.12 Output from PATTERN model for hill-slope biomass for three main matorral functional types for current CO_2 and 2 × current CO_2 greenhouse rainfall scenario. 2 × current CO_2 represents an increase of temperature of 2 °C in 50 years and a decrease in rainfall of 17.9 percent.

eco-hydrological change in drylands. These changes may, however, be important in bringing about structural readjustment in dryland ecosystems. Single rainfall events can bring about large changes in surface properties such as stone cover and meso-topography, while single extreme droughts can dramatically change the structure and composition of the plant community. Since extreme events are an important part of the dryland environment, change in the magnitude and frequency of these may be as important as secular changes in the average conditions. The next step in understanding the future of drylands will be to understand further these potential changes in magnitude and frequency and develop procedures and models that can reflect their impact on the structure and function of dryland systems.

CONCLUSIONS

Water is the critical ecological factor in drylands. The vegetation composition and structure is intimately controlled by it, and at the same time, exerts important feedbacks to it on a number of scales. At the largest spatial scales, the complexity of the links between soil–vegetation–atmosphere transfers are accentuated by the complexity of the dryland surface. Our knowledge of the details of boundary layer behaviour and the patterns of climate variability in drylands is weaker than for simpler grassland and forest systems. The importance of understanding the feedbacks between the climate and land-surface properties in drylands lies not only in the abstract understanding of the processes involved but also in the practical issues of desertification and land degradation and their feedbacks to climate at a regional and global scale. At smaller scales, dryland vegetation can be seen to respond in a complex series of ways to climatic variability on all timescales. Perhaps the most important processes in this case are the ways in which plants and communities respond to minimise the impact of such variability. A number of mechanisms have been identified, such as delayed responses and growth cycles and the development of clumped or banded structures. At a regional scale, the general models for runoff production and sediment transport in drylands developed over the last forty years seem to be applicable. However, at smaller scales, there are many details of the interrelationships between vegetation clumps, runoff production and erosion of sediments and nutrients that are only beginning to be understood.

Modelling can be seen as a useful way of developing our understanding of eco-hydrological processes in drylands, where data acquisition is often particularly difficult. Models with high spatial resolution and temporal resolutions that can cope with both rapidly occurring and longer-term processes are necessary. The PATTERN model developed by Mulligan (1996a, b) is used here to present some example results to illustrate how such models can be used to predict future changes in dryland environments.

On longer timescales, it can be seen that Mediterranean vegetation groupings are a relatively recent development. Although their origins can be seen in the later Pliocene, the types of association we see at the present time do not really stabilise until the last 600 ka. In this sense, they are a much later grouping than associations seen for example in Africa or North America. Over at least the last 6000 years, human activity can be seen to be possibly the most important control on Mediterranean eco-hydrology.

ACKNOWLEDGEMENTS

JW would like to acknowledge the role of Tony Parsons and Athol Abrahams in developing ideas and carrying out the work on overland flow and erosion in the American southwest. The field experiments described from Walnut Gulch were undertaken with the assistance and cooperation of the USDA-ARS Southwest Watershed Research Center, Tucson. Ken Renard, Gary Frasier, David Woolhiser, Leonard Lane and Mark Weltz are thanked for permission to use the facilities at the Tombstone field station, as are Roger Simanton and the technicians at Tombstone for their generous advice and assistance. The experiments described at Jornada were undertaken as part of the Jornada Long-Term Ecological Research project supported by the National Science foundation (Grant DEB 94-11971). None of the above-named, of course, is responsible for any errors and omissions contained herein.

MM would like to acknowledge UK NERC for the studentship GT4/91/AAPS/4 under which most of the modelling work was developed.

The modelling was carried out with data support from the EFEDA I and II projects and the Spanish National Meteorological Office.

Rainfall data mentioned in this chapter have been supplied courtesy of Météo-France, the Spanish National Meteorological Office, Kris Havstad of USDA-ARS Jornada Experimental Range and Reldon Beck of New Mexico State University.

Several of the figures were redrawn by Roma Beaumont of the King's College London Department of Geography drawing office. We would like to thank Andrew Baird and two anonymous referees for their useful comments in improving the chapter.

REFERENCES

Abrahams, A.D. and Parsons, A.J. (1991) Resistance to overland flow on desert pavement and its implications for sediment-transport modelling, *Water Resources Research* 27: 1827–1836.

Abrahams, A.D., Luk, S.-H. and Parsons, A.J. (1988) Threshold relations for the transport of sediment by overland flow on desert hillslopes, *Earth Surface Processes and Landforms* 13: 407–419.

Abrahams, A.D., Parsons, A.J. and Luk, S.-H. (1991) The effect of spatial variability in overland flow on the downslope pattern of soil loss on a semi-arid hillslope, southern Arizona, *Catena* 18: 255–270.

Abrahams, A.D., Parsons, A.J. and Wainwright, J. (1995) Effects of vegetation change on interrill runoff and erosion, Walnut Gulch, southern Arizona, *Geomorphology* 13: 37–48.

Abrahams, A.D., Parsons, A.J., Wainwright, J. and Schlesinger, W.H. (forthcoming) Biogeochemical transport, in W.H. Schlesinger, K.M. Havstad and L.F. Huenneke (eds) *Structure and Function of a Chihuahuan Desert Ecosystem: The Jornada LTER*, Oxford: Oxford University Press.

Acherar, M. and Rambal, S. (1992) Comparative water relations of four Mediterranean oak species, *Vegetatio* 99–100: 177–184.

Adam, D.P., Sarna-Wojcicki, A.M., Rieck, H.J., Bradbury, J.P., Dean, W.E. and Forester, R.M. (1989) Tulelake, California: the last 3 million years, *Palaeogeography, Palaeoclimatology, Palaeoecology* 72: 89–103.

Alexander, R.W. and Calvo, A. (1990) The influence of lichens on slope processes in some Spanish badlands, in J.B. Thornes (ed.) *Vegetation and Climate*, Chichester: Wiley, pp. 285–298.

Andrieu, V., Brugiapaglia, E., de Beaulieu, J.-L. and Reille, M. (1995) Enregistrement pollinique des modalités et de la chronologie de l'anthropisation des écosystèmes méditerranéens d'Europe et du Proche-Orient: bilan des connaissances, in M. Dubost (ed.) *MEDIMONT Second and Final Scientific Report, Volume II/III Ecological Regional Approaches*. Final Report on Contract EV5V-CT91-0045, Brussels: DGXII European Commission.

Baggioni, M., Suc, J.-P. and Vernet, J.-L. (1981) Le Plio-Pléistocène de Camerota (Italie Méridionale): géomorphologie et paléoflores, *Géobios* 14: 229–237.

Baird, A.J. (1997) Runoff generation and sediment mobilisation by water, in D.S.G. Thomas (ed.) *Arid Zone Geomorphology*, Chichester: Wiley, pp. 165–184.

Barbero, M., Bonin, G., Loisel, R. and Quézel, P. (1990) Changes and disturbances of forest ecosystems caused by human activities in the western part of the Mediterranean basin, *Vegetatio* 87: 151–173.

Barbero, M., Loisel, R. and Quézel, P. (1992) Biogeography, ecology and history of Mediterranean *Quercus ilex* ecosystems, *Vegetatio* 99–100: 19–34.

Beatley, J.C. (1974) Phenological events and their environmental triggers in Mojave Desert ecosystems, *Ecology* 55: 856–863.

Beatley, J.C. (1976) Vascular plants of the Nevada test site and central-southern Nevada: ecology and geographical distribution, National Technical Information Service TID-26881, Springfield, Va.: US Department of Commerce, 306 pp.

Bertoldi, R., Rio, D. and Thunell, R. (1989) Pliocene–Pleistocene vegetational and climatic evolution of the south-central Mediterranean, *Palaeogeography, Palaeoclimatology, Palaeoecology* 72: 263–275.

Betts, A., Sellers, P.J. and Hall, F.G. (1992) FIFE in 1992 – results, scientific gains, and future research directions, *Journal of Geophysical Research – Atmospheres* 97, D17: 19091–19109.

Blumler, M.A. (1993) Successional pattern and landscape sensitivity in Mediterranean and Near East, in D.S.G. Thomas and R.J. Allison (eds) *Landscape Sensitivity*, Chichester: Wiley, pp. 287–308.

Blyth, E.M., Dolman A.J. and Noilhan, J. (1994) The effect of forest on mesoscale rainfall – an example from HAPEX-MOBILHY, *Journal of Applied Meteorology* 33(4): 445–454.

Bolle, H.-J., Andre, J.C., Arrue, J.L., Barth, H.K., Bessemoulin, P., Brasa, A., de Bruin, H.A.R., Cruces, J., Dugdale, G., Engman, E.T., Evans, D.L., Fantechi, R., Fiedler, F., van de Griend, A., Imeson, A.C., Jochum, A., Kabat, P., Hratzsch, T., Lagouarde, J.-P., Langer, I., Llamas, R., Lopez-Baeza, E., Melia Miralles, J., Muniosguren, L.S., Nerry, F., Noilhan, J., Oliver, H.R., Roth, R., Saatchi, S.S., Sanchez Diaz, J., de Santa Olalla, M., Shuttleworth, W.J., Sogaard, H., Stricker, H., Thornes, J.B., Vauclin, M. and Wickland, D. (1993) European field experiment in desertification threatened areas, *Annales Geophysicae* 11: 173–189.

Brandt, C.J. (1986) Transformation of the kinetic energy of rainfall with variable tree canopies, unpublished PhD thesis, University of London.

Brandt, C.J. (1989) The size distribution of throughfall drops under vegetation canopies, *Catena* 16: 507–524.

Braud, I., Dantas-Antonino, A.C. and Vauclin, M. (1995) A stochastic approach to studying the influence of the spatial variability of soil hydraulic properties on surface fluxes, temperature and humidity, *Journal of Hydrology* 165(1–4): 283–310.

Carraro, F. (ed.) with contributions from 35 others (1996) Revisione del Villafranchiano nell'area-tipo di Villafranca d'Asti, *Il Quaternario* 9(1): 5–120.

Cerdà, A. (1997) The effect of patchy distribution of *Stipa tenacissima* L. on runoff and erosion, *Journal of Arid Environments* 36: 37–51.

Childs, S.W. and Flint, A.L. (eds) (1990) *Physical Properties of Forest Soils Containing Rock Fragments*, 7th North American Forest Soils Conference, Vancouver: Forestry Publications, University of British Columbia.

Christensen, N.L. (1994) The effects of fire on physical and chemical properties of soils in Mediterranean-climate shrublands, in J.M. Moreno and W.C. Oechel (eds) *The Role of Fire in Mediterranean-Type Ecosystems*, New York: Springer-Verlag, pp. 79–95.

Clemens, S.C. and Tiedemann, R. (1997) Eccentricity forcing of Pliocene–Early Pleistocene climate revealed in a marine oxygen-isotope record, *Nature* 385: 801–804.

COHMAP members (1988) Climatic changes of the last 18,000 years; observations and model predictions, *Science* 241: 1043–1052.

Combourieu-Nebout, N. (1993) Vegetation response to Upper Pliocene glacial/interglacial cyclicity in the central Mediterranean, *Quaternary Research* 40: 228–236.

Daget, P. (1980) Un élément actuel de la caractérisation du monde Méditerranéen: le climat, in *Colloque de la Fondation L. Emberger sur 'la mise en place, l'évolution et la caractérisation de la flore et de la végétation circomméditerranéennes'*, Montpellier 9–10 April 1980. Naturalia Monspeliensia, No. Hors Série, pp. 101–126.

Dale, V.H. (1997) The relationship between land use change and climatic change, *Ecological Applications* 7(3): 753–769.

Davis, F.W. and Burrows, D.A. (1994) Spatial simulation of fire regime in Mediterranean-climate landscapes, in J.M. Moreno and W.C. Oechel (eds) *The Role of Fire in Mediterranean-Type Ecosystems*, New York: Springer-Verlag, pp. 117–139.

de Lima, M.I.L.P. and de Lima, J.L.M.P. (1990) Water erosion of soils containing rock fragments, in U. Shamir and C.J. Aqi (eds) *The Hydrological Basis for Water Management* (Proceedings of the Beijing Symposium, October 1990), Wallingford: IASH Publication No. 197, pp. 141–147.

Dirmeyer, P.A. and Shukla, J. (1994) Albedo as a modulator of climate response to tropical deforestation, *Journal of Geophysical Research – Atmospheres* 99, D10: 20863–20877.

Douglas, I. (1967) Man, vegetation and the sediment yield of rivers, *Nature* 215: 925–928.

Dunkerley, D.L. (1997) Banded vegetation: survival under drought and grazing pressure based on a simple cellular automaton model, *Journal of Arid Environments* 35(3): 419–428.

Eltahir, E.A.B. and Bras, R.L. (1994) Sensitivity of regional climate to deforestation in the Amazon Basin, *Advances in Water Resources* 17(1–2): 101–115.

Elwell, H.A. and Stocking, M.A. (1976) Vegetal cover to estimate soil erosion hazard in Rhodesia, *Geoderma* 15: 61–70.

Evenari, M. (1985a) The desert environment, in M. Evenari, I. Noy-Meir and D.W.

Goodall (eds) *Hot Deserts and Arid Shrublands. Ecosystems of the World 12A*, Amsterdam: Elsevier, 1–19.

Evenari, M. (1985b) Adaptations of plants and animals to the desert environment, in M. Evenari, I. Noy-Meir and D.W. Goodall (eds) *Hot Deserts and Arid Shrublands. Ecosystems of the World 12A*, Amsterdam: Elsevier, 79–92.

Follieri, M., Magri, D. and Sadori, L. (1988) 250,000-year pollen record from Valle di Castiglione (Roma), *Pollen et Spores* XXX: 329–356.

Gasse, F., Fontes, J.C., Plaziat, J.C., Carbonel, P., Kaczmarska, I., De Deckker, P., Soulie-Marsche, I., Callot, Y. and Dupeuble, P.A. (1987) Biological remains, geochemistry and stable isotopes for the reconstruction of environmental and hydrological changes in the Holocene lakes from north Sahara. *Palaeogeography, Palaeoclimatology, Palaeoecology* 60: 1–46.

Goutorbe, J.P., Lebel, T., Tinga, A., Bessemoulin, P., Brouwer, J., Dolman, A.J., Engman, E.T., Gash, J.H.C., Hoepffner, M., Kabat, P., Kerr, Y.H., Monteny, B., Prince, S., Said, F., Sellers, P. and Wallace, J.S. (1994) HAPEX-Sahel – a large-scale study of land-atmosphere interactions in the semi-arid Tropics, *Annales Geophysicae – Atmospheres, Hydrospheres and Space Sciences* 12(1): 53–64.

Graf, W.L. (1988) *Fluvial Processes in Dryland Rivers*, Berlin: Springer Verlag.

Grove, A.T. (1957) Patterned ground in northern Nigeria, *Geographical Journal* 124: 528–533.

Guiot, J. (1987) Late Quaternary climate change in France estimated from multi-variate pollen time series, *Quaternary Research* 28: 100–118.

Hall, F.G. and Sellers, P.J. (1995) First international satellite land-surface climatology project (ISLSCP) field experiment (FIFE) in 1995, *Journal of Geophysical Research – Atmospheres* 100, D12: 25383–25395.

Henderson-Sellers, A. (1994) Land use change and climate, *Land Degradation and Rehabilitation* 5(2): 107–126.

Horowitz, A. (1989) Continuous pollen diagrams for the last 3.5 m.y. from Israel: vegetation, climate and correlation with the oxygen isotope record, *Palaeogeography, Palaeoclimatology, Palaeoecology* 72: 63–78.

Horowitz, A. and Horowitz, M. (1985) Subsurface late Cenozoic palynostratigraphy of the Hula Basin, Israel, *Pollen et Spores* XXVII(3–4): 365–390.

Imeson, A.C., Verstraten, J.M., van Mulligen, E.J. and Sevink, J. (1992) The effects of fire and water repellency on infiltration and runoff under Mediterranean type forest, *Catena* 19: 345–361.

Inbar, M. (1992) Rates of fluvial erosion in basins with a Mediterranean type climate, *Catena* 19: 393–409.

Jeltsch, F., Milton, S.J., Dean, W.R.J. and van Rooyen, N. (1996) Tree spacing and coexistence in semi-arid savannas, *Journal of Ecology* 84: 583–595.

Keigwin, L.D. and Thunell, R.C. (1979) Middle Pliocene climatic change in the western Mediterranean from faunal and oxygen isotope trends, *Nature* 282: 292–296.

Kirkby, M.J. (1990) A simulation model for desert runoff and erosion, in *Erosion,*

Transport and Deposition Processes, Proceedings of the Jerusalem Workshop, Wallingford: IAHS Publication No. 189, pp. 87–104.

Kirkby, M.J. and Neale, R.H. (1987) A soil-erosion model incorporating seasonal factors, in V.G. Gardiner (ed.) *International Geomorphology*, Proceedings of the First International Conference of Geomorphology, Manchester, UK, Chichester: Wiley, 189–210.

Kirkby, M.J., Baird, A.J., Diamond, S.M., Lockwood, J.G., McMahon, M.L., Mitchell, P.L., Shao, J., Sheehy, J.E., Thornes, J.B. and Woodward, F.I. (1996) The MEDALUS slope catena model: a physically based process model for hydrology, ecology and land degradation interactions, in C.J. Brandt and J.B. Thornes (eds) *Mediterranean Desertification and Land Use*, Chichester: Wiley, pp. 303–354.

Kosmas, C., Danalatos, N., Cammeraat, L.H., Chabart, M., Diamantopoulos, J., Farrand, R., Gutiérrez, M., Jacob, A., Marques, H., Martinez-Fernandez, J., Mizara, A., Moustakas, N., Nicolau, J.M., Oliveros, C., Pinna, G., Puddu, R., Puigdefábregas, J., Roxo, M., Simao, A., Stamou, G., Tomasi, N., Usai, D. and Vacca, A. (1997) The effect of landuse on runoff and soil erosion rates under Mediterranean conditions, *Catena* 29: 45–59.

Koster, R.D. and Suarez, M.J. (1994) The components of a SVAT scheme and their effects on a GCM's hydrological cycle, *Advances in Water Resources* 17(1–2): 61–78.

Kremer, R.G., Hunt, E.R., Running, S.W. and Coughlan, J.C. (1996) Simulating vegetational and hydrologic responses to natural climatic variation and GCM-predicted climate change in a semi-arid ecosystem in Washington, USA, *Journal of Arid Environments* 33: 23–38.

Kutiel, P. and Inbar, M. (1993) Fire impacts on soil nutrients and soil erosion in a Mediterranean pine forest plantation, *Catena* 20: 129–139.

Langbein, W.B. and Schumm, S.A. (1958) Yield of sediment in relation to mean annual precipitation, *Transactions of the American Geophysical Union* 39: 1076–1084.

Laval, H., Medus, J. and Roux, M. (1991) Palynological and sedimentological records of Holocene human impact from the Étang de Berre, southeastern France, *The Holocene* 1: 269–272.

Le Houérou, H.N. (1986) The desert and arid zones of northern Africa, in M. Evenari, I. Noy-Meir and D.W. Goodall (eds) Ecosystems of the World 12B, *Hot Deserts and Arid Shrublands B*. Amsterdam: Elsevier, pp. 101–147.

Le Houérou, H.N. (1997) Climate, flora and fauna changes in the Sahara over the past 500 million years, *Journal of Arid Environments* 37: 619–647.

Lean, J.J. and Rowntree, P.R. (1997) Understanding the sensitivity of a GCM simulation of Amazonian deforestation to the specification of vegetation and soil characteristics, *Journal of Climate* 10(6): 1216–1235.

Lefever R. and Lejeune, O. (1997) On the origin of tiger bush, *Bulletin of Mathematical Biology* 59(2): 263–294.

Levin, N. and Horowitz, A. (1987) Palynostratigraphy of the Early Pleistocene QI palynozone in the Jordan–Dead Sea Rift, Israel', *Israel Journal of Earth Sciences* 36: 45–58.

Mabbutt, J.A. and Fanning, P.C. (1987) Vegetation banding in arid Western Australia, *Journal of Arid Environments* 12: 41–59.

MacFadyen, W.A. (1950) Vegetation patterns in the semi-desert plains of British Somaliland, *Geographical Journal* 116: 199–211.

Maley, J. (1990) Les changements climatiques de la fin du Tertiaire en Afrique: leur conséquence sur l'apparition du Sahara et de sa végétation, in M.A.J. Williams and H. Faure (eds) *The Sahara and the Nile. Quaternary Environments and Prehistoric Occupation in Northern Africa*, Rotterdam: A.A. Balkema, pp. 63–86.

Marengo, J.A., Miller, J.R., Russell, G.L., Rosenzweig, C.E. and Abramopoulos, F. (1994) Calculations of river runoff in the GISS GCM – impact of a new land-surface parameterization and runoff routing model on the hydrology of the Amazon River, *Climate Dynamics* 10(6–7): 349–361.

Martinez-Mesa, E. and Whitford, W.G. (1996) Stemflow, throughfall and channel-ization of stemflow by roots in three Chihuahuan desert shrubs, *Journal of Arid Environments* 32: 271–287.

Meehl, G.A. and Washington, W.M. (1988) A comparison of soil moisture sensitivity in two global climate models, *Journal of Atmospheric Science* 45: 1476–1492.

Mommersteeg, H.J.P.M., Loutre, M.F., Young, R., Wijmstra, T.A. and Hooghiemstra, H. (1995) Orbital forced frequencies in the 975,000 year pollen record from Tenaghi Philippon (Greece), *Climate Dynamics* 11: 4–24.

Monteith, J. L. (1973) *Principles of Environmental Physics*, London: Edward Arnold.

Morgan, R.P.C. (1996) Verification of the European soil erosion model (EUROSEM) for varying slope and vegetation conditions, in M.G. Anderson and S.M. Brooks (eds) *Advances in Hillslope Processes*, Chichester: Wiley, pp. 657–668.

Mulligan, M. (1996a) Modelling hydrology and vegetation change in a degraded semi-arid environment, unpublished PhD thesis, King's College, University of London.

Mulligan, M. (1996b) Modelling the complexity of land surface response to climatic variability in Mediterranean environments, in M.G. Anderson and S.M. Brooks (eds) *Advances in Hillslope Processes*, Chichester: Wiley, pp. 1099–1149.

Naveh, Z. (1967) Mediterranean ecosystems and vegetation types in California and Israel, *Ecology* 48: 445–459.

Naveh, Z. (1975) The evolutionary significance of fire in the Mediterranean region, *Vegetatio* 29: 199–209.

Noilhan J., Lacarrere, P., Dolman, A.J. and Blyth, E.M. (1997) Defining area-average parameters in meteorological models for land surfaces with mesoscale heterogeneity, *Journal of Hydrology* 190(3–4): 302–316.

Nulsen, R.A., Bligh, K.J., Baxter, I.N., Solin, E.J. and Imrie, D.II. (1986) The fate of rainfall in a malle and heath vegetated catchment in southern Western Australia, *Australian Journal of Ecology* 11: 361–371.

O'Brien, K.L. (1996) Tropical deforestation and climate change, *Progress in Physical Geography* 20(3): 311–335.

Obando, J.A. (1996) Modelling the impact of land abandonment on runoff and soil erosion in a semi-arid catchment, unpublished PhD thesis, King's College, University of London.

Parsons, A.J., Abrahams, A.D. and Luk, S.-H. (1991) Size characteristics of sediment in interrill overland flow on a semi-arid hillslope, southern Arizona, *Earth Surface Processes and Landforms* 16: 143–152.

Parsons, A.J., Abrahams, A.D. and Simanton, J.R. (1992) Microtopography and soil-surface materials on semi-arid piedmont hillslopes, *Journal of Arid Environments* 22: 107–115.

Parsons, A.J., Abrahams, A.D. and Wainwright, J. (1994) Rainsplash and erosion rates in an interrill area on semi-arid grassland, southern Arizona, *Catena* 22: 215–226.

Parsons, A.J., Wainwright, J. and Abrahams, A.D. (1993) Tracing sediment movement in interrill overland flow on a semi-arid grassland hillslope using magnetic susceptibility, *Earth Surface Processes and Landforms* 18: 721–732.

Parsons, A.J., Wainwright, J. and Abrahams, A.D. (1996a) Runoff and erosion on semi-arid hillslopes, in M.G. Anderson and S.M. Brooks (eds) *Advances in Hillslope Processes*, Chichester: Wiley, pp. 1061–1078.

Parsons, A.J., Abrahams, A.D. and Wainwright, J. (1996b) Responses of interrill runoff and erosion rates to vegetation change in southern Arizona, *Geomorphology* 14: 311–317.

Parsons, A.J., Wainwright, J., Abrahams, A.D. and Simanton, J.R. (1997) Distributed dynamic modelling of interrill overland flow, *Hydrological Processes* 11: 1833–1859.

Pearson, G.W. and Stuiver, M. (1993) High-precision bidecadal calibration of the radiocarbon time scale, 500–2500 BC, *Radiocarbon* 35: 25–33.

Pelgrum, H. and Bastiaanssen, W.G.M. (1996) An intercomparison of techniques to determine the area-averaged latent-heat flux from individual in-situ observations – a remote-sensing approach using the European field experiment in a desertification-threatened area data, *Water Resources Research* 32(9): 2775–2786.

Poesen, J. (1986) Surface sealing as influenced by slope angle and position of simulated stones in the top layer of loose sediments, *Earth Surface Processes and Landforms* 11: 1–10.

Poesen, J. and Bunte, K. (1995) Effects of rock fragments on desertification processes in Mediterranean environments, in J.B. Thornes and C.J. Brandt (eds) *Mediterranean Desertification and Land Use*, Chichester: Wiley, pp. 247–269.

Poesen, J., Ingelmo-Sanchez, F. and Mucher, H. (1990) The hydrological response of soil surfaces to rainfall as affected by cover and position of rock fragments in the top layer, *Earth Surface Processes and Landforms* 15: 653–671.

Polcher, J. and Laval, K. (1994) A statistical study of the regional impact of deforestation on climate in the LMD GCM, *Climate Dynamics* 10(4–5): 205–219.

Pons, A. and Reille, M. (1988) The Holocene and Upper Pleistocene pollen record from Padul (Granada, Spain): a new study, *Palaeogeography, Palaeoclimatology, Palaeoecology* 66: 243–263.

Pons, A., Guiot, J., de Beaulieu, J.-L. and Reille, M. (1992) Recent contributions to the climatology of the last glacial–interglacial cycle based on French pollen sequences, *Quaternary Science Reviews* 11: 439–448.

Pope, K.O. and van Andel, T.H. (1984) Late Quaternary alluviation and soil formation in the southern Argolid: its history, causes and archaeological implications, *Journal of Archaeological Science* 11: 281–306.

Pressland, A.J. (1973) Rainfall partitioning by an arid woodland (*Acacia aneura* F. Meull.) in south-western Queensland, *Australian Journal of Botany* 21: 235–245.

Puigdefábrigas, J. and Sánchez, G. (1996) Geomorphological implications of vegetation patchiness on semi-arid slopes, in M.G. Anderson and S.M. Brooks (eds) *Advances in Hillslope Processes*, Chichester: Wiley, pp. 1027–1060.

Quézel, P. (1978) Analysis of the flora of Mediterranean and Saharan Africa, *Annals Missouri Botanical Garden* 65: 479–534.

Reynolds, J.F., Virginia, R.A., Kemp, P.R., de Soyza, A.G. and Tremmel, D.C., (in press) Impact of simulated drought on resource islands of shrubs in the Chihuahuan Desert: effects of species, season, and degree of island development, *Ecology*.

Ritchie, J.C., Eyles, C.H. and Haynes, C.V. (1985) (sic) Sediment and pollen evidence for an early to mid-Holocene humid period in the eastern Sahara. *Nature* 314: 352–355.

Rogers, R.D. and Schumm, S.A. (1991) The effect of sparse vegetative cover on erosion and sediment yield, *Journal of Hydrology* 123: 19–24.

Rosa Attolini, M., Galli, M., Nanni, T., Ruggiero, L. and Zuanni, F. (1988) Preliminary observations of the fossil forest of Dunarobba (Italy) as a potential archive of paleoclimatic information, *Dendrochronologia* 6: 33–48.

Rostagno, C.M. and del Valle, H.F. (1988) Mounds associated with shrubs in aridic soils of northeastern Patagonia: characteristics and probable genesis, *Catena* 15: 347–359.

Rumney, G.R. (1968) *Climatology and the World's Climates*. New York: Macmillan.

Sánchez, G. and Puigdefábregas, J. (1994) Interactions of plant growth and sediment movement on slopes in a semi-arid environment, *Geomorphology* 9: 243–260.

Savat, J. and Poesen, J. (1981) Detachment and transportation of loose sediment by raindrop splash. Part I: The calculation of absolute data on detachability and transportability, *Catena* 8: 1–17.

Schlesinger, W.H., Abrahams, A.D., Parsons, A.J. and Wainwright, J. (in press) Nutrient losses in runoff from grassland and shrubland habitats in southern New Mexico: I. Rainfall simulation experiments, *Biogeochemistry*.

Schlesinger, W.H., Reynolds, J.F., Cunningham, G.L., Huenneke, L.F., Jarrell, W.M., Virginia, R.A. and Whitford, W.G. (1990) Biological feedbacks in global desertification, *Science* 247: 1043–1048.

Schumm, S.A. (1965) Quaternary palaeohydrology, in H.E. Wright and D.G. Frey (eds) *The Quaternary of the United States*, Princeton, NJ: Princeton University Press, pp. 783–794.

Scoging, H.M. (1982) Spatial variations in infiltration, runoff and erosion on hillslopes in semi-arid Spain, in R.B. Bryan and A. Yair (eds) *Badland Geomorphology and Piping*, Norwich: GeoBooks, pp. 89–112.

Scoging, H.M. and Thornes, J.B. (1979) Infiltration characteristics in a semi-arid environment, in *Proceedings of the Canberra Symposium on the Hydrology of Areas of Low Precipitation*, Wallingford: IASH Publication No. 128, 159–168.

Sene, K.J. (1996) Meteorological estimates for the water balance of a sparse vine crop growing in semiarid conditions, *Journal of Hydrology* 179(1–4): 259–280.

Shackleton, N.J. and Opdyke, N.D. (1977) Oxygen isotope and palaeomagnetic evidence for early Northern Hemisphere glaciation, *Nature* 270: 216–219.

Shmida, A. (1972) The vegetation of Gebel Maghara, North Sanai, unpublished PhD thesis, The Hebrew University, Jerusalem.

Shmida, A. (1985) Biogeography of the desert flora, in M. Evenari, I. Noy-Meir, and D.W. Goodall (eds) *Hot Deserts and Arid Shrublands. Ecosystems of the World 12A*, pp. 23–72, Amsterdam: Elsevier.

Shukla, J. and Mintz, Y. (1982) The influence of land-surface evapo-transpiration on the Earth's climate, *Science* 215: 1498–1501.

Shuttleworth, W.M. and Wallace, J.S. (1985) Evaporation from sparse crops – an energy combination theory, *Quarterly Journal of the Royal Meteorological Society* 111: 839–856.

Shuttleworth, W.J, Gash, J.H.C., Roberts, J.M., Nobre, C.A., Molion, L.C.B. and Ribeiro, M.D.G. (1991) Post-deforestation Amazonian climate – Anglo-Brazilian research to improve prediction, *Journal of Hydrology* 129(1–4): 71–85.

Smith, E.A., Wai, M.M.K., Cooper, H.J. and Rubes, M.T. (1994) Linking boundary-layer circulations and surface processes during FIFE-89. 1. Observational analysis, *Journal of the Atmospheric Sciences* 51(11): 1497–1529.

Soler, M. and Sala, M. (1992) Effects of fire and of clearing in a Mediterranean *Quercus ilex* woodland: an experimental approach, *Catena* 19: 321–332.

Specht, R.L. (1972) Water use by perennial evergreen plant communities in Australia and Papua New Guinea, *Australian Journal of Botany* 20: 273–299.

Starkel, L. (1983) The reflection of hydrologic changes in the fluvial environment of the temperate zone during the last 15,000 years, in K.J. Gregory (ed.) *Background to Palaeohydrology*, Chichester: Wiley, pp. 213–235.

Starkel, L. (1987) The evolution of European rivers – a complex response, in K.J. Gregory, J. Lewin and J.B. Thornes (eds) *Palaeohydrology in Practice*, Chichester: Wiley, pp. 333–339.

Stockdale, T., Latif, M., Burgers, G. and Wolff, J.O. (1994) Some sensitivities of a coupled ocean–atmosphere GCM, *Tellus Series A – Dynamic Meteorology and Oceanography* 46(4): 367–380.

Stuiver, M. and Pearson, G.W. (1993) High-precision bidecadal calibration of the radiocarbon time scale, AD 1950–500 BC and 2500–6000 BC, *Radiocarbon* 35: 1–23.

Stuiver, M. and Reimer, P.J. (1993) Extended C^{14} data-base and revised Calib 3.0 C^{14} age calibration program, *Radiocarbon* 35: 215–230.

Sturdy, D., Webley, D. and Bailey, G.N. (1994) Prehistoric land use, in S.E. van der

Leeuw (ed.) *Understanding the Natural and Anthropogenic Causes of Soil Degradation and Desertification in the Mediterranean Basin. Volume 1: Land Degradation in Epirus*, Final Report on Contract EV5V-CT91-0021, Brussels: DGXII European Commission, pp. 81–142.

Suc, J.-P. (1984) Origin and evolution of the Mediterranean vegetation and climate in Europe, *Nature* 307: 429–432.

Suc, J.-P. and Bessais, E. (1990) Pérennité d'un climat thermo-xériquie en Sicile avant, pendant, après la crise de salinité messinienne, *Comptes Rendus de l'Académie des Sciences de Paris, Série II* 310: 1701–1707.

Suc, J.-P. and Zagwijn, W.H. (1983) Plio–Pleistocene correlations between the northwestern Mediterranean region and northwestern Europe according to recent biostratigraphic and palaeoclimatic data, *Boreas* 12: 153–166.

Sud, Y.C., Shukla, J. and Mintz, Y. (1985) *Influence of Land Surface Roughness on Atmospheric Circulation and Rainfall: A Sensitivity Experiment with a GCM.* NASA Technical Memo 86219, 233 pp.

Sud, Y.C., Walker, G.K., Kim, J.-H., Liston, G.E., Sellers, P.J. and Lau, K.M. (1996) Biogeophysical consequences of a tropical deforestation scenario: a GCM simulation study, *Journal of Geophysical Research* 101, D3: 7095–7109.

Terradas, J. and Savé, R. (1992) The influence of summer and winter stress and water relationships on the distribution of *Quercus ilex* L., *Vegetatio* 99–100: 137–145.

Tilman, D. (1994) Competition and biodiversity in spatially structured habitats, *Ecology*, 75: 2–16.

Thornes, J.B. (1988) Erosional equilibria under grazing, in J. Bintliff, D. Davidson and E. Grant (eds) *Conceptual Issues in Environmental Archæology*. Edinburgh: Edinburgh University Press, 193–210.

Thornes, J.B. (1990) The interaction of erosional and vegetational dynamics in land degradation: spatial outcomes, in J.B. Thornes (ed.) *Vegetation and Erosion*, Chichester: Wiley, 41–53.

Thunell, R.C. (1979) Climatic evolution of the Mediterranean Sea during the last 5.0 million years, *Sedimentary Geology* 23: 67–79.

Thunell, R.C. and Williams, D.F. (1983) Paleotemperature and paleosalinity history of the eastern Mediterranean during the late Quaternary, *Palaeogeography, Palaeoclimatology, Palaeoecology* 44: 23–39.

Torri, D., Sfalanga, M. and del Sette, M. (1981) Splash detachment: runoff depth and soil cohesion, *Catena* 14: 149–155.

Turner, J. and Greig, J.R.A. (1975) Some Holocene pollen diagrams from Greece, *Review of Palaeobotany and Palynology* 20: 171–204.

Tzedakis, P.C. (1994) Vegetation change through glacial–interglacial cycles: a long-term pollen-sequence perspective, *Philosophical Transactions of the Royal Society of London* 345B: 403–432.

UNEP (1977) *Desertification: its Causes and Consequences*, Nairobi: First United Nations Conference on Desertification.

UNEP (1991) *Status of Desertification and Implementation of the United Nations Plan of Action to Combat Desertification*, Nairobi: UNEP.

UNEP/ISRIC (1992) *World Atlas of Desertification*, London: Edward Arnold.

Uso-Domenech, J.L., Villacampa-Esteve, Y., Stubing-Martinez, G., Karjalainen, T. and Ramo, M.P. (1995) MARIOLA: a model for calculating the response of Mediterranean bush ecosystem to climatic variations, *Ecological Modelling* 80: 113–129.

Valentine, I. and Nagorcka, B.N. (1979) Contour patterning in *Atriplex vesicaria* communities, in D. Graetz and K.M.W. Howes (eds) *Studies of the Australian Arid Zone. Part 4*, Melbourne: CSIRO, pp. 121–146.

van Andel, T.H. and Runnels, C. (1987) *Beyond the Acropolis. A Rural Greek Past*, Stanford, Calif: Stanford University Press.

van Andel, T., Runnels, C.N. and Pope, K.O. (1986) Five thousand years of land use and abuse in the southern Argolid, Greece, *Hesperia* 55: 103–128.

van Andel, T.H., Zangger, E. and Demitrack, A. (1990) Land use and soil erosion in prehistoric and historical Greece, *Journal of Field Archaeology* 17: 379–396.

van der Maarel, E. (1988) Vegetation dynamics: patterns of change in time and space, *Vegetatio* 77: 7–19.

van der Wiel, A.M. and Wijmstra, T.A. (1987a) Palynology of the lower part (78–120 m) of the core Tenaghi Philippon II, Middle Pleistocene of Macedonia, Greece, *Review of Palaeobotany and Palynology* 52: 73–88.

van der Wiel, A.M. and Wijmstra, T.A. (1987b) Palynology of the 112.8–197.8 m interval of the core Tenaghi Philippon III, Middle Pleistocene of Macedonia, *Review of Palaeobotany and Palynology* 52: 89–117.

van Zinderen Bakker, E.M. (1983) Elements for the chronology of Late Cainozoic African climates, in W.C. Mahaney (ed.) *Correlation of Quaternary Chronologies*, Norwich: GeoBooks, pp. 23–38.

Vega, J.A. and Diaz Fiernos, F. (1987) Wildfire effects on soil erosion, *Ecologia Mediterranea* XIII: 119–125.

Vergnaud-Grazzini, C., Saliège, J.F., Urrutiager, M.J. and Iannace, A. (1990) Oxygen and carbon isotope stratigraphy of ODP Hole 653A and site 654: the Pliocene–Pleistocene glacial history recorded in the Tyrrhenian Basin (West Mediterranean), in K.A. Kastens, J. Mascle *et al.* (eds) *Proceedings of the Ocean Drilling Program. Volume 107 Scientific Results. Tyrrhenian Sea*, College Station TX: Ocean Drilling Program, pp. 361–386.

Vernet, J.-L. and Thiébault, S. (1987) An approach to northwestern Mediterranean recent prehistoric vegetation and ecologic implications, *Journal of Biogeography* 14: 117–127.

Wainwright, J. (1992) Assessing the impact of erosion on semi-arid archaeological sites, in M. Bell and J. Boardman (eds) *Past and Present Soil Erosion*, Oxford: Oxbow Books, pp. 228–41.

Wainwright, J. (1996a) Climats actuels et passés: implications humaines, in J. Gascó, L. Carozza, S. Fry, R. Fry, J.-D. Vigne and J. Wainwright (eds) *Le Laouret et la Montagne d'Alaric à la Fin de l'Âge du Bronze, un hameau abandonné entre Floure et Monze (Aude)*, Toulouse: Centre d'Anthropologie, pp. 45–56.

Wainwright, J. (1996b) Infiltration, runoff and erosion characteristics of agricultural land in extreme storm events, SE France, *Catena* 26: 27–47.

Wainwright, J. (1996c) A comparison of the infiltration, runoff and erosion characteristics of two contrasting 'badland' areas in S. France, *Zeitschrift für Geomorphologie Supplementband* 106: 183–198.

Wainwright, J. (1997) *History and Evolution of Mediterranean Desertification*, King's College London: Concerted Action on Mediterranean Desertification Thematic Report.

Wainwright, J. (in press, a) Climate and climatological variation, in W.H. Schlesinger, K.M. Havstad and L.F. Huenneke (eds) *Structure and Function of a Chihuahuan Desert Ecosystem: The Jornada LTER*. Oxford: Oxford University Press.

Wainwright, J. (in press, b) Contextes géomorphologiques et géoarchéologiques des habitats de l'Âge du Bronze en Méditerranée Occidentale, in *Proceedings, XXIVe Congrès Préhistorique de France*.

Wainwright, J. and Thornes, J.B. (forthcoming) *Environmental Issues in the Mediterranean*, Routledge: London.

Wainwright, J., Parsons, A.J. and Abrahams, A.D. (1995) Simulation of raindrop erosion and the development of desert pavements, *Earth Surface Processes and Landforms* 20: 277–291.

Wainwright, J., Parsons, A.J. and Abrahams, A.D. (in press) Field and computer simulation experiments on the formation of desert pavement, *Earth Surface Processes and Landforms*.

Walker, J. and Rowntree, P.R. (1987) The effect of soil moisture on circulation and rainfall in a tropical model, *Quarterly Journal of the Royal Meteorological Society* 103: 29–46.

Wallace, J.S. and Holwill, C.J. (1997) Soil evaporation from tiger-bush in south-west Niger, *Journal of Hydrology* 189(1–4): 426–442.

Walling, D.E. and Webb, B.W. (1983) Patterns of sediment yield, in K.J. Gregory (ed.) *Background to Palaeohydrology*, Chichester: Wiley, pp. 69–100.

Warren, A. and Khogali, M. (1992) *Assessment of Desertification and Drought in the Sudano-Sahelian Region 1985–1991*, New York: United Nations Sudano-Sahelian Office.

Watts, W.A., Allen, J.R.M. and Huntley, B. (1996) Vegetation history and palaeo-climate of the last glacial period at Lago Grande di Monticchio, southern Italy, *Quaternary Science Reviews* 15: 133–153.

Whitford, W.G. (1997) Desertification and animal biodiversity in the desert grass-lands of North America, *Journal of Arid Environments* 37: 709–720.

Wigley, T.M.L. (1992) Future climate of the Mediterranean Basin with particular emphasis on changes in precipitation, in T Jeftic, J. D. Milliman and G. Sestini (eds) *Climatic Change and the Mediterranean: Environmental and Societal Impacts of Climatic Change and Sea-Level Rise in the Mediterranean Region*, London: Edward Arnold, pp. 15–44.

Wijmstra, T.A. (1969) Palynology of the first 30 metres of a 120 m deep section in northern Greece, *Acta Botanica Neerlandica* 18: 511–527.

Wijmstra, T.A. and Smit, A. (1976) Palynology of the middle part (30–78 metres) of the 120 m deep section in northern Greece (Macedonia), *Acta Botanica Neerlandica* 25: 297–312.

Willis, K.J. (1994) The vegetational history of the Balkans, *Quaternary Science Reviews* 13: 769–788.

Wilson, M.F., Henderson-Sellers, A., Dickinson, R.E. and Kennedy, P.J. (1987) Investigation of the sensitivity of the land surface parameterisation of the NCAR community climate model in regions of tundra vegetation, *Journal of Climatology* 7: 319–343.

Wood, E.F. (1997) Effects of soil moisture aggregation on surface evaporative fluxes, *Journal of Hydrology* 190(3–4): 397–412.

Woodward, F.I. (1994) Vegetation growth modelling, in J.B. Thornes (ed.) *Third Interim Report for Medalus II*, pp. 61–65, EV5V-CT92-0128, Brussels: DGXII European Commission.

Worral, G.A. (1960) Patchiness in vegetation in the northern Sudan, *Journal of Ecology* 48: 107–115.

Wright, H.E., Kutzbach, J.E., Webb III, T., Ruddiman, W.F., Street-Perrott, F.A. and Bartlein, P.J. (eds) (1993) *Global Climates Since the Last Glacial Maximum*, Minneapolis: University of Minnesota Press.

Xue, Y.K. and Shukla, J. (1996) The influence of land surface properties on Sahel climate. Part II: afforestation, *Journal of Climate* 9(12): 3360–3275.

Zhang, X., Drake, N.A., Wainwright, J. and Mulligan, M. (1997) Global scale overland flow and soil erosion modelling using remote sensing and GIS techniques: model implementation and scaling, in *Proceedings of RSS '97 Remote Sensing in Action*, 2–4 September, Reading, Nottingham: Remote Sensing Society, pp. 379–384.

5

WATER AND PLANTS IN
FRESHWATER WETLANDS

Bryan D. Wheeler

INTRODUCTION

The most characteristic feature of wetlands is that they are wet. The effects of excess water dominate their formation and considerably control their processes and characteristics. The fundamental importance of water availability to the character of all wetlands, and the fact that certain types of wetland (peatlands) are largely based upon the remains of plants preserved by waterlogging, mean that any consideration of 'wetland plants and water' could easily become a review of much of wetland ecology. However, this review will be more selective and focus particularly upon (1) mechanisms of adaptation of wetland plants to waterlogged environments; (2) quantitative effects of water availability and regimes upon the distribution of wetland plants and upon the composition of wetland vegetation; (3) some effects of wetland plants upon the hydrological characteristics of the wetlands in which they grow; and (4) outline effects of water quality (chemical composition) upon the distribution of wetland plants. Consideration is restricted to freshwater wetlands.

The concept of 'wetland'

The term 'wetland' is used widely, but not consistently. It has been variously applied to a range of watery habitats, with differences of definition relating mainly to the position of the upper and lower limits (Mitsch and Gosselink, 1986). The Ramsar Convention on Wetlands of International Importance (1971) adopted a particularly broad concept, including habitats that are only periodically wet and with low summer water tables (e.g. wet meadows, pond margins) and those that are permanently inundated with up to 6 m of water. It is not difficult to find reasons why aquatic and wet terrestrial systems should be grouped together under a generic title of 'wetlands' (not least their frequent hydraulic and ontogenic links), but it is equally clear that many

127

ecological processes operate differently in the two types. From many (though not all) viewpoints, 'wet land' is more usefully seen as a particularly wet terrestrial ecosystem than an especially dry aquatic one: compare, for example, the relative ease of walking upon dry land, wet land and 6 m depth of water!

One way of maintaining compatibility with existing usage of the term 'wetland' is to make a primary subdivision into *aquatic wetlands* and *telmatic* (alternatively *'paludic'*) *wetlands*, these representing the broad categories of shallow-water and wet terrestrial wetland, respectively. This paper is concerned exclusively with telmatic freshwater wetlands, and the term 'wetland' is used here as shorthand for this. The main terms used are outlined in Figure 5.1.

Water sources and wetland types

The occurrence of any wetland ecosystem requires that the substratum is kept in a suitably wet condition for all, or part, of the year. Such saturation results from an interaction between landscape topography and sources of water. Broadly, wet conditions occur primarily because of water *detention* (impeded drainage) or because of high rates of water *supply*, or both. Supply may consist of *telluric* water (i.e. water that has had some contact with the mineral ground, such as river flooding, surface runoff and ground water discharge) or *meteoric* water (precipitation).

The behaviour of the water table in wetlands shows much variation. Some wetlands (e.g. some spring-fed sites) may have an almost constant water table, but usually the position of the water table shows some – often considerable – variation, induced by varying rates of water loss and recharge. Water level fluctuations in some wetlands are sometimes sufficient to cause periodic flooding or drying (or both) in whole or in part. This may occur on a daily

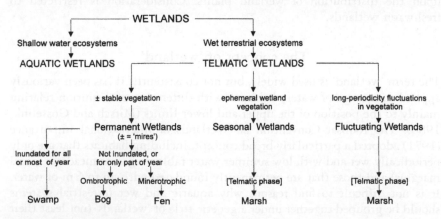

Figure 5.1 Wetland categories and terms as used in this chapter.

(tidal wetlands) or annual basis to produce episodically dry (or flooded) wetlands. Sometimes the periodicity is longer and irregular, as exemplified by the 'prairie glacial marshes' described by Van der Valk and Davies (1978).

The magnitude and period of water level fluctuation has a profound impact upon the character of wetlands. From the point of view of their plant ecology, three broad, intergrading types can be recognised:

1 *permanent wetlands*: wetlands where the normal amplitude of water level fluctuation is relatively small (or extremes are of short duration) and does not drive floristic change. Vegetation is relatively stable, with perennial plant species.
2 *seasonal wetlands*: wetlands in which the seasonal amplitude of water level is too great to support perennial wetland species; wetland vegetation, if it develops at all, is represented by opportunist, ephemeral species that temporarily colonise exposed, moist substrata.
3 *fluctuating wetlands*: sites with long-term water level fluctuations of sufficient magnitude and duration (several years) to cause either a phased change in the composition of perennial wetland vegetation or episodes when perennial wetland plants cannot grow.

The key component of this categorisation is whether water table fluctuations drive changes in vegetation composition or whether they are in equilibrium with a more or less stable vegetation. Some 'permanent wetlands' experience very considerable changes in the position of their water tables but nonetheless support 'stable' vegetation. The mean position of the water table varies considerably between examples of permanent wetlands.

There can be considerable variation in hydrodynamics within a single wetland site, and this is often responsible for the development of different vegetation types and contrasting habitat conditions (e.g. Yabe and Onimaru, 1996). Where the water table remains close to the surface for much of the year, wetlands often accumulate peat, especially in temperate and boreal regions. By contrast, sites with strongly fluctuating water tables (e.g. on river floodplains) may have a mineral substratum because periods of low water levels are inimical to peat accumulation. Such wetlands may also be subject to alluvial deposition. The term *marsh* is sometimes used for wetlands upon mineral soils.

The term *fen* is often applied to permanent wetlands receiving some telluric water input (du Rietz, 1949), particularly (but not exclusively) to peat-producing systems. Von Post and Granlund (1926) distinguished *topogenous* ('topography made') fens as sites that occupy flattish ground, shallow hollows, kettle-holes etc., where water naturally collects or where drainage is otherwise impeded because of the topography of the landscape and where the water surface is more or less horizontal. They also recognised *soligenous* ('soil-made') wetlands as those occurring (usually) on sloping ground and kept wet primarily by water supply (surface runoff or groundwater discharge). However, these categories do intergrade, not least because some topogenous fens experience considerable lateral water flow, sometimes derived from soligenous supply. A broad distinction can be made between *rheo-topogenous* fens (which have an ecologically important lateral flow component) and *stagno-topogenous* fens.

Von Post and Granlund further identified *ombrogenous* ('rain-made') wetlands (or *bogs*), which are irrigated directly and exclusively by meteoric water. All bogs depend on quite high rates of water supply (high precipitation-to-evaporation ratio), but they nonetheless show variation in their dependency on its amount and constancy. At the lower end of the precipitation range (*c.* 1200 mm year^{-1} in Britain), development of ombrogenous wetlands tends to be restricted to hollows or flattish surfaces where drainage of meteoric water is impeded, leading to the formation of shallow domes of ombrogenous peat (Ingram, 1982). Cooler, wetter climates permit ombrogenous peat formation to occur also on quite strongly sloping ground, where drainage impedance is smaller. Thus in a sense bogs form a series comparable with the topogenous–soligenous distinction made for fens, and they may be subdivided into *topogenous bog* and *hill bog** on this basis (Wheeler, 1995). This subdivision corresponds roughly to the long-recognised, though ill-defined, categories of *raised bog* and *blanket bog*. Topogenous bogs have often developed from a preceding phase of topogenous fen.

WETLAND PLANTS

Plants of wetlands

In aggregate, wetlands support a rich assemblage of plant species (a total of some 650 plant species has been recorded from fens and bogs in Britain). All of the main phyla are well represented. Herbaceous vascular plants are important constituents of most wetlands, with graminoid monocots often being especially prominent. Bryophytes are important in some wetlands and can be the main peat-forming species. Trees can also be dominant components in some wetlands, including some frequently flooded examples.

Not all the plants that grow in wetlands are restricted to this habitat. In Britain, fewer than half the species recorded from telmatic wetlands (*c.* 250) are largely confined to such sites (Wheeler, 1988; 1996; Wheeler and Shaw, 1995). However, the recognition of 'characteristic' wetland species can be

---

* The term 'hill bog' is used rather than 'soligenous bog' as the root meaning of soligenous ('soil-made') is inappropriate for a system dependent on meteoric water.

obfuscated by geographical variation in habitat range. For example, in England *Herminium monorchis* is a plant of dry calcareous grassland and *Schoenus nigricans* is a fen species (Wheeler, 1988), but the first grows widely in calcareous fens in mainland Europe, while the second is a constituent of summer-dry *Rosmarinus* heaths in southern France (Zwillenberg and de Wit, 1951).

The species comprising the broad assemblage of 'plants of wetlands' show very considerable variation in their tolerance of, and adaptation to, different water regimes. Even 'characteristic' wetland species do not necessarily require, or grow best in, sites with very high water tables (various species show greater growth when drainage is slightly improved, e.g. *Narthecium ossifragum*; Miles, 1976). Nor are 'non-characteristic' species necessarily restricted to drier sites (*Agrostis stolonifera* can grow in inundated conditions). In consequence, as a generic category 'wetland plants' can be misleading.

Tolerance of plants to waterlogged conditions in wetlands

Any assessment of the mechanisms that permit plants to survive in wetlands is complicated by the broad range of water conditions encompassed within the concept of wetland and by the variability of response of different plant species. The exclusion of species from permanently flooded swamps may have a different basis from their exclusion from fen meadows with a subsurface summer water table (Braendle and Crawford, 1987), or from sites with strongly fluctuating water levels, or those scoured by wave action, current flow or ice.

The principal 'problem' for the growth of plant species in many wetlands, at least when they are waterlogged, is the development of anoxia in the substratum, and the occurrence of low oxidation–reduction ('redox') potentials. The intensity of reducing conditions, their duration and distribution in the soil profile shows considerable variation, depending partly on the hydro-dynamics of the wetland (e.g. Pearsall, 1938; Urquhart and Gore, 1973; Armstrong and Boatman, 1967; Giller and Wheeler, 1986a; de Mars, 1996). Redox potentials typically decrease with depth in waterlogged soils (even fully saturated soils can have an oxidised surface skin: Patrick and DeLaune, 1972). Although the water levels of many temperate wetlands are higher in winter than in summer, redox potentials beneath the upper unsaturated zone are generally lower in summer than in winter, presumably reflecting increased microbial activity.

The chemical transformations that take place in waterlogged soils are, in gross terms, well known and have been reviewed by Ponnamperuma (1972), Armstrong (1982) and Ross (1995) among others. They can have a number of important repercussions upon plant growth. In particular, because oxygen diffusion rates in water are 10,000 times slower than rates in the atmosphere, waterlogging typically restricts the availability of oxygen to the underground organs of plants. Low redox potentials are associated with a series of chemical transformations, some of which (e.g. increase in availability of soluble, reduced phytotoxins, such as Fe^{2+}, Mn^{2+} and S^-) can be detrimental to plant growth. The magnitude and significance of this effect depends upon the initial concentration of the oxidised ions and pH as well as redox potentials; it may also be influenced by other soil chemicals (for example, in iron-rich soils reduced iron and sulphur can co-precipitate as pyrite: Bloomfield and Zahari, 1982).

The reviews of Armstrong (1982) and Crawford (1983) summarise the main ideas about waterlogging tolerance. Crawford (1996) has emphasised that survival in such habitats may depend on several linked attributes. Investigations on waterlogging have focused particularly upon mechanisms that exclude dryland* species from wetlands, but wetland species also vary in their sensitivity to waterlogging (Tanaka *et al.*, 1966, Ingold and Havill, 1984; Schat and Beckhoven, 1991; Sellars, 1991; Snowden and Wheeler, 1993; 1995).

Waterlogging avoidance

Some wetland plants effectively avoid the adverse effects of sustained high water tables. In sites with strongly fluctuating water levels, some plants avoid the damaging effects of periodic flooding (or drought) by surviving as seeds, tubers or other perennating organs (Squires and van der Valk, 1992). In wetlands with more stable water tables, occupancy of relatively well-aerated, elevated microsites (provided by microtopographical variation or hummock-forming plants) is probably an important avoidance mechanism. It sometimes has interesting ecological consequences. For example, precocious colonisation of upstanding tussocks of *Carex paniculata*, growing in swamp conditions, by shrub species can lead to the establishment of wooded vegetation much earlier in the hydrosere than would otherwise be the case (Lambert, 1951).

A related form of waterlogging avoidance is provided by shallow rooting. The formation of 'plate roots' by some tree species growing in wetlands is especially well known (e.g. Armstrong *et al.*, 1976), but many herbaceous plants, including some characteristic wetland species, also root in the uppermost, better-aerated layers of wetland soils (e.g. Orchidaceae, *Drosera anglica*, *Parnassia palustris*, *Viola palustris*) (e.g. Metsävainio, 1931; Schat, 1982). Armstrong and Boatman (1967) observed that roots of *Molinia caerulea* grew closer to the surface in stagnant conditions than in a better-aerated water

* 'Dryland' is used throughout this chapter to mean land that is not wetland. It is not used in the same sense as in Chapter 4.

track. Shallow rooting sometimes reflects the capacity of various wetland plants to form adventitious roots (Gill, 1970; 1975; Keeley, 1979; Etherington, 1984). Metsävainio (1931) observed that shallow-rooted plants in wetlands tended to have little development of aerenchyma, whereas deep-rooted species invariably possessed this tissue. More recently Justin and Armstrong (1987) have demonstrated, in experimental conditions, that rooting depth in flooded soils increases with fractional root porosity.

Oxygen transfer

It has long been recognised that many wetland species (and some dryland taxa) possess, or have the capacity to develop, anatomical structures (lacunae, aerenchyma and respiratory roots) that assist diffusion of atmospheric oxygen to the subterranean parts of the plants, where it helps to maintain aerobic respiration (e.g. Warming, 1909; Armstrong, 1980; Crawford, 1983; Justin and Armstrong, 1987). Wetland plants also often have some capacity to produce oxidising conditions in the rhizosphere of actively growing roots (e.g. Armstrong, 1964). This phenomenon is known as *radial oxygen loss* (ROL) and can be induced by the radial diffusion of oxygen through the cortex or by enzymatic oxidation on the root surface (Armstrong, 1967). The establishment of oxidising conditions in the rhizosphere is thought to provide a mechanism that reduces the uptake of the soluble reduced toxins, by their oxidative precipitation around or within the roots (Bartlett, 1961; Jones and Etherington, 1970; Jones, 1972) although in some circumstances, by steepening the concentration gradient of reduced toxins, this process may increase their uptake (Tessenow and Baynes, 1978). Oxidising conditions in the rhizosphere can also promote nitrification (Laan *et al.*, 1989), and this may contribute to the vigorous growth of such species as *Glyceria maxima* in some wetland soils (Engelaar *et al.*, 1995).

There can be little doubt that oxygen transfer provides the key to the survival of vascular plant species in waterlogged conditions. Indeed, one adaptation of some plants to periodic submergence is accelerated shoot growth, to extend above the level of flood (Squires and van der Valk, 1992; Blom *et al.*, 1994; Yabe and Onimaru, 1996). Doubts about the capacity of gas transfer to supply sufficient oxygen to plant roots in waterlogged soils (e.g. Crawford, 1983) have been largely answered by a better understanding of aeration mechanisms (Armstrong *et al.* 1992; W. Armstrong *et al.*, 1996), although even well-ventilated plants may still experience hypoxia or anoxia when growing in strongly reducing substrata, or if connection with the atmosphere is severed. Root porosity of wetland plants is also positively correlated with their tolerance to Fe^{2+} (Snowden and Wheeler, 1993; 1995), and capacity for oxygen transport and ROL seems to be critical to iron tolerance. ROL is unlikely, however, to be the only mechanism by which wetland plants accommodate reduced toxins: in iron-rich environments, the

Fe shoot content of some wetland species can be much greater than in low-iron examples (Wheeler *et al.*, 1985), and uptake of large amounts of iron sometimes occurs (e.g. *Eriophorum angustifolium*; Mansfield, 1990). Likewise, certain species (e.g. *Acorus calamus*, *Phragmites australis*) have some capacity to detoxify sulphide by formation of thiols within the rhizomes (Weber and Brändle, 1996; Fürtig *et al.*, 1996) – although exposure to high sulphide concentrations can still be injurious to *Phragmites* (J. Armstrong *et al.*, 1996).

The dynamics of oxygen transfer from shoots to roots are strongly dependent on shoot architecture and are related to overall shoot porosity (Justin and Armstrong, 1987), but even shoots with much pore space can have specific blockages to oxygen transfer. Conway (1937) concluded that the small cells forming the intercalary meristems at the base of the growing leaves of *Cladium mariscus* provided an effective barrier to gaseous diffusion and proposed that persistent dead leaves were critical to aeration of the subterranean parts of this plant. Likewise, Armstrong *et al.* (1992) point to the importance of old or broken culms of *Phragmites* for rhizome ventilation.

Anaerobic metabolism

Wetland plants, and some flood-tolerant dryland species, may display various metabolic adaptations to survival of anoxia, some of which may act by stimulating adventitious root formation, aerenchyma development or stem elongation (e.g. ethylene production: Osborne, 1984; van der Sman *et al.*, 1991). Roots of some, perhaps many, wetland plants can endure periods of anoxia, but early ideas that wetland plants may be preferentially able to accumulate low-toxicity metabolites of anaerobic respiration (e.g. malic acid) (MacMannon and Crawford, 1971; Crawford, 1983) seem to have been discarded in favour of their capacity to tolerate high concentrations of ethanol, or to vent toxic metabolites into their surroundings. There seems little doubt that anaerobic metabolism may provide tolerance to short periods of water-logging, but it is less clear what part it plays in long-term survival. Capacity for sustained anaerobic respiration is related to the magnitude of carbohydrate reserves (Schlüter *et al.*, 1966; Weber and Brändle, 1966). *Acorus calamus* has an unusually large capacity to endure anoxia and can tolerate this for about three months (Weber and Brändle, 1966). Various species may depend on anaerobic respiration for over-wintering in anoxic sediments and, particularly, for regenerating new leaves and shoots (e.g. Kausch *et al.*, 1981; Crawford, 1996). However, the reported ability of detached rhizomes of *Typha latifolia* to tolerate anoxia (Barclay and Crawford, 1982) does not seem to eliminate the dependency of this species on emergent dead shoots for over-wintering (Murkin and Ward, 1980). On balance, it seems likely that a substantial capacity for anaerobic metabolism may be restricted to a few wetland species, and it may provide a mechanism to endure short unfavourable periods rather than prolonged tolerance.

Seed survival and establishment in wetlands

In some wetlands, periodic drought or deep flooding can exceed the tolerances of some or all established wetland plants and lead to their death. With the return of appropriate conditions, re-establishment of wetland plants depends upon persistent propagules in the substratum or upon inward dispersal. Seasonal wetlands typically have a well-developed seed bank, often dominated by annual species (Haukos and Smith, 1993), which can survive many years of flooding (Salisbury, 1970). Sites with longer periods of flood or drought can have seed banks with a wider range of plant types (van der Valk and Davis, 1978; Keddy and Reznicek, 1982; Poiani and Johnson, 1989).

While re-establishment from seed is clearly important in seasonal and fluctuating wetlands, or in other sites subject to periodic damage (wave or current scour), its significance in 'undisturbed' wetlands, which are often dominated by long-lived phanaerophytes, is much less certain and is often assumed to be small. Long-term, persistent seed banks are only formed by relatively few perennial phanaerophytes – particularly, although not exclusively, species subject to frequent disturbance (Keddy and Reznicek, 1982; Voesenek and Blom, 1992) – although more have transient or short-term persistent reserves (Poschlod, 1995; Maas and Schopp-Guth, 1995). In particular, the rarer dicots, or those not very tolerant of waterlogged conditions (e.g. *Primula farinosa*), often do not have persistent seed banks (Maas, 1989), perhaps partly because their seeds have limited capacity to survive in a wetland environment (e.g. *Caltha palustris* (Skoglund, 1990); *Betula pubescens* (Skoglund and Verwijst, 1989)).

Ephemeral species are critically dependent upon detecting suitable germination conditions. Some annual mudflat species germinate best in shallowly submerged soils (e.g. *Elatine* spp.), but many species may require a virtual absence of standing water (Van der Valk and Davis, 1978). Wide temperature fluctuation can be indicative of an absence of deep water cover, or of gaps within the vegetation canopy, and is the most important cue to the germination of both ephemeral species (Baskin *et al.*, 1993; van der Valk and Davis, 1978) and many perennials (Thompson and Grime, 1983; Maas, 1989).

Tolerance of wetland plants to dry soils

In natural moisture gradients, the upper limit of the distribution of particular wetland species may be controlled by sensitivity to moisture stress, but surprisingly little attention has been paid to the tolerance of wetland plants to dry conditions. Nonetheless some, perhaps many, wetland plants can grow well in conditions drier than those in which they normally occur (e.g. Keddy, 1989), sometimes with reduced vigour (e.g. Conway, 1938; Wheeler *et al.*, 1985). This may help explain why it can be difficult to demonstrate clear relationships between experimental estimates of drought susceptibility and the field distribution of some wetland species (Schat and Beckhoven, 1991).

The absence of wetland species from drier habitats may have several causes: desiccation damage induced through water stress; constraints upon establishment; and competitive interactions with dryland species, in some cases conditioned by a reduction of vigour of wetland species in drier situations.

Water stress in wetland plants

Some wetland plants experience leaf water deficits, even in waterlogged soils (Bradbury and Grace, 1983), due both to high rates of water loss from the shoots and to constraints on water acquisition. Various wetland phanaerophytes have higher rates of 'cuticular' water loss from their leaves than do dryland plants (Hygen, 1953; Willis and Jefferies, 1963), and some insectivorous species have particularly limited control of transpirative water loss (Hygen, 1953), probably because water is readily lost from the leaf surface through the thin-walled secretory glands and tentacles. Other species may have inefficient stomatal control of transpiration. Conway (1940) reported that leaves of *Cladium mariscus* developed water deficits in hot, bright conditions, even when the plants were growing in water, and attributed this to restricted vascularisation. Many wetland species have shallow and meagre root systems (Metsävainio, 1931), adapted to avoid anoxia in waterlogged soils, and are little able to 'forage' for water in drier conditions. Such constraints may partly explain the occurrence of xeromorphic traits in some wetland plants (Schimper, 1898; Yapp, 1912), although some workers have suggested alternative possibilities. For example, Small (1973) concluded that xeromorphy had little to do with drought stress in bog plants but might be an incidental feature associated with slow growth in nutrient-deficient or cold-stressed environments. Jones and Etherington (1970) suggest that reduction of transpiration may help to reduce the uptake of soluble toxins from waterlogged soils.

Bryophytes are poikilohydric and have little ability to regulate water gain or loss but vary considerably in their capacity to acquire water and to withstand sustained desiccation. A number of species, including taxa that normally grow in very wet conditions, can survive some droughting (e.g. *Scorpidium scorpioides* (Kooijman, 1993); *Sphagnum cuspidatum* (Luken, 1985; Heikkilä and Löytönen, 1987)), although they may take longer to recover than do dryland species (Lee and Stewart, 1971; Bewley *et al.*, 1978).

Constraints on seedling establishment and species interactions

Seedlings tend to be more drought-sensitive than mature plants, and the absence of some wetland plants from dry soils may relate more to constraints on seedling establishment than to the tolerances of older plants (Ernst and Van der Ham, 1988). There is evidence that seedlings of some wetland plants establish less well in dry soils than do those of some dryland species (Evans and Etherington, 1991), in some cases, perhaps, because of an inability to produce questing roots that forage for a diminishing water supply (Reader et al., 1993).

There is also some evidence that some wetland species are restricted to wet habitats because potential competitor dryland species cannot grow so well in wet soils (Ellenberg, 1954; Keddy, 1988; Keddy et al. 1994).

WATER REGIMES AND PLANTS OF WETLANDS

Relationships between species, vegetation and water tables

The position of the water table in wetland soils undoubtedly exerts a major influence upon the distribution and performance of plant species and the composition of vegetation. This is especially obvious where there is a strong zonation of vegetation along a water level gradient, such as around the margins of lakes, where there may be a transition from swamp into fen and wet grassland, with woody vegetation (if present at all) often being furthest upslope (Spence, 1964). On a smaller scale, microtopographical variation is often associated with contrasting species distributions, as exemplified in the patterned surfaces of some ombrogenous bogs (Ratcliffe and Walker, 1958; Ivanov, 1981; see below).

Changing water tables also demonstrate the importance of water levels to wetland vegetation. Some wetlands oscillate naturally between 'dry' and 'wet' states and show compositional changes in time as well as space (e.g. van der Valk and Davis, 1978). Others are subject to long-term persistent changes, sometimes induced artificially (by drainage, etc.). Substantial, progressive change in water tables usually leads to some form of vegetation response, whether it be by flooding (e.g. Kadlec, 1962; Harris and Marshall, 1963; Rumberg and Sawyer, 1965; Sjöberg and Danell, 1983) or drying (Grootjans and Ten Klooster, 1980; van der Valk and Davis, 1980; Fojt and Harding, 1995; Grootjans et al., 1996). Observed responses of established plants to hydrological change depend upon the species and the magnitude of change and may include truncation of range, or complete disappearance, of species incapable of growing in the new conditions; persistence in sub-optimal conditions, often with reduced vigour; change in position of species along the gradient (if the topography permits); and continued, or even enhanced, growth in the new water depth range (Sjöberg and Danell, 1983). In addition, a range of new species may also colonise the new hydrological environment.

Numerous studies have examined the relationship between the distribution of wetland plant species (or plant communities) and water levels (e.g. Tüxen, 1954; Willis *et al.*, 1959a, b; Spence, 1964; Balátová-Tuláčková, 1968; Klötzli, 1969; Mörnsjö, 1969; Jeglum, 1971; Niemann, 1963; Grootjans and Ten Klooster, 1980; Egloff and Naef, 1982; Succow, 1988; Mooney and O'Connell, 1990; Scholle and Schrautzer, 1993). However, while it is possible to identify broad trends in water table–plant distributions in wetlands, the range of conditions occupied by particular species or communities can be wide and may be inconsistent between, or even within, sites. For example, the water level ranges of species and communities in shoreline zonations can differ considerably between lakes (Spence, 1964). There is some evidence that groups of species may show clearer relationships to water table behaviour than do individual species (Wierda *et al.*, 1997), but equally, because plant 'communities' are abstract, sometimes arbitrary and variable units, the use of syntaxa in exploring plant–environment interrelationships may sometimes be part of the problem rather than a road to its solution (van Wirdum, 1986). Willis *et al.* (1959a, b) were unable to demonstrate a sharp distinction between some communities in a dune slack toposequence based on their 'flooding index'. Thus, while it would be extremely desirable to establish robust hydrological limits for species and communities, this is often problematic. For example, Grootjans (1980) identified some water table limits for certain communities but, recognising the uncertainties involved, counselled against concluding that 'drainage within the indicated limits can be done without changing the floristic composition of the communities' – which suggests that the limits proposed have little practical value.

Difficulties in detecting clear relationships between water tables and species or community distributions in wetlands may stem from various causes, not least that other environmental (or biotic) variables may help to determine distributions, either independently of water tables or by modifying species' responses to water conditions. However, difficulties also arise because water regimes are often complex and difficult to characterise and relate to species distribution. Consideration of how variables influence species distributions in wetlands independently of water tables and water quality is outside the scope of this review, but attention will be given to some of the problems of relating species and vegetation to water table behaviour and to some of the ways in which other environmental and biotic variables can modify the interrelationships between plants and water conditions.

Relating wetland plants and vegetation to water table behaviour

There are several reasons why it may be difficult to relate species distributions and vegetation composition to water table behaviour:

1 Most studies are based on correlative procedures that assume equilibrium between the species (or vegetation) and the measured water regimes.
2 The behaviour of the water table in wetlands is often complex; it can be difficult to characterise and to identify hydrological variables salient to the distribution of plants; moreover, 'controlling variables' may differ for different species and hydrological situations.
3 In some circumstances, soil moisture content may be more meaningful to plant growth than the position of the water table.
4 Most studies have been concerned with established plants, whereas in some circumstances distributions may be determined by conditions that affect recruitment.
5 Other variables may modify plant–water table interrelationships.

These are considered further below.

Equilibrium conditions

The majority of field studies on plant–water table relationships in wetlands have been based upon some form of correlation (e.g. Nilsson and Keddy, 1988). Recently, some workers have used optimum species response curves (Huisman et al., 1993) to explore these relationships (Spieksma et al., 1995; Wierda et al., 1997). Such approaches require some form of stable equilibrium between water regime and vegetation, but this is more often assumed than investigated.

When considering equilibrium conditions, a broad distinction must be made between situations where water level flux is of sufficient magnitude or duration to drive changes in vegetation composition and those in which changing water levels occur within, and may be characteristic of, a more or less stable plant community. The first of these conditions is exemplified by the periodic drawdown cycle, and associated vegetation change, described for some prairie glacial marshes by van der Valk and Davis (1978). Here, any attempt to identify vegetation–water regime interrelationships would need to be made for each phase within the cycle – overall measurement would be meaningless. However, even within individual phases the vegetation is not necessarily in equilibrium with the behaviour of the water table (van der Valk and Davis, 1980).

Even in more stable wetlands, assumption of equilibrium between water table behaviour and vegetation may be invalid, either because species distributions within a given water regime are in a state of change, or because inertia against change permits the persistence of some species in sub-optimal conditions (Grootjans, 1980; van Diggelen et al., 1991b; Spieksma et al., 1995). Ironically, many, perhaps most, hydrological studies are made within regions subject to considerable recent, and often ongoing, water management (drainage by water engineers or re-wetting by conservation managers), where assumptions of equilibrium may be questionable. Disequilibria are also likely to be evident in systems where major community boundaries (e.g. extent of woody vegetation) are determined more by occasional extreme events (e.g. periodic deep flooding) than by persistent 'average' conditions (Hill and Keddy, 1992).

Identification and characterisation of salient components of water table behaviour

At many sites, water tables show substantial, and sometimes erratic, variation, which may be difficult to characterise simply. Each location has an 'average' water table (or 'wetness'), around which fluctuations occur. These vary in their magnitude, duration, frequency and timing (seasonality). The importance of these components in relation to plant distribution (inundation versus deficit; duration versus magnitude; winter flooding versus summer flooding; average values versus extremes; average extremes versus exceptional extremes) remains to be established fully and varies between species and situations. For example, the critical variables in wetlands subjected to protracted periods of inundation are likely to be different from those important in wetlands with high water tables but limited flooding, or in sites that experience periods of low water levels.

Various studies have drawn attention to the importance of different components of water table behaviour in regulating species distributions and vegetation composition, but it is quite difficult to extract generalities from these. This is because there are a relatively small number of studies, and not all of these have considered every component of water table behaviour; some studies are concerned with controls on species distribution or performance, others with vegetation composition or the position of community boundaries; individual hydrological variables may show strong cross-correlation (Nilsson and Keddy, 1988; Wierda et al., 1997); different components may be important in different wetland types; and variables may act in complex combination, which may be site-specific. For example, Johnson (1994) assessed controls on the expansion of marginal Populus–Salix woodland into river channels in response to reduced flow and concluded that the process was regulated by three main variables: low June river flow (which coincided with the main tranche of tree seed germination), summer drought, and ice coupled with high winter flows (the biggest impact on seedling survivorship).

Amplitude, frequency and duration of water table fluctuations

Amplitude of water level fluctuation is important both to vegetation composition and to community boundaries. Its impact depends upon its magnitude and duration and also upon the identity and stature of the plant species concerned. A 'high' water level may just saturate the soil or it may completely submerge some or all of the wetland plants; a 'low' water table may be just below the surface or well below the rooting zone. The impact of a given water level change depends upon its magnitude relative to the dimensions of the species growing in a wetland, not just upon its absolute value.

Water table fluctuations of large magnitude and duration can cause catastrophic change to some or all components of the established vegetation. The effect of such fluctuations depends much upon their frequency. Wetlands with frequent large fluctuations may be characterised mainly, or exclusively, by opportunistic annuals and ephemerals capable of growing and reproducing on moist muds for the duration of their exposure (Poiani and Johnson, 1989) (see 'Seed survival and establishment in wetlands'). With less frequent, or lower-amplitude, fluctuations, perennial vegetation can survive, with community composition and limits determined partly by the water level range. In some shoreline situations, regularly flooded zones at lower elevations may be composed mainly of ruderal species that can effectively avoid the effects of flooding, while the higher levels are composed of flood-tolerant perennials.

Wide amplitude fluctuations are particularly associated with some lake shoreline and river margin environments and can have considerable repercussions on vegetation composition and zonation. Hill and Keddy (1992) observed that, for some lake shoreline vegetation, greater amplitude of water levels was associated with reduced development of woody plants, a wider zone of wet meadow and greater species diversity. In this instance, constraints on scrub invasion – perhaps caused more by periodic flooding episodes than by 'average' water conditions – may have promoted high diversity and a large number of rare species, but in other situations a high amplitude of water table fluctuation is often associated with reduced species richness, as Weiher and Keddy (1995) have demonstrated experimentally.

The duration of hydrological events is of much importance to the character of wetland vegetation. In sites subjected to flooding episodes, species distributions can be determined by the duration, as well as the depth, of inundation (Noest, 1994). Tree species vary considerably in the duration of flooding that they can tolerate (Gill, 1970; Kozlowski, 1984), and, in wetlands subjected to regular water level change, such as freshwater tidal wetlands, the lower limits of tree species can be related to both the duration and frequency of periods of inundation (e.g. Zonneveld, 1959), although these effects are difficult to separate.

Long-period fluctuations can have particularly complex effects. Keddy and Reznicek (1982) describe shoreline vegetation that varies from being species-poor in high-water phases to species-rich in low-water phases. They further suggest that the species richness of the low-water phase is itself partly a product of the unstable water environment, because the periodic high-water phases prevent the establishment of dominance, and concomitant reduction of diversity, during the low-water phases.

In somewhat less variable hydrological contexts, Spieksma *et al.* (1995) suggest that the most discriminating variables (with regard to the occurrence of plant communities) are the mean, highest and lowest groundwater levels together with the possibility of inundation during the growing season. Wierda *et al.* (1997) concluded that mean highest water level was particularly important in controlling the occurrence of plant species in some types of wetland, along with amplitude of fluctuation. Scholle and Schrautzer (1993) used the combination of 'average' water level, a 'groundwater fluctuation index' (which assessed the fluctuation pattern) and the duration of inundation to characterise six water regime types, each of which could be related quite closely to vegetation composition. Various workers have concluded that the cumulative period of time for which a particular water level is exceeded can provide a sensitive characterisation of hydrological regimes with regard to vegetation composition (Willis *et al.*, 1959a, b; Klötzli, 1969; Niemann, 1973; Grootjans and Ten Klooster, 1980). This concept takes into account elements of both the magnitude and duration of water level fluctuation. Plots of water level against the length of time that each value is exceeded are known as *duration lines*, and importance is attached not just to their absolute values but also to the shape of the curves, which can sometimes be used, *inter alia*, to deduce water sources. Nonetheless, duration lines constructed from different examples of one vegetation type can show considerable variation, limiting their predictive value (van Diggelen *et al.*, 1991b). Moreover, they are most appropriate for situations in which periods of low water levels control vegetation composition.

Effects of timing and seasonality

Rather little is known in detail about the effects of timing and seasonality of water table fluctuations upon wetland plants and vegetation. Reported examples sometimes have limited general applicability, but there is no doubt that plant species show different seasonal responses to water regimes. In experimental conditions, *Urtica dioica* has been shown to be intolerant of summer flooding but can survive autumn flooding, while *Phalaris arundinacea* can tolerate both (Klimesova, 1994). In general, flooding during the growing period seems to have the greatest impact on plant distributions (Noest, 1994).

In an early study at Wicken Fen (England), Godwin and Bharucha (1932) concluded that the main hydrological control on the distribution of vegetation, and particularly the lower limit of bush growth, was determined by winter water excess rather than summer water levels. However, it seems unlikely that these observations have general applicability. Many examples of well-established fen woodland are regularly inundated during the winter, and, in general, trees seem better able to survive flooding in winter than during the growing season (Kramer, 1969; Gill, 1970). The duration of summer flooding is negatively associated with the abundance of trees in wetland vegetation (e.g. Gill (1970) suggested that colonisation by woody plants could be prevented by flooding for more than about 40 percent of the growing season). Toner and Keddy (1997), using regression models, likewise conclude that the distribution of wooded vegetation in a riparian wetland was best accounted for by a combination of two variables: the last day of the first flood and the time of the second flood. They suggest that the longer the period between the two floods, the greater the chance that trees would recover sufficiently to withstand the second flood and hence the more likely the occurrence of wooded vegetation.

There is thus evidence that species distribution and the composition, and community limits, of wetland vegetation can be influenced by occasional extreme hydrological events, by average minima and maxima, by 'average' water levels, and by the frequency and duration of fluctuations and the timing of these events. Since the importance of these terms can differ both with type of wetland and species, it is scarcely surprising that clear and consistent relationships between plants and water table behaviour have sometimes proved difficult to detect!

Water tables versus soil moisture content

Most investigations of plant–water interrelationships in wetlands have concentrated on the behaviour of water tables. The measurement of water tables may well be appropriate when they are high, but when they sink substantially below the surface (or, perhaps more critically, well below the main rooting zone), soil moisture content may be more important to plant growth than the position of the water table. At any given low water level, the moisture content of the upper soil layers can vary considerably between soil types (von Müller, 1956) due to variation in soil capillary properties and moisture removal via evapotranspiration. In 'dry' conditions, the behaviour of water tables may be uncoupled from the soil moisture conditions experienced by the vegetation. The significance of this, in relation to understanding the plant ecology of wetlands, has been little explored, but in such conditions measurements of soil moisture, or redox potential, in the rooting zone may be more meaningful than water depth (de Mars, 1996) (see below).

Other water variables

Additional to the behaviour of the water table, it is important to recognise that in some circumstances water may have a further impact upon the composition and community limits of wetland vegetation, in terms of wave action and water (or ice) scour. These can influence the overall character of zonations and can modify the distributional limits of wetland plants along a shoreline water level gradient (Keddy, 1983; Shipley, Keddy and Lefkovitch, 1991).

Relationship to recruitment conditions

Many investigations have focused particularly upon the relationship of water table behaviour to the distribution of established plants in vegetation, but in some instances conditions prevailing during the regenerative phase may determine distributions. This effect is particularly important for ephemeral species but can also apply to perennials (see above, 'Seed survival and establishment in wetlands'). Ernst and van der Hamm (1988) reported that exceptional flooding permitted the germination of *Schoenus nigricans* above the normal limits of seedling recruitment of the plant in dune slack systems and, because the mature plants could survive at this elevation in 'normal' conditions, this led to a local expansion in the water level range of the plant.

The timing of hydrological events may be important to the regeneration of wetland plants from seeds, especially to those that germinate in response to a falling water table near the start of the growing season and that have short-lived seeds (such as *Salix* spp.) (Johnson, 1994; Klimesova, 1994). Van Splunder *et al*. (1995) concluded that the initial zonation of some tree species in a riverine forest could be related to the timing of seed dispersal and water level fluctuation, but in general the importance of such processes to the composition of wetland vegetation is not well known.

Various workers have examined the importance of recruitment conditions to the formation of species zonations on lake and river shorelines. There is considerable evidence that the germination and early life history character-istics of some species are related to the position that they occupy in shoreline zonations (e.g. Keddy and Ellis, 1985; Keddy and Constabel, 1986; Welling *et al*., 1988; Squires and van der Valk, 1992; Blom *et al*., 1994; Coops *et al*., 1996; van Splunder *et al*., 1995), but, in general, little relationship has been found between the distribution of seeds in the soil and zonations of established plants (Keddy and Reznicek, 1982; Welling *et al*., 1988). In general, it appears that post-recruitment processes are considerably more critical to the development of shoreline zonations than are seed distributions (Welling *et al*., 1988).

Effects of other environmental variables on plant–water table relationships

Relationship of water table to oxidation–reduction potentials

There is a general negative correlation between water table height and oxidation–reduction potentials in wetlands (Giller and Wheeler, 1986a; Shaw and Wheeler, 1990; de Mars, 1996), but the relationship is far from exact, and important discrepancies have been reported. De Mars (1996) observed that the relationship between depth of water below the surface and redox potentials (measured at 15 cm) varied with the type of wetland: redox potentials at 15 cm depth were consistently quite low (<300 mV) in some rich fens, even when water levels were well below the surface, whereas in poor fens and bogs there was a greater tendency for higher redox potentials to develop at 15 cm depth when water levels were low. He attributed this to greater microbial activity in the rich fens.

Likewise, water table fluctuations are not necessarily accompanied by a corresponding change in redox potential. Giller and Wheeler (1986a) observed that a large (>10 ha) area of solid peat had a lower water table and higher redox potentials in summer than did an adjoining area of fen with a semi-floating, saturated vegetation raft, but they also found that the higher redox potentials persisted throughout the winter, when the area of solid peat was itself inundated. De Mars (1996) observed a similar 'redox blockade' in a once drained, since re-wetted, fen and attributed it to a scarcity of readily decomposable organic material.

Since available concentrations of some phytotoxic elements are related to redox potentials rather than water levels, redox measurements may give a better estimate of some environmental conditions experienced by plant roots than do water levels *per se*. Redox potentials are also generally less labile than are water levels, and de Mars suggests that they may be better for investigating plant–water regime relationships. He proposes the use of redox duration lines (in conjunction with water level measurements). Although these arguments are persuasive, Snowden and Wheeler (1995) found a strong and highly significant correlation between the iron tolerance index of plant species and the summer water table of the soils in which they grew, but not with soil redox potential.

It remains to be established whether redox potentials provide a better predictor of plant distributions than do water levels. At present, few relevant synoptic redox data have been published. However, it seems likely that the uncoupling of redox potentials from water levels in certain wetland soils, and the varying relationships in different wetland types, almost certainly contributes to the difficulties of detecting clear relationships between plant distribution and water table behaviour.

Influences of lateral water flow

Some lateral water movement occurs in many wetlands, but its importance to plant growth and distribution has received rather limited study. Potentially complicating factors include those of scale and rate: slow lateral flow within wetland soils or affecting an entire wetland may have different, and probably less obvious, floristic impacts from more rapid flow concentrated into narrow water tracks.

Kulzyncski (1949) emphasised the importance of lateral water flow in his classic study of the peatlands of Polesye (Prypet Marshes), and water movement characteristics formed the basis of his classification of peatlands. He interpreted many aspects of the types of mire that occurred, and the characteristics of their peat and vegetation, in terms of the effects (direct and indirect) of water flow. For example, he suggested that peatlands fed by strong fluvial flow tend to form a 'heavy' peat across which water flows, whereas those subjected to smaller flows may form a 'light', partly floating peat (which, when developed, may help to reduce flow rates yet further). Various other workers have also recognised that water movement can have a significant effect on the character of wetland environments and the composition of wetland vegetation (e.g. Malmer, 1962a; Heinselman, 1970; Ingram, 1967; Daniels and Pearson, 1974).

Several studies have shown that zones of moving water within wetlands tend to have higher redox potentials than do more stagnant examples (Sparling, 1966; Armstrong and Boatman, 1967; Ingram, 1967); that this may be associated with a lower availability of phytotoxins with redox-related solubilities (Fe^{2+}, Mn^{2+}, S^-); and that some species (e.g. *Molinia caerulea*) can grow better in water tracks than in stagnant, waterlogged soils. Shaw and Wheeler (1991) found that the mean (±SE) redox potential (at 10 cm depth) in soils of soligenous fens was 297 (±32) mV while that for topogenous fens was 223 (±26) mV. This may help to explain the occurrence of some typical dryland plants (e.g. *Briza media*, *Linum catharticum*, *Trifolium pratense*) in spring-fed fens. However, even in soligenous sites the wetland plant *Parnassia palustris* still often grows in elevated microniches (Eades, 1997).

There is no doubt that water movement can increase the availability of nutrients to plant roots, in terms of both greater import of allochthonous solids and increased rates of solute supply (Kulczynski, 1949; Gorham, 1950; Ingram, 1967). Chapin *et al.* (1988), examining the influence of flow in water tracks in the Alaskan tundra, concluded that the observed flow rates were six and eight times as rapid as the diffusion of phosphate and ammonium in water and probably explained the ten times greater productivity of *Eriophorum vaginatum* in the water track compared with the adjacent tundra. They also found slightly higher soil temperatures, deeper thaw, higher soil phosphatase and protease activity and quicker N mineralisation, although some of these trends relate specifically to the location of the study within the tundra. Sparling (1966) has also shown that increased flow rates tend to be associated with an increase in pH and, in some cases, bicarbonate concentration, together with a diminution in aluminium concentration in mire waters.

One problem in assessing the impact of water flow on solute availability and floristic composition is that of distinguishing truly flow-related phenomena from those determined by differences in water source. For example, Bellamy (1968) combined Kulczynski's approach to peatland classification with a large number of chemical analyses to distinguish major types of peatland, but his examples of rheophilous (= 'flow-loving') mire were mostly also fed by base-rich water sources. He did not examine the floristic trade-off between flow rate and solute concentration, nor have most other authors, although Sjörs (1950b) suggested that variation in vegetation composition along the rich-fen–poor-fen gradient (see below) within individual mires may be partly a product of varying flow rate. Variation of flow rates in endotelmic water tracks in ombrogenous peatlands would provide an ideal context in which to study the effects of water flow *per se* on aspects of hydrochemistry and vegetation composition, but even here care would be needed to ensure that these tracks receive no telluric water (Malmer, 1962a).

Interactions with chemical variables

The reducing conditions found in wetland soils can increase the availability of some phytotoxins, especially Fe^{2+}, Mn^{2+} and S^-. The precise part that these play in determining the distribution of plant species within wetlands is not fully known, but Snowden and Wheeler (1993) showed that species that were intolerant of high concentrations of iron supplied in laboratory experiments were also restricted to wetland sites that had small concentrations of extractable iron. Some iron-sensitive species, such as *Epilobium hirsutum* and *Filipendula ulmaria*, can grow widely in the drier parts of many wetlands, with relatively high redox potentials in the rooting zone; they can also grow in wetter conditions when the soils contain small concentrations of reduced toxins (Wheeler *et al.*, 1985). It is not certain whether the damaging effects of high iron concentrations on these species are a form of direct toxicity, or that they result from an interference with the plant's phosphorus metabolism (Snowden and Wheeler, 1995), but the availability of reduced iron seems to regulate the water table ranges occupied by some plants.

Soil nutrient status may also modify species' responses to waterlogging, although rather little seems to be known about this. Ellenberg (1988) suggested that the representation of dryland species in wet grasslands can be increased without drainage by nitrogen fertilisation. Recently, Eades (1997) has observed a similar effect experimentally using the wetland species *Anagallis tenella*. When grown on its native soil (a nutrient-poor calcareous peat, with very small total concentrations of Fe and Mn), optimum growth was found where the water level was kept at 10 and 20 cm below the surface; most plants died with surface water levels. However, when grown in a nutrient-enriched peat there was good growth at all three water levels, with no significant differences between the means. The basis for these responses is not known (they may be related to observations that nitrate application can reduce symptoms of waterlogging damage in young shoots of barley (Drew *et al.*, 1979), but such observations suggest that the water regime tolerances of some wetland plants may be partly determined by the nutrient status of the substratum in which they grow. If so, this raises the possibility that nutrient enrichment of wetlands may facilitate colonisation by species that might have been excluded from nutrient-poor examples on account of their sensitivity to waterlogging. However, there is also some evidence that in some situations excessive nutrient enrichment may affect the ability of wetland plants to tolerate anoxia. Koncalová *et al.* (1993) presented evidence that high nitrogen supply can reduce aerenchyma formation in *Carex* spp., although the significance of this to the survival and distribution of the plants is not known.

Effects of species interactions upon plant–water table relationships

Competition and water table ranges

The water level range occupied by individual species can be modified by competitive interactions. One of clearest demonstrations of this was provided by Grace and Wetzel (1981), who examined the distribution of the closely related species *Typha angustifolia* and *T. latifolia* along the shoreline gradient of a small experimental pond. When both species were present, *T. angustifolia* usually grew in the deeper water, below *T. latifolia*, but in the absence of the latter *T. angustifolia* not only grew throughout the water level range otherwise occupied by *T. latifolia* but grew best where the water level was lower than its limits when growing with *T. latifolia*. This suggests that in natural situations *T. angustifolia* may be excluded from parts of its potential water level range when competitors such as *T. latifolia* are present. By contrast Van der Valk and Welling (1988) found little evidence that zonation patterns were much influenced by species competition.

Keddy and co-workers have also examined the role of competition in regulating the distribution of shoreline species, and in some instances were unable to show that it had much importance (Shipley *et al.*, 1991). However, Keddy (1989) reported a field experiment that examined whether the restriction of herbaceous vegetation to the lower shore, and its replacement by woody vegetation higher up, was a result of water level preferences or competitive interactions. He showed that up-shore patches cleared of scrub developed more lower-shore species than did uncleared examples, although within the four-year period examined the response was rather slow and involved less than one-quarter of the herbaceous flora (mainly those species with a large buried seed bank such as *Drosera intermedia*). Locations with frequent disturbance and low fertility showed no evidence for such competitive release. More recently, Keddy *et al.* (1994) have shown that flooding can exert a significant influence upon the competitive response of seedlings of some wetland plants grown in combination experimentally, although the relevance of this to field distribution has yet to be established fully.

Kotowski *et al.* (in press) have recently compared mean water tables and water table amplitude associated with selected plant species in two hydrologically comparable, but widely separated, fens (Drentse A, the Netherlands; Peene, eastern Germany). They observed that the majority of species examined were associated with wetter conditions and greater water level flux in the Peene valley. The Peene plants also tended to have narrower water table ranges, which the authors attributed to greater species interactions, possibly caused by a larger species pool.

Much more work is needed to assess fully the impact of competition in modifying the relationship of species to water tables, but existing evidence suggests that it may be of considerable importance. If so, for some species at least, it may be more meaningful to consider their water table relationships within specific communities, or species pools, than to identify overall trends.

Facilitative oxygenation

The recognition of the occurrence of radial oxygen loss (ROL) from the roots of certain wetland plants had led to a recurrent suggestion that some well-adapted wetland species (i.e. species with high rates of ROL) may oxygenate wetland substrata sufficiently to permit some more waterlogging-sensitive species to grow in situations that would otherwise be too wet for them. However, effective oxygenation is likely to require a high root density (Justin and Armstrong, 1987), and while there is no doubt that some species (e.g. *Juncus subnodulosus*, *Schoenus nigricans*) can greatly increase the redox potential in experimental waterlogged root chambers (Eades, 1997), their capacity to oxygenate field substrata is less certain and is doubted by some workers (e.g. Bedford *et al.*, 1991). Nonetheless, Hansen and Andersen (1981) and Armstrong *et al.* (1992) have demonstrated substratum oxidation by *Phragmites australis*, and Schat and Beckhoven (1991) present some evidence to suggest that the growth of *Plantago coronopus* in wet sites in Dutch dune slacks may be facilitated by ROL from *Juncus maritimus*. This interaction has also been reported for salt marsh plants (Hacker and Berness, 1995). The topic has been the focus of debate between Sorrel and Armstrong (1994) and Bedford and Bouldin (1994), but with particular reference to methodological constraints.

A related form of facilitative oxygenation may be provided by photosynthesis of wet bryophyte mats, but again existing evidence is largely inconclusive. Some semi-submerged mats of *Scorpidium scorpioides* and *Drepanocladus* spp. have much higher redox potentials than those measured only a few centimetres beneath them, and free S^- detected below the mats is not present within them (Giller and Wheeler, 1986a; Sellars, 1991). This may help to explain why some waterlogging-sensitive species are sometimes found associated with bryophyte mats in fens (Clapham, 1940), but critical evidence is lacking.

More work is required to establish the importance of facilitative oxygenation in field conditions. If it turns out to be significant, it has obvious implications for assessing the water regime tolerances of waterlogging-sensitive species, because their actual field range could be strongly influenced by the presence or absence of strong oxygenators.

The hummock-hollow gradient of Sphagnum species in ombrogenous bogs

Some important insights into the complexity of water table relationships of plants in wetlands have come from examination of the microtopographical distribution of *Sphagnum* species. Many ombrogenous bogs have patterned surfaces, created by the growth of peat-forming plants, within which various topographical components can be recognised (e.g. pools, hollows, lawns and hummocks). Different plant species tend to occupy different parts of the microtopographical gradient (e.g. Tansley, 1949; Ratcliffe and Walker, 1958; Vitt *et al.*, 1975; Ivanov, 1981; Andrus *et al.*, 1983; Luken 1985; Andrus, 1986). Ivanov (1981) provides details of water levels associated with different species of *Sphagnum*.

Water relations are considered to be important in determining the microtopographical zonation of *Sphagnum* species and have been the subject of quite detailed research, but the relationships are not simple. For example, in experimental conditions the hummock species *S. capillifolium* survives desiccation less well than does the pool species *S. auriculatum* (Clymo, 1973; Clymo and Hayward, 1982). However, the hummock species do appear better able to acquire and retain water in low-water conditions.

When water tables are relatively low (10 cm below the surface) hummock species (especially *S. capillifolium*) have a greater tissue moisture content and growth rates than do lawn (*S. papillosum*) and hollow (*S. cuspidatum*) species grown in similar conditions (Luken, 1985; Heikkilä and Löytönen, 1987). This seems to be because they have a greater capacity for water acquisition, being better able to transport water along a continuous capillary path to the capitula on account of small capillary spaces between the leaves (Clymo, 1973; Clymo and Hayward, 1982; Hayward and Clymo, 1983; Wagner and Titus, 1984; Titus and Wagner, 1984). Moreover, high stem densities generally retard evaporative water loss (Luken, 1985), and stem dichotomies occur more often among *Sphagnum* plants growing in drier habitats (Clymo, 1970; Lane, 1977). It thus appears that *Sphagnum* species of lawns and hollows are less well adapted for growth at higher elevations than are hummock species.

Although their field distribution is skewed away from hollows, hummock species of *Sphagnum* are not intrinsically intolerant of wet conditions: *S. capillifolium* can grow better in hollow conditions than on hummocks, even though it is a hummock former (Clymo and Reddaway, 1971), and measurements of the photosynthetic capacity of *Sphagnum* transplants have shown that hummock species can readily tolerate the environmental conditions of hollows (Rydin and McDonald, 1985). However, in very wet conditions, hollow and pool species seem able to grow more rapidly than hummock species (Clymo and Reddaway, 1971; Pedersen, 1975; Luken, 1985) and may consistently out-compete hummock species (Andrus, 1986). Pakarinen (1978) studied the production of three *Sphagnum* species in southern Finnish raised bogs and showed that annual increment in length and net productivity per unit area was smaller in *S. fuscum* than in the two hollow species *S. balticum* and *S. majus*. The greater productivity of pool species may be partly due to a longer growing period (Hulme and Blyth, 1982).

There is also some evidence that the distribution of some *Sphagnum* species may be influenced by positive inter-specific interactions. Under dry conditions, *S. magellanicum* appears to have a greater capacity for water acquisition than does *S. papillosum*. It has a better water transport ability and greater water content, correlating with greater stem diameter, greater pore number and smaller leaf size (Li *et al.*, 1992). However, at higher hummock positions, *S. magellanicum* may permit *S. papillosum* to occupy higher elevations than would be possible if it were a single-species stand. This proto-cooperation is thought to occur through lateral transport of water along interconnecting fascicular branches (*ibid.*). Similar interactions have also been observed between *S. fuscum* and *S. balticum* (Rydin, 1985) and between *S. tenellum* and *S. capillifolium* (Heikkilä and Löytönen, 1987).

Some approaches to the assessment of plant–water table relationships

It is clear that not only are the water table relationships of wetland plants potentially difficult to identify but they are also susceptible to modification by other influences. In consequence, while trends can be identified, informed workers are likely to hesitate before making exact statements of the water regimes appropriate for individual species or communities, either in terms of optimal or tolerable conditions, because these may well be simplistic or inaccurate (Newbould and Mountford, 1997).

Intuitive assessments

A comprehensive attempt to assess the water conditions associated with particular plant species was made by Ellenberg (1974) for 2000 Central European plant species. The character, value and limitations of Ellenberg's subjective twelve-point scale of 'moisture value numbers' has been discussed by Wheeler and Shaw (1995). The coarseness of the scale (many wetland plants belong to just points 8 or 9) means that while there is little reason to dispute the values assigned to most species, they do not provide a sensitive indicator of differences within the wetland habitat. Mountford and Chapman (1993), examining plant species distribution on ditch banks, concluded that Ellenberg values were correlated with measured height of species above ditch water level, but inspection of their data suggests that species with values between 6 and 10 (which includes all wetland species) either occupied similar ditch bank ranges or showed inconsistent variation.

Models

Various workers have explored the manipulation of Ellenberg values in the calculation of indices and the development of descriptive and predictive models (Ter Braak and Gremmen, 1987; Gremmen et al., 1990; Mountford and Chapman, 1993). While such developments are of interest, they can be limited by their intrinsic assumptions. For example, they may take little or no account of the varying amplitude of response of a given species to water tables (Gremmen et al., 1990). Moreover, such numerical indices and models are unlikely to be more reliable than the crude scale on which they are based, yet they can generate a numerical precision absent from the original categorisation.

To be realistic, the assessment of species–water table relationships needs to take account of abiotic and biotic interactions affecting species' response to water conditions as well as water table behaviour itself. Modelling may provide one route to achieving this. Van Wirdum (1986) and Olff et al. (1995) have reviewed the properties of some models used in the Netherlands. Broadly, there have been two main approaches: (1) based on 'expert judgement' in which indicator values of the individual species are used to derive an estimate of the site values for particular variables (e.g. the WAFLO model: Gremmen et al., 1990 and the MOVE model: Latour et al., 1993); and (2) based on measurements of environmental variables at a large number of sites supporting particular species (e.g. the ICHORS model: Barendregt et al., 1986). The first approach has the obvious limitation of subjectivity, the second the frequent lack of truly comprehensive data sets coupled with the difficulties of identifying meaningful determinative relationships between environmental variables and species distributions using purely correlative procedures (Van Wirdum, 1986). In addition, some 'mechanistic' models have also been derived that include the description of fluxes of carbon and nutrients. These models describe the interrelation between plant composition and soil conditions and permit some predictions on competition between plant species. However, the value of any model can be no better than the data and assumptions on which it is based, and it is clear that there is little conclusive information on many plant–water table relationships.

Experimental approaches

Targeted, controlled experiments can generate important insights about plant–water table relationships, but the limitations of experimentation must also be recognised. Some of the simplest studies have examined species growth with respect to controlled variation in water levels and, while some report that growth in experimental conditions shows a clear relationship to field behaviour (e.g. Coops *et al.*, 1996), this is not always so. It seems likely that some, or all, of the variables that modify species' responses to water tables in the field may be equally important in determining the outcome of growth experiments, perhaps to the extent that the concept of a determinable 'physiological optimum', with specific respect to water level, has limited value – as does simple screening of species' responses along water gradients. Ellenberg (1954) was far-sighted in demonstrating that, in experimental conditions, the water table range occupied by plant species is influenced by species' interactions, but he considered only a very small range of species. The most ambitious comparisons of interactions, reported by Keddy *et al.* (1994) and Weiher and Keddy (1995), have involved prodigious experimentation yet still included only twenty species and a limited range of 'environments', whereas the true complexity of species–environment interactions is potentially much greater than this.

IMPACT OF VEGETATION ON WATER DYNAMICS

Hydroregulation by vegetation

Wetlands may contain various structures that help to regulate their hydrological environment, both to retain and to dissipate water. These include the occurrence, area and disposition of hollows and pools (which provide surface water storage) and the distribution of water tracks, etc. (Sjörs, 1950b; Ivanov, 1981; Beets, 1992). In peatlands, insofar as such structures have been formed by the variable growth of plants and the differential accumulation or decomposition of peat, they can be regarded as some form of hydroregulation generated by the (past) vegetation. But there are also more immediate and direct forms of hydroregulation produced by the present vegetation. One important topic (rates of evapotranspiration from wetland vegetation) is considered by Wetzel (see Chapter 8, particularly Table 8.8) and is not discussed further here, but some consideration will be given to two vegetation-produced hydroregulatory structures: the *acrotelm* and *vegetation rafts*.

The acrotelm

As outlined by Ivanov (1981), the *acrotelm* is the thin (<1 m) 'active' layer that forms the surface skin of peatlands, superimposed upon a less active *catotelm* (which usually represents the bulk of the deposit). The key differences between the two layers are listed in Table 5.1. Some of these attributes have been questioned by subsequent workers (e.g. the catotelm may not be as 'inactive' as Ivanov implies, not least because substantial peat decomposition

Table 5.1 Differences between the layers of a peat deposit (after Ivanov, 1981).

Acrotelm	Catotelm
Living plant cover + protopeat	Peat
Intensive exchange of moisture with the atmosphere and the surrounding area	Very slow exchange of water with the underlying mineral strata and surrounding area
Frequent fluctuations in water level and moisture content	Constant, or little changing, water content
High hydraulic conductivity and water yield (rapidly declining with depth)	Very low hydraulic conductivity (3–5 orders of magnitude lower than the acrotelm)
Access of air to pore space	No of atmospheric oxygen to pore s
Large aerobic microbial population, facilitation of decomposition of plants and transformation into peat	Reduced microbial population, lacking aerobes

can occur within it: Clymo, 1984a). Ivanov suggests that the key feature of the acrotelm is its fluctuating water table, which enables it to function as both an aerated layer and a peat-forming layer, and proposes that the thickness of the acrotelm is 'equal to the distance from the surface of the mire to the average minimum level of water in the warm season.' Because of this, and perhaps also because of its convenience to hydrologists interested in modelling the hydrodynamics of peatlands, the catotelm is often defined as being permanently saturated, with the acrotelm as an unsaturated layer (e.g. Kirkby *et al.*, 1995). However, such usage has led to some confusion, because, whereas most peatlands have a surface unsaturated layer, this does not always have all of the properties of the acrotelm proposed by Ivanov. For example, the unsaturated layer of peatlands damaged by drainage or peat extraction does not necessarily have high hydraulic conductivity and water yield (Schouwenaars and Vink, 1992), nor does it usually form 'protopeat'. It can be argued that the residual massif of a cut-over peatland is more of a catotelm with an unsaturated upper layer than a truly diplotelmic deposit, as is implicit in the use of the term *haplotelmic* for such sites (Ingram and Bragg, 1984). Nonetheless, other workers (interestingly, apparently including Ivanov*) have generalised the acrotelm–catotelm concepts to apply to badly damaged peatlands as well as intact ones.

* Ivanov (1981), p. 45: 'Draining adds to the acrotelm a significant part, and sometimes even the whole thickness of the peat deposit.'

The scope of the acrotelm concept is not just a matter of terminological niceties but relates to the eco-hydrological function of this layer in peatlands. In a natural ombrogenous bog, the acrotelm layer may provide some hydrological regulation, dependent specifically upon properties such as high hydraulic conductivity and water yield (Ivanov, 1981; Ingram, 1992; Joosten, 1993; Schouwenaars, 1996) that are not features of the aerated upper layer of a damaged catotelm. The combination of high hydraulic conductivity and high specific yield means that the acrotelm allows rapid dissipation of water excess (e.g. heavy rainfall, snowmelt) without a significant rise of water level. During drought the acrotelm may shrink, bringing the mire surface closer to the water table. Ivanov (1981) suggests that these functions are hydroregulatory and thus provide a positive feedback in which the plants that form the acrotelm help to produce conditions appropriate for their continued growth. Such regulation may be especially important in ombrogenous peatlands, particularly for examples in regions subjected to periods of drought (Joosten, 1993), although the magnitude and exact mechanisms of such postulated hydroregulation have yet to be established critically. *Sphagnum* is particularly important to acrotelm hydroregulation on account of the loosely woven, expansible surface it often creates (Bragg, 1989), its capacity to store water (Romanov, 1968; Schouwenaars and Vink, 1992) and reduction of evapotranspiration by 'mulching' and albedo of bleached *Sphagnum* (Schouwenaars, 1990). Such hydroregulatory properties may be a particular feature of spongy *Sphagnum* surfaces (Ingram, 1992; Joosten, 1993; Schouwenaars, 1996); it is not clear whether the acrotelm of bogs naturally dominated by graminoids (e.g. *Eriophorum* spp.) or tree species has a comparable capacity.

The hydrological importance of the acrotelm in fens is not well known. Fens have a periodically unsaturated surface layer, but the depth of this can vary very considerably and its hydroregulatory function remains to be established. It is possible that *Sphagnum*-dominated poor fens may have a capacity for hydroregulation comparable to that of bogs (Ingram, 1992), but the hydrodynamics of many fens are controlled by external events independent of any properties of their surface layers. For example, the stable water regimes of some soligenous fens may be imposed more by the constancy of spring flow than by internal mechanisms, while in many other fens seasonal variation in recharge generates strongly fluctuating water levels. However, the upper layers of some little-modified rheo-topogenous fens may provide the main water flow paths through the systems (Kulczynski, 1949; Succow, 1988).

Although the status and role of the acrotelm in fens is mostly uncertain, some fens do possess structures that provide some measure of hydrological self-regulation, created by vertically mobile mats of vegetation. These undoubtedly respond to variation in water supply and in some circumstances may help to regulate the hydrodynamics of the wetland.

Vegetation rafts

While hydrologists are often particularly concerned with absolute water levels, the growth and distribution of wetland plants is related more to relative water levels (position of the water table relative to the soil surface, or to the rooting zone). This distinction becomes important when the vegetation surface has some capacity for vertical mobility, as in vegetation rafts.

Vegetation rafts are perhaps most obviously associated with some types of hydroseral colonisation of (usually small) water bodies, both in natural wetlands and in reflooded turbaries ('turf ponds'). However, raft-like surfaces have also developed in a range of other situations, and quaking, tremulous surfaces are a feature of many fens. Rafts vary enormously in their character and thickness, from thin skins of hydroseral vegetation formed from the entangled rhizomes of hydrophytes growing over deep water and muds, to thick accumulations of relatively solid peat that have 'broken away' from lower layers of peat. The thicker examples are generally the least mobile, and some rafts are grounded except at high water levels (e.g. Yabe and Onimaru, 1996).

Kulczynski (1949) paid particular attention to vegetation rafts and observed that buoyant layers of peat frequently occur on top of heavier, solid peats, with varying degrees of independence, sometimes separating only in high water conditions. His so-called *dysaptic* structures consist of an admixture of tall plants (immersive perennials) rooting into a solid underlying layer of peat topped by a buoyant mat of semi-floating vegetation (Mörnsjö (1969) also provides some good examples of this). The two layers may experience very different environmental conditions. However, Kulczynski observed that the upper profiles of many fens did not show the contrasting layering of a dysaptic structure but had a more uniform, unseparated profile in which the uppermost layers were generally less humified and which 'undergo the biggest changes in their volume when the . . . water level oscillates.' This *cryptodysaptic* structure – which is widespread in many wet, topogenous fens – is more of an expandable peat mass than a true raft and has some obvious affinities with the acrotelm. Kulczynski (1949) emphasised that both crypto-dysaptic and, especially, dysaptic structures provide an important water storage role that helps to reduce water level change. However, a buoyant raft also damps water level fluctuations relative to the vegetation on account of its mobility (Buell and Buell, 1941; Giller and Wheeler, 1988; van Wirdum, 1991), and the key difference between a raft and an acrotelm is that whereas water levels fluctuate within an acrotelm, a mobile raft retains a fairly stable position relative to the surface of a fluctuating water level. Of course, vegetation rafts may also have an acrotelm in terms of a surface unsaturated layer that can presumably supplement or, as the raft thickens and solidifies, supervene raft-based hydroregulation.

The hydroregulatory properties of rafts may explain their frequent association with a range of plant species that are either uncommon or unexpected
(Giller and Wheeler, 1986a, 1988; Den Held *et al.*, 1992; Van Wirdum *et al.*,
1992). For example, some mobile rafts permit acidophiles such as *Sphagnum*
spp. to grow over very base-rich water, because their surface is rarely flooded
(Giller and Wheeler, 1988; van Wirdum, 1991). Yet other examples (in
topogenous fens) support plant species that are particularly characteristic
of soligenous situations (so-called 'seepage indicator species': Van Wirdum,
1991). Some workers have attributed the occurrence of such species on
vegetation rafts to the influence of groundwater discharge (Segal, 1966), but
by no means all examples are in discharge areas (van Wirdum, 1991). The
environmental features associated with vegetation rafts supporting 'seepage
indicator species' in Britain and the Netherlands are base-rich, low-fertility
water and a damped amplitude of water level fluctuation (Giller and Wheeler,
1986a, b; van Wirdum, 1991). In these respects, such vegetation rafts may
mimic some salient characteristics of base-rich, soligenous fens without being
subject to groundwater discharge.

WATER CHEMICAL COMPOSITION

In many wetlands, the inflowing water provides a major mechanism for
the import of chemical elements. It helps to determine the availability
of essential plant nutrients as well as regulating other components of the
chemical environment of the rooting zone, such as pH. Variation in the
chemical composition of water sources helps to control species distribution
and vegetation composition in wetlands, and water *quality* can be of equal
ecological importance to water *quantity*. This should be recognised by water
engineers and others who may be tempted to remedy water deficit in a
wetland by introducing supplementary water of contrasting chemical composition to that of the original sources.

A large number of chemical studies have been made in wetlands (e.g.
Malmer, 1962a, b). Among their main foci have been characterisation of
hydrochemical conditions; identification of sources of water supply; examination and quantification of hydrochemical processes; and exploration of links
between hydrochemistry and wetland vegetation. Various aspects of wetland
hydrochemistry have been reviewed by Shotyk (1988) and Ross (1995),
among others. Here consideration will be restricted to some aspects of water
chemistry that are particularly pertinent to the distribution of plants and the
composition of wetland vegetation.

Vegetation-related gradients in the chemical composition of water in wetlands

Trends in the chemical composition of water in wetlands

Gibbs (1970) has pointed out that the gross chemical composition of the world's surface waters is controlled by three main mechanisms: atmospheric precipitation, supply from rocks and evaporation–crystallisation processes. He considers that it is possible to characterise the variation in surface waters using just two axes, the concentration of total dissolved salts and the 'weight ratio' $[Na]/([Na] + [Ca])$, or $[Cl]/([Cl] + [HCO_3])$.

Independently of Gibbs, van Wirdum (1982, 1991) developed a comparable approach for representing the composition of water samples from wetlands. He proposed that much of the variation in the chemical composition of wetland water can be summarised as a two-dimensiona' ordination, where one axis is a measure of overall concentration as estimate by electrical conductivity (EC_{25}) and the second is an estimate of ioni composition based on the quotient 'IR', where:

$$IR = 100[\frac{1}{2}Ca^{2+}]/([\frac{1}{2}Ca^{2+}] + [Cl^-]);$$

Van Wirdum (1991) found that most water samples plotted within a discrete, sub-triangular portion of the EC–IR ordination and that certain combinations of EC and IR were rare or absent. He concluded that the clusters of samples could be characterised by reference to three more or less extreme points. These effectively form the apices of the sub-triangle and have been distinguished as:

• *Lithotrophic*: character derived from much contact with rocks (high IR, medium EC)
• *Atmotrophic*: character based on atmospheric water (low IR, low EC)
• *Thalassotrophic*: character based on oceanic water (low IR, high EC)

This 'LAT framework' much simplifies the actual variation found in wetland water chemistry and provides a way of handling water samples for which complete chemical analyses are not available (van Wirdum, 1991). It can provide valuable information about water sources and can correlate well with the distribution of plant communities (Koerselman et al., 1990), suggesting that for some purposes more detailed chemical determinations may be superfluous.

Working from a rather different perspective, Wheeler and Shaw (1995) made a synoptic comparison of the environmental conditions associated with fen vegetation in Britain and identified two primary gradients of vegetation-related chemical variation. The principal gradient was one of base richness

158

and included several closely correlated variables (positively related to base richness: Ca, Mg, HCO_3; negatively related: pH, Fe, Mn, Al). The second gradient, orthogonal to the first, was one of fertility.

The capacity of van Wirdum's LAT framework to represent the distinctive character of nutrient-enriched water is not fully established. Since ions containing N and P generally contribute rather little to electrical conductivity, it may be suspected that nutrient-rich water may be difficult to detect in his ordination. Nonetheless, many forms of enrichment are associated with concentration changes in solutes additional to growth-limiting nutrients, and van Wirdum (1991) has shown that polluted (*molunotrophic*) water occupies a distinctive position on the EC–IR plot, presumably on account of its overall ionic signature.

The base-richness gradient

Variation in base richness of wetland waters corresponds broadly to the atmotrophic–lithotrophic gradient of the LAT framework and has long been considered to be the primary gradient in relation to the floristic composition of mire vegetation (e.g. du Rietz, 1949; Sjörs, 1950a; Malmer, 1962a, 1986; Heinselman, 1970; Wheeler and Shaw, 1995). The acidity of wetlands depends upon the composition of their water sources and upon their capacity to buffer acidity produced endogenously by decomposition processes and protonation by plants, especially *Sphagnum* spp. (Clymo, 1984b; Shotyk, 1988).

Reflecting its importance to floristic composition, the base-richness gradient has provided the basis for some of the main hydrochemical categories of wetlands. Du Rietz (1949, 1954) recognised the two main categories of bog and fen, with fen further subdivided into poor fen and rich fen. This classification is still quite widely used, with its categories usually regarded as units of base status (in the sequence bog < poor fen < rich fen) (Sjörs, 1950a), although Du Rietz's units were actually based on floristics, not on hydrological or hydrochemical definitions. Bog and poor fen vegetation often has a prevalence of *Sphagnum* and, in general, is less species-rich than rich-fen vegetation of comparable low productivity.

The split between bog and fen corresponds – by definition – to an important difference in water source (exclusively meteoric versus telluric water-fed systems), and the two categories are widely considered to be 'fundamental units'. Nonetheless, they do not seem to represent the primary hydrochemical or floristic subdivision of mires. The pH value of the water of natural mires ranges between the limits of about pH 2.5 and 9, but various workers have observed a bimodal frequency distribution of water pH (Gorham *et al.*, 1985; Proctor, 1995). One mode (pH < 5.0) appears to represent water buffered by humic material; the other (pH > 6.0) represents water buffered by the bicarbonate system. The separating antimode (the point of 'natural' subdivision?) approximates quite closely to the split between rich fen and poor fen plus bog, not to the split between bog and fen (Malmer, 1986). Likewise, multivariate floristic classifications of peatland vegetation do not necessarily identify bog versus fen as the primary split (Daniels, 1978; Gignac *et al.*, 1991) but tend to support the approach of many European phytosociologists, who have generally grouped bog vegetation together with much poor fen vegetation into a single class (Oxycocco-Sphagnetea Br.-Bl. et Tx. 1943). Thus, if there is any single 'natural' subdivision within mires, it seems more likely to be fixed at rich fen versus the remainder rather than at bog versus fen.

In some peatlands, the base-rich to base-poor gradient is linked to ontogenic processes, especially to a diminishing influence of base-rich groundwater upon the surface, often created by growth of plants (especially *Sphagnum* species) above the influence of base-rich water. Surface acidification can also be induced by reduction of groundwater flow and the accumulation of lenses of rainwater in the upper peat, and thus drive floristic change rather than be the product of it. The importance of acid rain inputs to the base status and floristic composition of wetlands is not at all well known. Although the effects of this might perhaps be expected to be greatest in ombrogenous mires, which not only have a complete dependence on precipitation inputs but which sometimes also occur in regions of particularly high total acid deposition, such systems are naturally of quite low pH (Proctor and Maltby, 1998). The most pervasive floristic consequences of acid rain are more likely to be upon high pH, but weakly buffered, wetlands (Kooijman, 1993; Haesebroeck *et al.*, 1997).

The fertility gradient

The productivity of wetlands shows enormous variation (Bradbury and Grace, 1983), much of which is probably in response to variation in soil 'fertility'. This is primarily controlled by the availability of growth-limiting nutrients, mainly nitrogen and phosphorus, sometimes potassium. In general, the most naturally fertile wetlands occur alongside rivers, in areas subject to alluvial deposition – a process that may be encouraged by entrapment by vegetation. Fertile wetlands, at least when unmanaged, usually support robust, productive, species-poor vegetation dominated by one or a few vigorous, monopolistic species (e.g. Wheeler and Giller, 1982; Vermeer and Beredense, 1983; Wheeler and Shaw, 1991). Very high concentrations of nutrients may be directly detrimental to the metabolism and growth of some wetland species, including potential dominants such as *Phragmites australis* (Kubín and Melzer, 1996; Votrubová and Pecháčková, 1996).

Studies suggest that N, P and K availability can all limit plant growth in wetlands, although K limitation seems to be the least common. Some workers have been particularly interested in identifying limiting nutrients for individual sites and vegetation types. Koerselman and Verhoeven (1995) tabulated examples of various European studies and concluded that bogs were mostly P-limited and fens mostly N-limited. However, this assessment may partly reflect a preponderance of studies made in the Netherlands, because there is plenty of evidence for P limitation in fens (e.g. Egloff, 1983; Richardson and Marshall, 1986; Boyer and Wheeler, 1989; Kooijman, 1993; Boeye et al., 1997). Moreover, the search for general patterns in the identity of limiting nutrients in wetlands may have only limited value. Hayati and Proctor (1991) present evidence that the performance of different species can be influenced by different nutrients, even on the same soil or site, with perhaps several nutrients being close to their limiting concentrations; and that uptake of specific ions may be influenced by the presence of others. There is also some evidence that availability of N and P can influence the uptake of both elements (O'Connell and Grove, 1985; Perez-Corona and Verhoeven, 1996). If the availability of N helps to control root phosphatase activity, so that increased N can relieve P deficiency by stimulating hydrolysis of organic P sources, which element is then 'limiting'?

Major nutrients may be imported into wetlands by inflowing water, from naturally or artificially enriched sources, but their availability can also be regulated by the position of the water table. Mineralisation processes are widely considered to be much reduced by anoxia (although substantial N mineralisation has been reported from some waterlogged soils: Williams and Wheatley, 1988). Conversely, denitrification occurs mostly in low-redox environments ($< c.$ 300 mV). It can, however, be rate-limited by insufficient nitrate flux, due inter alia to slow rates of nitrification and mineralisation or to limited penetration, or insufficient residence, of nitrate-rich inflows in wetlands (Warwick and Hill, 1988; Koerselman et al., 1989; Lowrance, 1992). Nonetheless, in appropriate circumstances, especially given base-rich conditions and fluctuating water tables, denitrification can lead to loss of much N from wetlands (e.g. Guthrie and Duxbury, 1978; Reddy et al., 1980).

A variety of hydrochemical processes can help to regulate phosphorus availability in wetlands. High water tables can be associated with an increase in P availability, probably largely on account of desorption processes (Koerselman and Verhoeven, 1995). Conversely, in highly calcareous fens P can be adsorbed onto calcite precipitated in response to degassing of supersaturated groundwater (Boyer and Wheeler, 1989; Boeye et al., 1995). Some workers also point to Ca–P immobilisation through formation of hydroxyapatites (Kemmers, 1986), but this mechanism has sometimes been invoked for relatively low Ca concentrations, where its occurrence is far from certain (Wilson and Fitter, 1984; Wassen et al., 1989; Wassen and Barendregt, 1992).

Salinity gradients

In freshwater wetlands, outside the influence of marine or estuarine waters, salinity gradients are generally of little consequence. However, there can be a greater thalassotrophic influence in coastal freshwater systems, derived from storm surges that send periodic pulses of oligohaline water upstream and into wetlands (Giller and Wheeler, 1986b); from penetration of brackish groundwater (Wassen *et al.*, 1989); and from 'fossil salinity' derived from former estuarine sediments now covered by freshwater deposits, especially where the surface of estuarine clays has been exposed by peat removal (Giller and Wheeler, 1986a, b).

Relating water chemical composition to vegetation composition

While the vegetation composition of wetlands is undoubtedly strongly influenced by the chemical composition of water, the identification of exact chemical limits for particular plant species or communities has generally proved elusive. Some early analyses of vegetation–hydrochemical relationships remain instructive. Gorham (1950) examined the hydrochemical boundary conditions of *Carex lasiocarpa* and concluded that they were very variable. Sjörs (1950a) observed a considerable inconsistency of association between measured chemical concentrations and plant distributions in between-site comparisons and noted that, while some species showed the same rank order along hydrochemical gradients at different sites, there could be great differences in actual concentrations associated with them at each individual site. More recent workers, optimistically trying to define coherent chemical conditions associated with species and communities, often do little more than confirm Gorham's and Sjörs's observations about variability. Even when clear relationships are observed at individual sites, they do not always have wider relevance. For example, Wassen *et al.* (1989) identified six broad water types in the fens in the Naardermeer (Netherlands), with clear correlations with species distribution, but there is little reason to suppose that the species necessarily show the same hydrochemical 'preferences' and limits elsewhere. For example, in the UK such species as *Eupatorium cannabinum*, *Berula erecta* and *Lycopus europaeus* do not seem to be preferentially associated with water with conductivities in excess of $700 \, \mu S \, cm^{-1}$ as observed by Wassen *et al.* (B.D. Wheeler and S.C. Shaw, unpublished data).

Difficulties of defining clear vegetation–hydrochemical relationships stem partly from the considerable spatial and temporal variation in water composition, even within the same stand (Malmer, 1962a; Summerfield, 1974; Proctor, 1994), sometimes associated with water level fluctuations (Giller and Wheeler, 1986b; Proctor, 1994). De Mars *et al.* (1997) report that the best correlation between water composition in the root zones and vegetation types

162

was found during very dry or very wet weather. Concentrations of ions in interstitial water do not necessarily well reflect their bio-availability, especially those of growth-limiting nutrients, whose measured concentrations may be much determined by biological uptake (e.g. Urban *et al.*, 1988; Wheeler *et al.*, 1992), or which occur as hydrolysable organic complexes.

Water sources, water chemistry and wetland vegetation

Since different water sources can have strikingly different chemical composition, changes in the proportionate contribution of contrasting sources may be expected to affect the floristic composition of wetlands, even when there is no net change in water levels. In the Netherlands, some workers consider that flow of calcareous groundwater, derived from adjoining upland recharge areas, is critical to the occurrence of some of the more 'desirable' species and communities in wetlands. They also suggest that reduction of groundwater flow, or changing flow paths (induced by groundwater abstraction or drainage), can lead to a deterioration of floristic quality, even without a strong water level decrease, on account of an increased proportion of alternative water sources, such as nutrient-enriched land-drainage water or rainwater (Bakker *et al.*, 1987; Grootjans *et al.*, 1988; Van Diggelen *et al.*, 1991a), and, in some situations, brackish groundwater (Schot *et al.*, 1988; Wassen *et al.*, 1989). Because of this, there is much interest among Dutch workers in identifying and maintaining (or reinstating) the regional ground water flow systems responsible for the production of the desired hydrochemical conditions (Grootjans, 1985; Wassen *et al.*, 1988; 1990; van Diggelen *et al.*, 1995).

Nonetheless, it is important to recognise that, from the point of view of plant growth and distribution, the chemical environment in the immediate vicinity of the plant is more critical than its mechanism of supply. Calcium-rich, mesotrophic water, in the Dutch fens and elsewhere, can originate from sources other than direct discharge of regional groundwater (van Wirdum, 1982; 1991; Wassen *et al.*, 1990; Schot and Wassen, 1993). It can even have an artificial origin (Boeye *et al.*, 1995). Providing its quality is appropriate, plants may not 'mind' too much from whence their water comes, or how it arrives! Thus, while reinstatement of former regional flow patterns provides one important mechanism for restoring appropriate hydrochemical conditions to wetland sites, it is not the only approach and should not obscure the possibility that similar, or equally desirable, floristic outcomes may be achieved by alternative supply mechanisms. For example, the fact that some rare 'seepage indicator species' can grow on semi-floating vegetation rafts in reflooded peat workings that do not receive any water 'seepage' (groundwater discharge) points to the possibilities of maintaining species populations in hydrological and hydrochemical environments that do not correspond exactly to any former 'natural condition'. Such opportunities need to be identified and

ECO-HYDROLOGY

grasped, because in some highly developed landscapes provision of 'new' environments and hydrological mechanisms may offer the only realistic prospect for wetland restoration. Such an approach does, however, demand a more thorough understanding of the interrelationships between hydrology and hydrochemistry and vegetation composition than currently exists.

ACKNOWLEDGEMENTS

I thank H.A.P. Ingram, P.A. Keddy, R.P. Money, M.C.F. Proctor, S.C. Shaw and G. van Wirdum for helpful comments on an early draft of this paper. N. Malmer made some valuable comments on the importance of the acrotelm in fens. J. Schouwenaars contributed material on eco-hydrological models used in the Netherlands.

REFERENCES

Andrus, R.E. (1986) Some aspects of *Sphagnum* ecology, *Canadian Journal of Botany* 64: 416–426.

Andrus, R.E., Wagner, D.J. and Titus, J.E. (1983) Vertical distribution of *Sphagnum* mosses along hummock–hollow gradients, *Canadian Journal of Botany* 61: 3128–3189.

Armstrong, J., Armstrong, W. and Beckett, P.M. (1992) *Phragmites australis*: venturi- and humidity-induced pressure flows enhance rhizome aeration and rhizosphere oxidation, *New Phytologist* 120: 197–202.

Armstrong, J., Armstrong, W., Wu, Z. and Afreen-Zobayed, F. (1996) A role for phytotoxins in the *Phragmites* die-back syndrome? *Folia Geobotanica and Phytotaxonomica* 31: 127–142.

Armstrong, W. (1964) Oxygen diffusion from the roots of some British bog plants, *Nature* 204: 801–802.

Armstrong, W. (1967) The oxidising activity of roots in waterlogged soils, *Physiologia Plantarum* 20: 920–926.

Armstrong, W. (1980) Aeration in higher plants, *Advances in Botanical Research* 7: 226–332.

Armstrong, W. (1982) Waterlogged soils, in J.R. Etherington (ed.) *Environment and Plant Ecology*, second edition, Chichester: John Wiley, pp. 290–330.

Armstrong, W. and Boatman, D.J. (1967) Some field observations relating the growth of bog plants to conditions of soil aeration, *Journal of Ecology* 55: 101–110.

Armstrong, W., Armstrong, J. and Beckett, P.J. (1996) Pressurised aeration in wetland macrophytes: some theoretical aspects of humidity–induced convection and thermal transpiration, *Folia Geobotanica and Phytotaxonomica* 31: 25–36.

Armstrong, W., Booth, T.C., Priestley, P. and Read, D.J. (1976) The relationship between aeration, stability and growth of Sitka spruce (*Picea sitchensis* (Bong.) Carr.) on upland peaty gleys, *Journal of Applied Ecology* 13: 585–591.

Bakker, J.P., Brouwer, C., van den Hof, L. and Jansen, A. (1987) Vegetation succession, management and hydrology in a brookland (The Netherlands), *Acta Botanica Neerlandica* 36: 39–58.

Balátová-Tuláčková, E. (1968). Grundwasserganglinien und Wissengesellschaften (Vergleichende Studie der Wiesen aus Südmähren und der Südwestslowakei), *Acta Scientiarum Naturalium Academiae Scientiarum Bohemoslovacae Brno* 2: 1–37.

Barclay, A.M. and Crawford, R.M.M. (1982) Plant growth and survival under strict anaerobiosis, *Journal of Experimental Botany* 33: 541–549.

Barendregt, A., Wassen, M.J., de Smidt J.T. and Lippe, E. (1986) Ingreep effect voorspelling voor waterbeheer, *Landschap* 3: 41–55.

Bartlett, R.J. (1961) Iron oxidation proximate to plant roots, *Soil Science* 92: 372–379.

Baskin, C.C., Baskin, J.M. and Chester, E.W. (1993) Seed germination ecophysiology of four summer annual mudflat species of Cyperaceae, *Aquatic Botany* 45: 41–52.

Bedford, B.L. and Bouldin, D.R. (1994) 'On the difficulties of measuring oxygen release by root systems of wetland plants', by B.K. Sorrell and W. Armstrong, *Journal of Ecology* 82: 185–186.

Bedford, B.L., Bouldin, D.R. and Beliveau, B.D. (1991) Net oxygen and carbon dioxide balances in solutions bathing roots of wetland plants, *Journal of Ecology* 79: 943–959.

Beets, C.P. (1992) The relation between the area of open water in bog remnants and storage capacity with resulting guidelines for bog restoration, in O.M. Bragg, P.D. Hulme, H.A.P. Ingram and R.A. Roberstson (eds) *Peatland Ecosystems and Man: An Impact Assessment*, University of Dundee: Dundee/International Peat Society: Jyväskylä, pp. 133–140.

Bellamy, D.J. (1968) An ecological approach to the classification of European mires, *Third International Peat Congress, Quebec 1968*, pp. 74–79.

Bellamy, D.J. and Bellamy, S.R. (1966) An ecological approach to the classification of the lowland mires of Ireland, *Proceedings of the Royal Irish Academy* B65: 237–251.

Bewley, J.D., Halmer, P., Krochko, J. and Winner, W.E. (1978). Metabolism of a drought-tolerant and a drought-sensitive moss. Respiration, ATP synthesis and carbohydrate status, in J.H. Crowe and J.S. Clegg (eds) *Dried biological systems*, New York: Academic Press, pp. 185–203.

Blom, C.P.W.M., Voesenek, L.A.C.J., Banga, M., Engelaar, W.M.H.G., Rijnders, J.H.M.G., van Steeg, H.M. and Visser, E.J.W. (1994) Physiological ecology of riverside species, adaptive responses upon submergence, *Annals of Botany* 74: 253–263.

Bloomfield, C. and Zahari, A.B. (1982) Acid sulphate soils, *Outlook on Agriculture* 11: 48–54.

Boeye, D., van Straaten, D. and Verheyen, F. (1995) A recent transformation from poor to rich fen caused by artificial ground water recharge, *Journal of Hydrology* 169: 111–129.

Boeye, D., Verhagen, B., van Haeseboroeck, V. and Verheyen, R.F. (1997) Nutrient limitation in species-rich lowland fens, *Journal of Vegetation Science* 8: 415–424.

Boyer, M.H.L. and Wheeler, B.D. (1989) Vegetation patterns in spring-fed calcareous fens: calcite precipitation and constraints on fertility, *Journal of Ecology* 77: 597–609.

Bradbury, I.K. and Grace, J. (1983) Primary production in wetlands, in A.J.P. Gore (ed.) *Ecosystems of the World. 4A. Mires: Swamp, Bog, Fen and Moor. General Studies*, Amsterdam: Elsevier, pp. 285–310.

Braendle, R. and Crawford, R.M.M. (1987) Rhizome anoxia tolerance and habitat specialization in wetland plants, in R.M.M. Crawford (ed.) *Plant Life in Aquatic and Amphibious Habitats*, Oxford: Blackwell Scientific Publications, pp. 397–410.

Bragg, O.M. (1989). The importance of water in mire ecosystems, in W. Fojt and R. Meade (eds) *Cut-over Lowland Raised Mires*, Peterborough, Research and Survey in Nature Conservation 24, Nature Conservancy Council, pp. 61–82.

Buell, M.F. and Buell, H.F. (1941) Surface level fluctuations in Cedar Creek Bog, Minnesota, *Ecology* 22: 314–321.

Chapin, F.S., Fetcher, N., Kielland, K., Everett, K.R. and Linkins, A.E. (1988) Productivity and nutrient cycling of Alaskan tundra: enhancement by flowing soil water, *Ecology* 69: 693–702.

Clapham, A.R. (1940) The role of bryophytes in the calcareous fens of the Oxford district, *Journal of Ecology* 28: 71–80.

Clausen, E. (1952) Hepatics and humidity, a study on the occurrence of hepatics in a Danish tract and the influence of relative humidity on their distribution, *Dan. Bot. Ark.* 15: 5–80.

Clymo, R.S. (1970) The growth of *Sphagnum*: methods of measurement, *Journal of Ecology* 58: 13–49.

Clymo, R.S. (1973) The growth of *Sphagnum*: some effects of environment, *Journal of Ecology* 61: 849–869.

Clymo, R.S. (1984a) The limits to peat bog growth, *Philosophical Transactions of the Royal Society of London* B303: 605–654.

Clymo, R.S. (1984b) *Sphagnum*-dominated peat bog: a naturally acid ecosystem, *Philosophical Transactions of the Royal Society of London* B305: 487–499.

Clymo, R.S. and Reddaway, E.J.F. (1971) Productivity of *Sphagnum* (bog moss) and peat accumulation, *Hidrobiologia* 12: 181–192.

Clymo, R.S. and Hayward P.M. (1982) The ecology of *Sphagnum*, in A.J.E. Smith (ed.) *Bryophyte Ecology*, London: Chapman and Hall, pp. 229–291.

Conway, V.M. (1937) Studies in the autecology of *Cladium mariscus* R. Br. Part III. The aeration of the subterranean parts of the plant, *New Phytologist* 36: 64–96.

Conway, V.M. (1938) Studies in the autecology of *Cladium mariscus* R. Br. IV. Growth rates of the leaves, *New Phytologist* 37: 254–278.

Conway, V.M. (1940) Growth rates and water loss in *Cladium mariscus* R. Br., *Annals of Botany* 4: 151–164.

Coops, H., van den Brink, F.W.B. and van der Velde, G. (1996) Growth and morphological response of four helophyte species in an experimental water depth gradient, *Aquatic Botany* 54: 11–24.

Coops, H. and van der Velde, G. (1995) Dispersal, germination and seedling growth of six helophyte species in relation to waterlevel zonation, *Freshwater Biology* 34: 13–20.

Crawford, R.M.M. (1977) Tolerance of anoxia and ethanol metabolism in germinating seeds, *New Phytologist* 79: 511–517.

Crawford, R.M.M. (1983) Root survival in flooded soils, in A.J.P. Gore (ed.) *Ecosystems of the World. 4A. Mires: Swamp, Bog, Fen and Moor. General Studies*, Amsterdam: Elsevier, pp. 257–283.

Crawford, R.M.M. (1996) Whole plant adaptations to fluctuating water tables, *Folia Geobotanica Phytotaxonomia* 31: 7–24.

Daniels, R.E. (1978) Floristic analyses of British mires and mire communities, *Journal of Ecology* 66: 773–802.

Daniels, R.E. and Pearson, M.C. (1974) Ecological studies at Roydon Common, Norfolk, *Journal of Ecology* 62: 127–150.

de Mars, H. (1996) *Chemical and Physical Dynamics of Fen Hydro-ecology*, Nederlanse Geografische Studies 203, Koninklijk Nederlands Aardrijkskundig Genootschap/ Faculteit Ruimtelijke Wetenschappen Universiteit Utrecht.

de Mars, H., Wassen, M.J. and Olde Venterink, H. (1997) Flooding and ground water dynamics in fens in the Biebrza Valley, N.E. Poland, *Journal of Vegetation Science* 8: 319–328.

den Held, A.J., Schmitz, M. and van Wirdum, G. (1992) Types of terrestrializing fen vegetation in the Netherlands, in J.T.A. Verhoeven (ed.) *Fens and Bogs in the Netherlands* Dordrecht: Kluwer, pp. 237–322.

Drew, M.C., Sisworo, E.J. and Saker, L.R. (1979) Alleviation of waterlogging damage to young barley plants by application of nitrate and synthetic cytokinin and comparison between the effects of waterlogging, nitrogen deficiency and root excision, *New Phytologist* 82: 315–329.

du Rietz, G.E. (1949) Huvudenheter och granser i Svensk myrvegetation, *Svensk Botanisk Tidskrift* 43: 299–309.

du Rietz, G.E. (1954) Die Mineralbodenwasserzeigergrenze als Grundlage einer Natürlichen Zwiegleiderung der Nord- und Mitteleuropäischen Moore, *Vegetatio* 5–6: 571–585.

Eades, P.A. (1997) Experimental studies into the effects of water level changes upon the vegetation and fertility of calcareous spring-fed fens, unpublished PhD thesis, University of Sheffield.

Egloff, T. (1983) Der Phosphor als primär limitierender Nährstoff im Streuwiesen (Molinion). Düngungsexperiment im unteren Reusstal, *Ber. Geobot. Inst. Eidg. Tech. Hochsch. Stuft. Rübel Zürich* 50: 119–148.

Egloff, T. and Naef, E. (1982) Grundwasserstandsmessungung in Streuwiesen des unteren Reußtales, *Ber. Geobot. Inst. ETH, Stiftung Rübel* 49: 154–194.

Ellenberg, H. (1954) Über einige Fortschritte der kausalen Vegetationskunde, *Vegetatio* 5–6: 199–211.

Ellenberg, H. (1974) Zeigerwerte der Gefäßpflanzen Mitteleuropas, *Scripta Geobotanica* 9: 1–97.

Ellenberg, H. (1988) *Vegetation Ecology of Central Europe*, fourth edition, Cambridge: Cambridge University Press.

Engelaar, W.M.H.G., Symens, J.C., Laanbroeck, H.J. and Blom, C.W.P.M. (1995) Preservation of nitrifying capacity and nitrate availability in waterlogged soils by radial oxygen loss from the roots of wetland plants, *Biology and Fertility of Soils* 20: 243–248.

Ernst, W.H.O. and van der Ham, N.F. (1988) Population structure and rejuvenation potential of *Schoenus nigricans* in coastal dune slacks, *Acta Botanica Neerlandica* 37: 451–465.

Etherington, J.E. (1984) Comparative studies of plant growth and distribution in relation to waterlogging. X. Differential formation of adventitious roots and their experimental excision in *Epilobium hirsutum* and *Chamerion angustifolium*, *Journal of Ecology* 72: 389–404.

Evans, C.E. and Etherington, J.R. (1991) The effect of soil water potential on seedling growth of some British plants, *New Phytologist* 118: 571–580.

Fojt, W. and Harding, M. (1995) Thirty years of change in the vegetation communities of three valley mires in Suffolk, England, *Journal of Applied Ecology* 32: 561–577.

Fürtig, K., Rüegsegger, A., Brunold, C. and Brändle, R. (1996) Sulphide utilization and injuries in hypoxic roots and rhizomes of common reed (*Phragmites australis*), *Folia Geobotanica and Phytotaxonomica* 31: 143–151.

Gibbs, R.J. (1970) Mechanisms controlling world water chemistry, *Science* 170: 1088–1090.

Gignac, L.D., Vitt, D.H., Zoltai, S.C. and Bayley, S.E. (1991) Bryophyte response surfaces along climatic, chemical and physical gradients in peatlands of western Canada, *Nova Hedwigia* 53: 27–71.

Gill, C.J. (1970) The flooding tolerance of woody species – a review, *Forestry Abstracts* 31: 671–678.

Gill, C.J. (1975) The ecological significance of adventitious rooting as a response to flooding in woody species, with special reference to *Alnus glutinosa* (L.) Gaertn., *Flora* 164: 85–97.

Giller, K.E. and Wheeler, B.D. (1986a) Past peat cutting and present vegetation patterns in an undrained fen in the Norfolk Broadland, *Journal of Ecology* 74: 219–247.

Giller, K.E. and Wheeler, B.D. (1986b) Peat and peat water chemistry of a floodplain fen in Broadland, Norfolk, UK, *Freshwater Biology* 16: 99–114.

Giller, K.E. and Wheeler, B.D. (1988) Acidification and succession in a flood-plain mire in the Norfolk Broadland, UK, *Journal of Ecology* 76: 849–866.

Godwin, H. and Bharucha, F.R. (1932) Studies in the ecology of Wicken Fen. II. The fen water table and its control on plant communities, *Journal of Ecology* 20: 157–191.

Gorham, E. (1950) Variation in some chemical conditions along the borders of a *Carex lasiocarpa* fen community, *Oikos* 2: 217–240.

Gorham, E., Eisenreich, S.J., Ford, J. and Santelmann, M.V. (1985) The chemistry of bog waters, in W. Stumm (ed.) *Chemical Processes in Lakes*, New York: Wiley, pp. 339–363.

Grace, J.B. and Wetzel, R.G. (1981) Habitat partitioning and competitive displacement in cattails (*Typha*): experimental field studies, *The American Naturalist* 118: 463–474.

Gremmen, N.J.M., Reignen, M.J.S.M., Wiertz, J. and van Wirdum, G. (1990) A model to predict and assess the effects of ground-water withdrawal on vegetation in the Pleistocene areas of the Netherlands, *Journal of Environmental Management* 31: 143–155.

Grootjans, A.P. (1980) Distribution of plant communities along rivulets in relation to hydrology and management, in O. Wilmanns and R. Tüxen (eds) *Epharmonie. Berichte über die internationalen Symposien der Internationale Vereinigung für Vegetationskunde 1979*, Cramer Verlag, pp. 143–170.

Grootjans, A.P. (1985) Changes of ground water regime in wet meadows, PhD thesis, University of Groningen.

Grootjans, A.P. and Ten Klooster, W.P. (1980) Changes of ground water regime in wet meadows, *Acta Botanica Neerlandica* 29: 541–554.

Grootjans, A.P., Fresco, L.F.M., de Leeuw, C.C. and Schipper, P.C. (1996) Degeneration of species-rich *Calthion palustris* hay meadows; some considerations on the community concept, *Journal of Vegetation Science* 17: 185–194.

Grootjans, A.P., van Diggelen, R., Wassen, M.J. and Wiersinga, W.A. (1988) The effects of drainage upon ground water quality and species distribution in stream valley meadows, *Vegetatio* 75: 37–48.

Guthrie, T.F. and Duxbury, J.M. (1978) Nitrogen mineralisation and denitrification in organic soils, *Soil Science Society of America Journal* 42: 908–912.

Hacker, S.D. and Berness, M.D. (1995) Morphological and physiological consequences of a positive plant interaction, *Ecology* 76: 2165–2175.

Haesebroeck, V., Boeye, D., Verhagen, B. and Verheyen, R.F. (1997) Experimental investigation of drought induced acidification in a rich fen soil, *Biogeochemistry* 37: 15–32.

Hansen, J.I. and Andersen, F.O. (1981) Effects of *Phragmites australis* roots on redox potentials, nitrification and bacterial numbers in sediments, in A. Bromberg and T. Tiren (eds) *Ninth Nordic Symposium on Sediments*, pp. 72–78.

Harris, S.W. and Marshall, W.H. (1963) Ecology of water level manipulations on a northern marsh, *Ecology* 44: 331–343.

Haukos, D.A. and Smith, L.M. (1993) Seed-bank composition and predictive ability of field vegetation in playa lakes, *Wetlands* 13: 32–40.

Hayati, A.A. and Proctor, M.C.F. (1991) Limiting nutrients in acid-mire vegetation:

peat and plant analyses and experiments on plant responses to added nutrients, *Journal of Ecology* 79: 75–95.

Hayward, P.M. and Clymo, R.S. (1983) The growth of *Sphagnum* – experiments on, and simulation of, some effects of light flux and water table depth, *Journal of Ecology* 71: 845–863.

Heikkilä, R. and Löytönen, M. (1987) Observations on *Sphagnum* species and their relation to vegetation and ecological factors in Ostanberg Stormossen, South Finland, *Suo* 38: 63–70.

Heinselman, M.L. (1970) Landscape evolution, peatland types and the environment in the Lake Agassiz Peatlands Natural Area, Minnesota, *Ecological Monographs* 40: 236–261.

Hill, N.M. and Keddy, P.A. (1992) Prediction of rarities from habitat variables: coastal plain plants on Nova Scotian lakeshores, *Ecology* 73: 1852–1859.

Hulme P.D. and Blyth A.W. (1982) The annual growth period of some *Sphagnum* species on the Silver Flowe National Nature Reserve, south-west Scotland, *Journal of Bryology* 12: 287–291.

Huisman, J., Olff, H. and Fresco, L.F.M. (1993) A hierarchical set of models for species response analysis, *Journal of Vegetation Science* 4: 47–56.

Hygen, G. (1953) Studies in Plant Transpiration. II, *Physiologia Plantarum* 6: 106–133.

Ingold, A. and Havill, D.C. (1984) The effect of sulphide on the distribution of higher plants in salt marshes, *Journal of Ecology* 72: 1043–1054.

Ingram, H.A.P. (1967) Problems of hydrology and plant distribution in mires, *Journal of Ecology* 55: 711–724.

Ingram, H.A.P. (1982) Size and shape in raised mire ecosystems: a geophysical model, *Nature* 297: 300–303.

Ingram, H.A.P. (1992) Introduction to the ecohydrology of mires in the context of cultural perturbation, in O.M. Bragg, P.D. Hulme, H.A.P. Ingram and R.A. Roberstson (eds) *Peatland Ecosystems and Man: An Impact Assessment*, University of Dundee: Dundee/International Peat Society: Jyväskylä, pp. 67–93.

Ingram, H.A.P. and Bragg, O.M. (1984) The diplotelmic mire: some hydrological consequences reviewed, *Proceedings of the 7th International Peat Congress, Dublin, June, 1984*. Irish National Peat Committee, for the International Peat Society 1: 220–234.

Ivanov, K.E. (1981) *Water Movement in Mirelands*, trans. A. Thomson and H.A.P. Ingram, London: Academic Press.

Jeglum, J.K. (1971) Plant indicators of pH and water level in peatlands at Candle Lake, Saskatchewan, *Canadian Journal of Botany* 49: 1661–1676.

Johnson, W.C. (1994) Woodland expansion in the Platte River, Nebraska: patterns and causes, *Ecological Monographs* 64: 45–84.

Jones, H.E. (1972) Comparative studies of plant growth and distribution in relation to waterlogging. V. The uptake of iron and manganese by dune and dune slack plants, *Journal of Ecology* 60: 131–140.

Jones, H.E. and Etherington, J.R. (1970) Comparative studies of plant growth and distribution in relation to waterlogging. I. The survival of *Erica cinerea* L. and *Erica tetralix* L. and its apparent relationship to iron and manganese uptake in waterlogged soil, *Journal of Ecology* 58: 487–496.

Joosten, H.J. (1993) Denken wie ein Hochmoor: Hydrologische Selbsregulation von Hochmooren und deren Bedeutung für Wiedernässung und Restauration, *Telma* 23: 95–115.

Justin, S.H.F.W. and Armstrong, W. (1987) The anatomical characteristics of roots and plant response to soil flooding, *New Phytologist* 106: 465–495.

Kadlec, J.A. (1962) Effects of a drawdown on a waterfowl impoundment, *Ecology* 43: 1443–1457.

Kausch, A.P., Seago, J.L. and Marsh, L.C. (1981) Changes in starch distribution in the overwintering organs of *Typha latifolia* (Typhaceae), *American Journal of Botany* 68: 877–880.

Keddy, P.A. (1983) Shoreline vegetation in Axe Lake, Ontario: Effects of exposure on zonation patterns, *Ecology* 64: 331–344.

Keddy, P.A. (1989) Effects of competition from shrubs on herbaceous wetland plants: a 4-year experiment, *Canadian Journal of Botany* 67: 708–716.

Keddy, P.A. and Constabel, P. (1986) Germination of ten shoreline plants in relation to seed size, soil particle size and water level: an experimental study, *Journal of Ecology* 74: 133–141.

Keddy, P.A. and Ellis, T.H. (1985) Seedling recruitment of 11 plant species along a water level gradient: shared or distinct responses?, *Canadian Journal of Botany* 63: 1876–1879.

Keddy, P.A. and Reznicek, A.A. (1982) The role of seedbanks in the persistence of Ontario's coastal plain flora, *American Journal of Botany* 69: 13–22.

Keddy, P.A., Twolan-Strutt, L. and Wisheu, I.C. (1994) Competitive effect and response rankings in 20 wetland plants: are they consistent across three environments? *Journal of Ecology* 82: 635–643.

Keeley, J.E. (1979) Population differentiation along a flood frequency gradient: physiological adaptations to flooding in *Nyssa sylvatica*, *Ecologological Monographs* 49: 89–108.

Kemmers, R.H. (1986) Calcium as hydrochemical characteristic for ecological states, *Institute for Land and Water Management Research (ICW), Wageningen, the Netherlands. Technical Bulletin, ICW New Series* 47: 1–16.

Kirkby, M.J., Kneale, P.E., Lewis, S.L. and Smith, R.T. (1995) Modelling the form and distribution of peat mires, in J.M.R. Hughes and A.L. Heathwaite (eds) *Hydrology and Hydrochemistry of British Wetlands*. Chichester: John Wiley, pp. 83–93.

Klimesova, J. (1994) The effects of timing and duration of floods on growth of young plants of *Phalaris arundinacea* L. and *Urtica dioica* L.: an experimental study, *Aquatic Botany* 48: 21–29.

Klötzli, F. (1969) Die Grundwasserbeziehungen der Streu- und Moorwiesen im nördlichen Schweizer Mitteland, *Beitrag geobot. Landesaufnahme* 52: 1–296.

Koerselman, W. and Verhoeven, J.T.A. (1995) Eutrophication of fen ecosystems:

external and internal nutrient sources and restoration strategies, in B.D. Wheeler, S.C. Shaw, W.J. Fojt and R.A. Robertson (eds) *Restoration of Temperate Wetlands*, Chichester: Wiley, pp. 91–112.

Koerselman, W., de Caluwe, H. and Kieskamp, W. (1989) Denitrification and dinitrogen fixation in two quaking fens in the Vechtplassen area, the Netherlands, *Biogeochemistry* 8: 153–165.

Koerselman, W., Claessens, D., ten Den, P. and van Winden, E. (1990) Dynamic hydrochemical and vegetation gradients in fens, *Wetlands Ecology and Management* 1: 73–84.

Koncalová, H., Kvet, J. Pokorny, J. and Hauser, V. (1993) Effect of flooding with sewage water on three wetland sedges, *Wetlands Ecology and Management* 2: 199–211.

Kooijman, A. (1993) Changes in the bryophyte layer of rich fens as controlled by acidification and eutrophication, doctoral thesis, Universiteit Utrecht.

Kotowski, W., van Diggelen, R. and Kleinke, J. (in press) On the consistency of the realised niches of plants – comparison of wetland species behaviour in two geographically distant areas, *Acta Botanica Neerlandica*.

Kozlowski, T.T. (ed.) (1984) *Flooding and Plant Growth*, Orlando, Fla.: Academic Press.

Kramer, P.J. (1969) *Plant and Soil Water Relationships*, New York: McGraw-Hill.

Kubín, P. and Melzer, A. (1996) Does ammonium affect accumulation of starch in rhizomes of *Phragmites australis*? *Folia Geobotanica and Phytotaxonomica* 31: 99–109.

Kulczynski, S. (1949) Peat bogs of Polesie, *Mémoires de l'académie Polonaise des Sciences et des Lettres, Serie B: Science Naturelles* 15: 1–356.

Laan, P., Smolders, A., Blom, C.W.P.M. and Armstrong, W. (1989) The relative roles of internal aeration, radial oxygen loss, iron exclusion and nutrient balances in flood tolerance of *Rumex* species, *Acta Botanica Neerlandica* 38: 131–145.

Lambert, J.M. (1951) Alluvial stratigraphy and vegetational succession in the region of the Bure valley broads. III. Classification, status and distribution of communities, *Journal of Ecology* 39: 149–170.

Lane, D.M. (1977) Extent of vegetative reproduction in eleven species of *Sphagnum* from northern Michigan, *Michigan Botanist* 16: 83–89.

Latour, J.B., Reiling, R.J. and Wiertz, J. (1993) MOVE; a multiple stress model for the vegetation, Proceedings of the CHO–TNO Symposium *Ecohydrologische voorspellingsmodellen*, 25 May 1993. CHO–TNO 47, Delft.

Lee, J.A. and Stewart, G.R. (1971) Desiccation injury in mosses. I. Intra-specific differences in the effect of moisture stress on photosynthesis, *New Phytologist* 70: 1061–1068.

Li, Y., Glime J.M. and Liao, C. (1992). Responses of two interacting *Sphagnum* species to water level, *Journal of Bryology* 17: 59–70.

Lowrance, R. (1992) Ground water nitrate and denitrification in a coastal plain riparian forest, *Journal of Environmental Quality* 21: 401–405.

Luken J.O. (1985) Zonation of *Sphagnum* mosses: interaction among shoot growth, growth form and water balance, *The Bryologist* 88: 374–379.

Maas, D. (1989) Germination characteristics of some plant species from calcareous fens in southern Germany and their implications for the seed bank, *Holarctic Ecology* 12: 337–344.

Maas, D. and Schopp–Guth, A. (1995) Seed banks in fen areas and their potential use in restoration ecology, in B.D. Wheeler, S.C. Shaw, W.J. Fojt and R.A. Robertson (eds) *Restoration of Temperate Wetlands*. Chichester: Wiley, pp. 189–206.

MacMannon, M. and Crawford, R.M.M. (1971) A metabolic theory of flooding tolerance: the significance of enzyme distribution and behaviour, *New Phytologist* 70: 299–306.

Malmer, N. (1962a) Studies on mire vegetation in the Archaean area of southwestern Götaland (south Sweden). I. Vegetation and habitat conditions on the Åkuhlt mire, *Opera Botanica* 7: 1–322.

Malmer, N. (1962b) Studies on mire vegetation in the Archaean area of southwestern Götaland (south Sweden). II. Distribution and seasonal variation in elementary constituents on some mire sites, *Opera Botanica* 7: 1–67.

Malmer, N. (1968) Über die Gliederung der Oxycocco-Sphagnetea und Scheuchzerio-Caricetea fuscae, in R. Tüxen (ed.) *Planzensoziologische Systematik. Bericht über das internationale Symposium in Stolenzau / Weser 1964 der Internationalen Vereinigung für Vegetationskunde*, Den Haag: Junk, pp. 293–305.

Malmer, N. (1986) Vegetation gradients in relation to environmental conditions in northwestern European mires, *Canadian Journal of Botany* 64: 3750383.

Mansfield, S.M. (1990) Ecophysiological effects of high concentrations of iron and other heavy metals on *Eriophorum angustifolium* Honck. and *Phragmites australis* (Cav.) Trin. ex Steudel, unpublished PhD thesis, University of Sheffield.

Metsävainio, K. (1931) Untersuchungen über das Wurzelsystem der Moorpflanzen, *Ann. bot. soc. zool.–bot. fenn. Vanamo* 1.

Miles, J. (1976) The growth of *Narthecium ossifragum* in some southern English mires, *Journal of Ecology* 64: 849–858.

Mitsch, W.J. and Gosselink, J.G. (1986) *Wetlands*, New York: Van Nostrand Reinhold.

Mooney, E.P. and O'Connell, M. (1990) The phytosociology and ecology of the aquatic and wetland plant communities of the lower Corrib basin, County Galway, *Proceedings of the Royal Irish Academy* 90B: 57–97.

Mörnsjö, T. (1969) Studies on vegetation and development of a peatland in Scania, South Sweden, *Opera Botanica* 24: 1–187.

Mountford, J.O. and Chapman, J.M. (1993) Water regime requirements of British wetland vegetation: using the moisture classifications of Ellenberg and Londo, *Journal of Environmental Management* 30: 275–288.

Murkin, H.R. and Ward, P. (1980) Early spring cutting to control cattail in a northern marsh, *Wildlife Society Bulletin* 8: 254–256.

Newbould, C. and Mountford, O. (1997) *Water Level Requirements of Wetland Plants and Animals*, English Nature Freshwater Series No. 5, Peterborough: English Nature.

Niemann, E. (1963) Beziehungen zwischen Vegetation and Grundwasser, *Archiv für Naturschutz und Landschaftsforschung* 3: 3–37.

Niemann, E. (1973) Grundwasser und Vegetationsgefüge, *Nova Acta Leopold* 6: 1–147.

Nilsson, C. and Keddy, P.A. (1988) Predictability of change in shoreline vegetation in a hydroelectric reservoir, northern Sweden, *Canadian Journal of Fisheries and Aquatic Sciences* 45: 1896–1904.

Noest, V. (1994) A hydrology–vegetation interaction model for prediction of the occurrence of plant species in dune slacks, *Journal of Environmental Management* 40: 119–128.

O'Connell, A.M. and Grove, T.S. (1985) Acid phosphatase activity in karri in relation to soil phosphate and nitrogen supply, *Journal of Experimental Botany* 36: 1359–1372.

Olff, H., Berendse, F., Verkaar, D. and van Wirdum, G. (1995) Modelling vegetation succession, *Landschap* 12: 69–82.

Osborne, D.J. (1984). Ethylene and plants of aquatic and semi-aquatic environments. A review, *Plant Growth Regulation* 2: 167–185.

Pakarinen, P. (1978) Production and nutrient ecology of three *Sphagnum* species in southern Finnish raised bogs, *Annales Botanici Fennici* 15: 15–26.

Patrick, W.H. and DeLaune, R.D. (1972) Characterisation of the oxidised and reduced zones in flooded soil, *Soil Science Society of America, Proceedings* 36: 573–576.

Pearsall, W.H. (1938) The soil complex in relation to plant communities. I. Oxidation–reduction potentials in soils, *Journal of Ecology* 26: 180–193.

Pedersen, A. (1975) Growth measurements of five *Sphagnum* species in south Norway, *Norwegian Journal of Botany* 22: 277–284.

Perez–Corona, M.E. and Verhoeven, J.T.A. (1996) Effects of soil P status on growth and P and N uptake of *Carex* species from fens differing in P availability, *Acta Botanica Neerlandica* 45: 381–392.

Poiani, K.A. and Johnson, W.C. (1989) Effect of hydroperiod on seedbank composition in semipermanent prairie wetlands, *Canadian Journal of Botany* 67: 856–864.

Ponnamperuma, F.N. (1972) The chemistry of submerged soils, *Advances in Agronomy* 24: 29–96.

Poschlod, P. (1995) Diaspore rain and diaspore bank in raised bogs and implications for the restoration of peat-mined sites, in B.D. Wheeler, S.C. Shaw, W.J. Fojt and R.A. Robertson (eds) *Restoration of Temperate Wetlands*, Chichester: J. Wiley, pp. 471–494.

Proctor, M.C.F. (1994) Seasonal and shorter-term changes in surface-water chemistry on four English ombrogenous bogs, *Journal of Ecology* 82: 597–610.

Proctor, M.C.F. (1995) Hydrochemistry of the raised bog and fens at Malham Tarn National Nature Reserve, Yorkshire, UK, in J.M.R. Hughes and A.L. Heathwaite (eds) *Hydrology and Hydrochemistry of British Wetlands*, Chichester: J. Wiley, pp. 273–289.

Proctor, M.C.F. and Maltby, E. (1998) Relations between acid atmospheric deposition and the surface pH of some ombrotrophic bogs in Britain. *Journal of Ecology* 86: 329–340.

Ratcliffe D.A. and Walker, D. (1958) The Silver Flowe, Galloway, Scotland, *Journal of Ecology* 46: 407–445.

Reader, R.J., Jalili, A., Grime, J.P., Spencer, R.E. and Matthews, N. (1993) A comparative study of plasticity in seedling rooting depth in drying soil, *Journal of Ecology* 81: 543–550.

Richardson, C.J. and Marshall, P.E. (1986) Processes controlling movement, storage and export of phosphorus in a fen peatland, *Ecological Monographs* 56: 279–302.

Romanov, V.V. (1968) *Hydrophysics of Bogs*, Jerusalem: Monson Bindery.

Ross, S.M. (1995) Overview of the hydrochemistry and solute processes in British wetlands, in J.M.R. Hughes and A.L. Heathwaite (eds) *Hydrology and Hydro-chemistry of British Wetlands*, Chichester: Wiley, pp. 133–181.

Rumberg, C.B. and Sawyer, W.A. (1965) Response of wet meadow vegetation to length and depth of surface water from wild-flood irrigation, *Agronomy Journal* 57: 245–247.

Rydin, H. (1985) Effect of water level on desiccation of *Sphagnum* in relation to surrounding *Sphagna*, *Oikos* 45: 374–379.

Rydin, H. and McDonald, A.J.S. (1985) Photosynthesis in *Sphagnum* at different water contents, *Journal of Bryology* 13: 579–584.

Salisbury, E.J. (1970) The pioneer vegetation of exposed muds and its biological features, *Philosophical Transactions of the Royal Society of London* B259: 207–255.

Schat, H. (1982) On the ecology of some Dutch dune slack plants, Doctoral thesis, Vrije Universiteit: Amsterdam.

Schat, H. and Beckhoven (1991) Water as a stress factor in the coastal dune system, in J. Rozema and J.A.C. Verkleij (eds) *Ecological Responses to Environmental Stresses*, Dordrecht: Kluwer Academic Publishers, pp. 76–89.

Schimper, A.F.W. (1898) *Plant Geography upon a Physiological Basis* (English translation), Oxford: Clarendon Press.

Schlüter, U., Albrecht, G. and Wiedenroth, E.M. (1996) Content of water soluble carbohydrates under oxygen deprivation in plants with different flooding tolerance, *Folia Geobotanica and Phytotaxonomica* 31: 57–64.

Scholle, D. and Schrautzer, J. (1993) Zur Grundwasserdynamik unterschiedlicher Niedermoor-Gesellschafter Schleswig-Holsteins, *Zeitschrift für Ökologie und Naturschutz* 2: 87–98.

Schot, P.P. and Wassen, M.J. (1993) Calcium concentrations in wetland ground water in relation to water sources and soil conditions in the recharge area, *Journal of Hydrology* 141: 197–217.

Schot, P.P., Barendrecht, A. and Wassen, M. (1988) Hydrology of the wetland Narderneer: influence of the surrounding area and impact on vegetation, *Agric. Water Manag.* 14: 459–470.

Schouwenaars, J.M. (1990) A study on the evapotranspiration of *Molinia caerulea* and *Sphagnum papillosum*, using small weighable lysimeters, in Schouwenaars, J.M., Problem oriented research on plant–soil–water relationships, PhD thesis, Agricultural University: Wageningen.

175

Schouwenaars, J.M. (1996) The restoration of water storage facilities in the upper peat layer as a temporary substitute for acrotelm functions, in G.W. Lüttig (ed.) *Peatlands Use – Present, Past and Future*. 10th International Peat Congress, Bremen, Vol 2: Proceedings pp. 475–487.

Schouwenaars, J.M. and Vluk, J.P.M (1992) Hydrophysical properties of peat relicts in a former bog and perspectives for *Sphagnum* regrowth, *International Peat Journal* 5: 15–28.

Segal, S. (1966) Ecological studies of peat-bog vegetation in the north-western part of the province of Overijsel (the Netherlands), *Wentia* 15: 109–141.

Sellars, B. (1991) The response and tolerance of wetland plants to sulphide, unpublished PhD thesis, University of Sheffield.

Shaw, S.C. and Wheeler, B.D. (1990) *Comparative Survey of Habitat Conditions and Management Characteristics of Herbaceous Poor-Fen Vegetation Types*, Survey Report 129, Peterborough: Nature Conservancy Council.

Shipley, B., Keddy, P.A. and Lefkovitch, L.P. (1991) Mechanisms producing plant zonation along a water depth gradient: a comparison with the exposure gradient, *Canadian Journal of Botany* 69: 1420–1424.

Shotyk, W. (1988) Review of the inorganic geochemistry of peats and peat waters, *Earth Science Reviews* 25: 95–176.

Sjöberg, K. and Danell, K. (1983) Effects of permanent flooding on *Carex Equisetum* wetlands in Northern Sweden, *Aquatic Botany* 15: 275–286.

Sjörs, H. (1950a) On the relation between vegetation and electrolytes in North Swedish mire waters, *Oikos* 2: 241–258.

Sjörs, H. (1950b) Regional studies in North Swedish mire vegetation, *Botaniska Notiser* 2: 173–222.

Skoglund, J. (1990) *Seed Banks, Seed Dispersal and Regenetation Processes in Wetland Areas*. Uppsala: Acta Universitatis Upsaliensis.

Skoglund, J. and Verwijst, T. (1989) Age structure of woody species populations in relation to seed rain, germination and establishment along the river Dalälven, Sweden, *Vegetatio* 82: 25–34.

Small, E. (1973) Water relations of plants in raised *Sphagnum* peat bogs, *Ecology* 53: 726–728.

Snowden, R.E.D. and Wheeler, B.D. (1993) Iron toxicity to fen plants, *Journal of Ecology* 81: 35–46.

Snowden, R.E.D. and Wheeler, B.D. (1995) Chemical changes in selected wetland plant species with increasing Fe supply, with specific references to root precipitates and Fe tolerance, *New Phytologist* 131: 503–520.

Sorrell, B.K. and Armstrong, W. (1994) On the difficulties of measuring oxygen release by root systems of wetland plants, *Journal of Ecology* 82: 177–183.

Sparling, J.H. (1966) Studies on the relationship between water movement and water chemistry in mires, *Canadian Journal of Botany* 44: 747–758.

Spence, D.N.H. (1964) The macrophytic vegetation of freshwater lochs, swamps and associated fens, in J.H. Burnett (ed.) *The Vegetation of Scotland*, Edinburgh: Oliver & Boyd, pp. 306–425.

Spieksma, J.F.M., Schouwenaars, J.M. and van Diggelen, R. (1995) Assessing the impact of water management options upon vegetation development in drained lake side wetlands, *Wetlands Ecology and Management* 3: 249–262.

Squires, L. and van der Valk, A.G. (1992) Water depth tolerances of the dominant emergent macrophytes of the Delta Marsh, Manitoba, *Canadian Journal of Botany* 70: 1860–1867.

Succow, M. (1988) *Landschaftsökologische Moorkunde*, Jena: Gustav Fischer Verlag.

Summerfield, R.J. (1974) The reliability of mire water chemical analyses as an index of plant nutrient availability, *Plant and Soil* 40: 97–106.

Tanaka, A., Loe, R. and Navasero, S.A. (1966) Some mechanisms involved in the development of iron toxicity symptoms in the rice plant, *Soil Science and Plant Nutrition* 12: 158(32)–164(38).

Tansley, A.G. (1949) *The British Islands and their Vegetation*, second edition, Cambridge: Cambridge University Press.

Ter Braak, C.J.F. and Gremmen, N.J.M. (1987) Ecological amplitudes of plant species and the internal consistency of Ellenberg's indicator values for moisture, *Vegetatio* 69: 79–87.

Tessenow, U. and Baynes, Y. (1978) Experimental effects of *Isoetes lacustris* L. on E_h, pH, Fe and Mn in lake sediments, *Verh. Int. Ver. Theor. Angew. Limnol.* 20: 2358–2362.

Thompson, K. and Grime, J.P. (1983) A comparative study of germination responses to diurnally-fluctuating temperatures, *Journal of Applied Ecology* 20: 141–156.

Titus, J.E. and Wagner D.J. (1984) Carbon balance for two *Sphagnum* mosses, *Ecology* 65: 1765–1774.

Toner, M. and Keddy, P.A. (1997) River hydrology and riparian wetlands: a predictive model for ecological assembly, *Ecological Applications* 7: 236–246.

Tüxen, R. (1954). Pflanzengesellschaften und Grundwasserganglinien, *Angewandte Pflanzensoziologie* 8: 64–96.

Urban, N.R., Eisenreich, S.J. and Bayley, S.E. (1988) The relative importance of denitrification and nitrate assimilation in midcontinental bogs, *Limnology and Oceanography* 33: 1611–1617.

Urquhart, C. and Gore, A.J.P. (1973) The redox characteristics of four peat profiles, *Soil Biology and Biochemistry* 5: 659–672.

van der Sman, A.J.M., Voesnek, L.A.C.J., Blom, C.W.P.M., Harren, F.J.M. and Reuss, J. (1991) The role of ethylene in shoot elongation with respect to survival and seed output of flooded *Rumex maritimus* L. plants, *Functional Ecology* 5: 304–313.

van der Valk, A.G. and Davis, C.B. (1978) The role of seed banks in the vegetation dynamics of prairie glacial marshes, *Ecology* 59: 322–335.

van der Valk, A.G. and Davis, C.B. (1980) The impact of a natural drawdown on the growth of four emergent species in a prairie glacial marsh, *Aquatic Botany* 9: 301–322.

van der Valk, A.G., Davis, C.B. and Welling, C.H. (1988) The development of zonation in freshwater wetlands – an experimental approach, in H.J. During,

M.J.A. Werger and J.H. Willems (eds) *Diversity and Pattern in Plant Communities*, Den Haag: SPB Publishing, pp. 145–158.

van Diggelen, R., Grootjans, A.P., Kemmers, R.H., Kooijman, A.M., Succow, M., de Vries, N.P.J. and van Wirdum, G. (1991a) Hydroecological analysis of the fen system Lieper Posse, eastern Germany, *Journal of Vegetation Science* 2: 465–476.

van Diggelen, R., Grootjans, A.P., Wierda, A., Berkunk, P. and Hoogendoorn, J. (1991b) Prediction of vegetation changes under different hydrological scenarios, in *Hydrological Basis of Ecologically Sound Management of Soil and Ground water*, IAHS Publication No, 202, pp. 71–80.

van Splunder, I., Coops, H., Voesenek, L.A.C.J. and Blom, C.W.P.M. (1995) Establishment of alluvial forest species in floodplains: the role of dispersal timing, germination characteristics and water level fluctuation, *Acta Botanica Neerlandica* 44: 269–278.

van Wirdum, G. (1982) The ecohydrological approach to nature protection, *Research Institute for Nature Management, Leersum, Annual Report 1981*: 60–74.

van Wirdum, G. (1986) Water related impacts on nature protection sites, *TNO Committee on Hydrological Research, Proceedings and Information* 34: 27–57.

van Wirdum, G. (1991) Vegetation and hydrology of floating rich-fens, doctoral thesis, University of Amsterdam.

van Wirdum, G., den Held, A.J. and Schmitz, M. (1992) Terrestrializing fen vegetation in former turbaries in the Netherlands, in J.T.A. Verhoeven (ed.) *Fens and Bogs in the Netherlands*, Dordrecht: Kluwer, pp. 323–360.

Vermeer, J.D. and Beredense, F. (1983) The relationship between nutrient availability, shoot biomass and species richness in grassland and wetland communities, *Vegetatio* 53: 121–126.

Vitt, D.H., Crum, H. and Snider, J.A. (1975) The vertical zonation of *Sphagnum* species in hummock–hollow complexes in northern Michigan, *Michigan Botanist* 14: 190–200.

Voesenek, L.A.C.J. and Blom, C.W.P.M. (1992) Germination and emergence of *Rumex* species in river flood-plains. I. Timing of germination and seedbank characteristics, *Acta Botanica Neerlandica* 41: 319–329.

von Müller, A. (1956) Über die Bodenwasser-Bewegung unter einigen Grünland Gesellschaften des mittleren Wesentales und seiner Randgebiet, *Angwandte Pflanzensoziologie* 12: 1–85.

von Post, L. and Granlund, E. (1926) Sodra Sveriges tortillangar I, *Sveriges Geol. Unders.*, C335, 127pp.

Votrubová, O. and Pecháčková, A. (1996) Effect of nitrogen over-supply on root structure of common reed. *Folia Geobotanica Phytotaxonomica* 31: 119–125.

Wagner, D.J. and Titus, J.E. (1984). Comparative desiccation tolerance of two *Sphagnum* mosses, *Oecologia* 62: 182–187.

Warming, E. (1909) *Oecology of Plants. An Introduction to the Study of Plant Communities*, Oxford: Clarendon Press.

Warwick, J. and Hill, A.R. (1988) Nitrate depletion in the riparian zone of a small woodland stream, *Hydrobiologia* 157: 231–240.

Wassen, M.J. and Barendregt, A. (1992) Topographic position and water chemistry of fens in a Dutch river plain, *Journal of Vegetation Science* 3: 447–456.

Wassen, M.J., Barendregt, A., Palcynski, A., de Smidt, J.T. and de Mars, H. (1990) The relationship between fen vegetation gradients, ground water flow and flooding in an undrained valley mire at Biebrza, Poland, *Journal of Ecology* 78: 1106–1122.

Wassen, M.J., Barendregt, A., Palcynski, A., de Smidt, J.T. and de Mars, H. (1992) Hydro-ecological analysis of the Biebrza mire (Poland), *Wetlands Ecology and Management* 2: 119–134.

Wassen, M.J., Barendregt, A. and de Smidt, J.T. (1988) Ground water flow as conditioning factor in fen ecosystems in the Kortenhoef area, the Netherlands, in M. Ruzicka, T. Hrnciarova and L. Miklos (eds) *Proceedings of the 8th Symposium on Problems of Landscape Ecological Research*, Institute of Experimental Biology and Ecology: Bratislava, pp. 241–251.

Wassen, M.J., Barendregt, A., Bootsma, M.C. and Schot, P.P. (1989) Ground water chemistry and vegetation of gradients from rich fen to poor fen in the Nardermeer (the Netherlands), *Vegetatio* 79: 117–132.

Wassen, M.J., Barendregt, A., Schot, P.P. and Beltman, B. (1990) Dependency of local mesotrophic fens on a regional ground water flow system in a poldered river plain in the Netherlands, *Landscape Ecology* 5: 21–38.

Weber, M. and Brändle, R. (1996) Some aspects of extreme anoxia tolerance of the sweet flag, *Acorus calamus* L., *Folia Geobotanica and Phytotaxonomica* 31: 37–46.

Weiher, E. and Keddy, P.A. (1995) The assembly of experimental wetland plant communities, *Oikos* 73: 323–335.

Welling, C.H., Pederson, R.L. and van der Valk, A.G. (1988) Recruitment from the seedbank and the development of zonation of emergent vegetation during a drawdown in a prairie wetland, *Journal of Ecology* 76: 483–496.

Wheeler, B.D. (1988) Species richness, species rarity and conservation evaluation of rich-fen vegetation in lowland England and Wales, *Journal of Applied Ecology* 25: 331–353.

Wheeler, B.D. (1993) Botanical diversity in British mires, *Biodiversity and Conservation* 2: 490–512.

Wheeler, B.D. (1995) Introduction: restoration and wetlands, in B.D. Wheeler, S.C. Shaw, W.J. Fojt and R.A. Robertson (eds) *Restoration of Temperate Wetlands*, Chichester: Wiley, pp. 1–18.

Wheeler, B.D. (1996) Conservation of peatlands, in E. Lappalainen (ed.) Jyskä: International Peat Society, *Global Peat Resources*, pp. 285–301.

Wheeler, B.D. and Giller, K.E. (1982) Species richness of herbaceous fen vegetation in Broadland, Norfolk in relation to the quantity of above ground plant material, *Journal of Ecology* 70: 179–200.

Wheeler, B.D. and Shaw, S.C. (1991) Above-ground crop mass and species-richness of the principal types of herbaceous rich fen vegetation of lowland England and Wales, *Journal of Ecology* 79: 285–301.

Wheeler, B.D. and Shaw, S.C. (1995) A focus on fens – controls on the composition of fen vegetation in relation to restoration, in B.D. Wheeler, S.C. Shaw, W.J. Fojt and R.A. Robertson (eds) *Restoration of Temperate Wetlands*, Chichester: J. Wiley, pp. 49–72.

Wheeler, B.D. and Shaw, S.C. (1995) Plants as hydrologists? An assessment of the value of plants as indicators of water conditions in fens, in J.M.R. Hughes and A.L. Heathwaite (eds) *Hydrology and Hydrochemistry of British Wetlands*, Chichester: Wiley, pp 63–93.

Wheeler, B.D., Al-Farraj, M.M. and Cook, R.E.D. (1985) Iron toxicity to plants in base-rich wetlands: comparative effects on the distribution and growth of *Epilobium hirsutum* L. and *Juncus subnodulosus* Schrank, *New Phytologist* 100: 653–669.

Wheeler, B.D., Shaw, S.C. and Cook, R.E.D. (1992) Phytometric assessment of the fertility of undrained rich-fen soils, *Journal of Applied Ecology* 29: 466–475.

Wierda, A., Fresco, L.F.M., Grootjans, A.P. and van Diggelen, R. (1997) Numerical assessment of plant species as indicators of the ground water regime, *Journal of Vegetation Science* 8: 707–716.

Williams, B.L. and Wheatley, R.E. (1988) Nitrogen mineralization and water table height in oligotrophic deep peat, *Biology and Fertility of Soils* 6: 141–147.

Willis, A.J. and Jefferies, R.L. (1963) Investigations on the water relations of sand-dune plants under natural conditions, in A.J. Rutter and F.H. Whitehead (eds) *The Water Relations of Plants*, London: Blackwell Scientific Publications, pp. 168–189.

Willis, A.J., Folkes, B.F., Hope-Simpson, J.F. and Yemm, E.W. (1959a) Braunton Burrows: the dune system and its vegetation. Part I, *Journal of Ecology* 47: 1–24.

Willis, A.J., Folkes, B.F., Hope-Simpson, J.F. and Yemm, E.W. (1959b) Braunton Burrows: the dune system and its vegetation. Part II, *Journal of Ecology* 47: 249–288.

Wilson, K.A. and Fitter, A.H. (1984) The role of phosphorus in vegetational differentiation in a small valley mire, *Journal of Ecology* 72: 463–473.

Wilson, S.D. and Keddy, P.A. (1986) Species competitive ability and position along a natural stress/disturbance gradient, *Ecology* 67: 1236–1242.

Yabe, K. and Onimaru, K. (1996) Key variables controlling the vegetation of a cool temperate mire in northern Japan, *Journal of Vegetation Science* 8: 29–36.

Yapp, R.H. (1912) *Spirea ulmaria* L. and its bearing on the problem of xeromorphy in marsh plants, *Annals of Botany* 26: 815–870.

Zonneveld, I.S. (1959) *De Brabantse Biesbosch. Een Studie van Bodem en Vegetatie van een Zoetwatergetijdendelta*, Verslagem van Landbouwkundige Onderzoekingen No. 65: Wageningen.

Zwillenberg, L.O. and de Wit, R.J. (1951) Observations sur le *Rosmarineto-Lithospermetum schoenetosum* du Bas-Languedoc, *Acta Botanica Neerlandica* 1: 310–323.

6

PLANTS AND WATER IN FORESTS AND WOODLANDS

John Roberts

INTRODUCTION

Trees often constitute the climax vegetation in a region. This is a consequence of the tall habit, which allows trees to capture the majority of radiation. They are also likely to have extensive and deep root systems exploiting water resources unavailable to vegetation with shallow roots, and they are able to survive extended periods in which no rain falls. Many land cover classes in most of the zones of the world are dominated by trees (Table 6.1). Trees and woodlands cover approximately 55×10^6 km^2 (*c.* 37 percent) of the Earth's land surface. This forest cover is a principal route by which water in the hydrological cycle enters and leaves soils and groundwater systems. Forests are regarded as having many important influences with respect to water: the common perception of forests is that they influence the quantity and timing of water released to streams, rivers and groundwater; forest cover is considered to reduce the likelihood of flooding; forests that are not being manipulated are considered to yield water of good quality in terms of low amounts of dissolved material and sediments. At the local scale of a few kilometres there is still debate about the role of forests in modifying rainfall and climate, but there is convincing evidence from general circulation models (GCMs) of the role of forests in determining climate and rainfall at the regional and continental scale. The common perception is that trees use more water than short vegetation, but this is not always the case. There are examples of very high water use by forests and examples where forest water use is modest and probably equivalent to that of short vegetation. The focus of this chapter is the water use by forests; it examines the important hydrological processes in forests and surveys the measurement techniques available to study them. Mechanisms leading to high or modest water use are examined in the context of the transpiration and interception processes. The impact of trees on forest soils is also discussed. Two case studies, a temperate coniferous forest and a tropical rain forest, are presented for comparison, and attention is drawn to similarities and differences in the evaporation processes between the two.

181

Table 6.1 The IGBP–DIS land cover classification (after Woodward and Steffen, 1996).

1. *Evergreen needleleaf forests*: Lands dominated by trees with greater than 60% cover and height exceeding 2 m. Almost all trees remain green. Canopy is never without green foliage.

2. *Evergreen broadleaf forests*: Lands dominated by trees with greater than 60% cover and height exceeding 2 m. Almost all trees remain green all year. Canopy is never without green foliage.

3. *Deciduous needleleaf forests*: Lands dominated by trees with greater than 60% cover and height exceeding 2 m. Consists of seasonal needleleaf tree communities with an annual cycle of leaf-on and leaf-off periods.

4. *Deciduous broadleaf forests*: Lands dominated by trees with greater than 60% cover and height exceeding 2 m. Consists of seasonal broadleaf tree communities with an annual cycle of leaf-on and leaf-off periods.

5. *Mixed forests*: Lands dominated by trees with greater than 60% cover and height exceeding 2 m. Consists of tree communities with interspersed mixtures or mosaics of the other four forest cover types. None of the forest types exceeds 60% of the landscape.

6. *Closed shrublands*: Lands with woody vegetation less than 2 m in height and with greater than 60% cover. The shrub foliage can be either evergreen or deciduous.

7. *Open shrublands*: Lands with woody vegetation less than 2 m in height and a cover between 10 and 60%. The shrub foliage can be either evergreen or deciduous.

8. *Woody Savannahs*: Lands with herbaceous and other understorey systems, and with forest canopy cover between 30 and 60%. The forest cover height exceeds 2 m.

9. *Savannahs*: Lands with herbaceous and other understorey systems, and with forest canopy cover between 10 and 30 percent. The forest cover height exceeds 2 m.

THE FOREST HYDROLOGICAL CYCLE

The hydrological cycle in a forest is shown in Figure 6.1. The magnitude of some of the components will differ to some degree from one forest type to another, but a greater difference will exist in the size of specific components between forests and short vegetation. Some rain falling on the forest canopy is held on the foliage and branches. Water that is evaporated directly from these surfaces, both during rain and after it has stopped, is the rainfall *interception* loss. Water reaches the forest floor either by direct *throughfall* or as drainage down the tree trunks (*stemflow*). Forests may have a shrubby or herbaceous understorey, which will also have interception and stemflow associated with

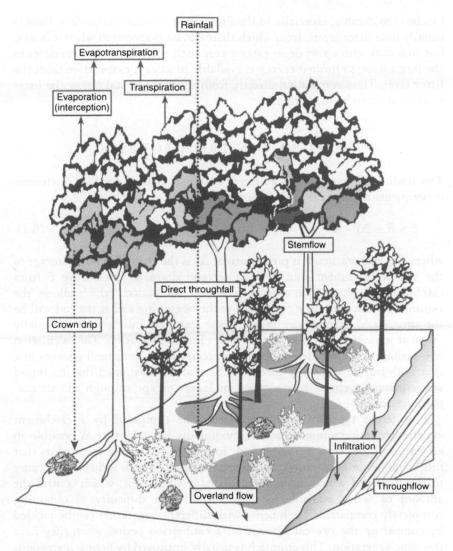

Figure 6.1 The hydrological cycle in a tropical rain forest. Redrawn from Bruijnzeel (1990) with permission from UNESCO and L.A. Bruijnzeel.

it. Water reaching the soil surface may run off directly (*overland flow*) but this would be unusual in a forest. Water enters the soil to replenish the soil water store, contributes to *subsurface flow* downslope if the forest is in a hilly area, or drains to deep layers in the soil (deep drainage or *recharge*).

As well as the interception loss, water returns to the atmosphere as transpiration. In some types of vegetation, particularly crops, there is likely to be bare soil, at least at some times of the year; evaporation from the bare soil can be considerable, especially in the first day or so after a rainstorm. Forests usually have litter layers, from which there can be evaporation when it is wet, but in forests with a very dense canopy very little solar radiation penetrates to the forest floor, so limited energy is available to sustain evaporation from the litter layer. Thus evaporation directly from the mineral soil below the litter layer is likely to be negligible.

EVAPORATION FROM FORESTS

The traditional way to measure forest evaporation is to conduct catchment experiments to solve the water balance equation:

$$E = P \pm \Delta S - R \tag{6.1}$$

where E is evaporation, P is precipitation, ΔS is the change in water storage of the soil or catchment, and R is runoff and drainage. Measuring E from catchments is often done over a twelve-month period, which allows the assumption that soil water conditions at the beginning and at the end will be the same, so ΔS becomes negligible. These start and finish points are usually taken at some time when the soil is at or close to saturation. The calculation of evaporation as the difference between precipitation and runoff assumes first that both can be accurately measured. Second, it is assumed that no liquid water enters except as precipitation or leaves except through the stream-gauging device.

If different types of vegetation are to be compared by a catchment experiment it is required that the catchments be as similar as possible in all respects other than the vegetation. Important features of catchments that might influence evaporation are slope and aspect, which influence exposure to radiation and wind, and soil and drainage conditions, which control the amount of water available for evaporation. The difficulty of obtaining completely comparable catchments with different vegetation can be tackled by comparing the two catchments for a calibration period when they have the same vegetation. This approach is usually improved by fitting regressions of runoff against climatic variables for each catchment and testing the significance of before- and after-treatment regressions (Kovner and Evans, 1954).

Catchment experiments are the singular means for investigating effects of catchment management on the stream's behaviour and its water quality, but understanding of evaporation and its controls can only proceed by using methods that separate the components of evaporation (interception and transpiration) and allow analysis of these processes in relation to seasonal and diurnal changes in environmental conditions. Measurements of interception and transpiration losses from forests are usually made at scales smaller than that of the catchment; i.e. at the leaf, tree or plot scale. Models would then be employed to exploit the information from these measurements at larger scales. The following sections look at the interception and transpiration processes in more detail.

INTERCEPTION

Measurement of rainfall interception

There is no reason why the evaporation of intercepted water cannot be measured over short periods of as little as an hour or less, using the micro-meteorological methods of Bowen ratio or eddy covariance (see 'Measurement of transpiration', below). Practically, though, these methods are troublesome to use in wet conditions because raindrops will interfere with the correct operation of the instruments or, because during and immediately after rainstorms, humidity deficits of the atmosphere are small and gradients in the atmosphere would be difficult to measure. Nevertheless, micrometeorological methods have been used successfully during and immediately after rainfall has ceased to investigate and understand processes controlling the rainfall evaporation process and to determine short-term rates of evaporation from canopies under wet conditions (Stewart, 1977; Jarvis, 1993). More usually, interception loss from vegetation is measured as the difference between gross rainfall, either measured above the forest or in a nearby clearing, and net rainfall measured inside the forest. There are two main approaches to measuring net rainfall.

Plastic sheet rainfall collectors

Calder and Rosier (1976) describe a gauge that has been used successfully in a number of forests in temperate and tropical regions. The gauge consists of a large plastic sheet arranged on the forest floor as a series of roofs and troughs created by joining rows of adjacent trees with ropes and laying the sheet over the rope and back to ground level, with several such configurations repeated. At each tree the sheet is cut and sealed around the trunk with mastic compounds. The whole sheet slopes downwards in one direction and all the net rainfall (no discrimination is made between throughfall and stemflow) is led to a tipping bucket gauge connected to a data logger. If the gauge is to be left in place for extended periods of months or years, it will be necessary to provide sub-irrigation to prevent droughting and leaf fall of the trees in the gauge. A plastic sheet net rainfall gauge was used very effectively by Calder (1977) to provide estimates of rainfall interception and net rainfall inputs to a natural lysimeter in spruce forest in mid-Wales (see also 'Measurement of transpiration'). This approach was also used successfully in a tropical rain forest in Java (Calder et al., 1986). However, a similar approach proved unsuccessful in tropical rain forest in Amazonia (C.R. Lloyd, Institute of Hydrology, UK, personal communication) because an inordinately large gauge would have been necessary to sample net rainfall adequately. In tropical rain forest, there are 'drip points' where there can be a considerable concentration of throughfall. A plastic sheet gauge that is too small can conceivably sample one or several drip points, and it is possible to measure more net rainfall than gross rainfall. Alternatively, there is clearly a serious risk of under-sampling net rainfall.

Throughfall and stemflow gauges

Another common approach to measuring throughfall is to use fixed or moving rain gauges located randomly on the forest floor. Recording rain gauges can be used in the random positions, but this is likely to be prohibitively expensive. It is more usual to employ bottle gauges, which can be read manually at regular intervals or after rainstorms and then replaced or located in new random positions. It will be necessary to measure stemflow separately, and this is usually done by attaching a guttering around the trunks of a number of trees and collecting the stemflow in a container. Usually a number of trees are fitted with gauges, trees in the sample spanning the girth range of trees in the forest. An alternative to rain gauges on the forest floor are randomly located but fixed throughfall troughs. There would also be a need for stemflow gauges as described above.

A major challenge in interception studies is to achieve adequate sampling such that confidence limits on throughfall estimates are acceptably low. A plastic sheet net rainfall gauge of around 100 m^2 would be adequate in a young, densely stocked coniferous or broadleaf plantation, but there are other situations where the plastic sheet approach is likely to be unsuitable. In forests with widely spaced trees the gauges would need to be excessively large to sample the variation in throughfall. The problem of localised concentration of net rainfall in drip points in tropical rain forest has already been referred to. In tropical forests there is likely to be a similar problem for fixed trough gauges (Bruijnzeel and Wiersum, 1987). When bottle gauges are used it is necessary to have a sufficiently large sample size; a further reduction in the error estimate of net rainfall will be achieved by randomly relocating gauges, preferably after individual rainstorms, but this may prove impractical in remote situations. The number of gauges for adequate sampling has been discussed for broadleaf woodlands by Czarnowski and Olszewski (1970), for conifers by Kimmins (1973), and for tropical forest by Lloyd and Marques Filho (1988). These last authors show that a better than 5 percent error in the estimate of mean throughfall can be achieved with forty gauges with twenty random arrangements. Based on concerns about inadequate sampling, Bruijnzeel (1990) was sceptical about many reports of rainfall interception losses for tropical forests.

Interception losses from forests

There are a considerable number of studies of rainfall interception in temperate broadleaf and conifer forests. Figure 6.2 shows a representative portion of the data that have come from studies in the UK. Because of the leafless nature of broadleaf forest canopies in winter, the interception fraction of gross rainfall is usually substantially less than for evergreen conifers in the same rainfall.

A feature of interception losses from temperate broadleaf forests (Figure 6.2) is that interception losses are far more variable at a given value of precipitation than for conifers. A number of factors may contribute to this greater variability. Broadleaf forests in which interception studies have been carried out may have contained a mixture of tree species and be of uneven age. Understoreys would be present in some forests but not in others. By contrast, the data for conifers would generally be from even-aged plantations of one species. Particularly in the data for conifers presented in Figure 6.2, there appears to be a greater percentage of interception loss with lower rainfall totals. In low-rainfall climates, there is often a greater percentage of small storms, out of which the interception loss may constitute a considerable portion of the storm. This effect should not be limited to conifers. In a study of oak interception on the Dutch coast, where annual gross rainfall is low

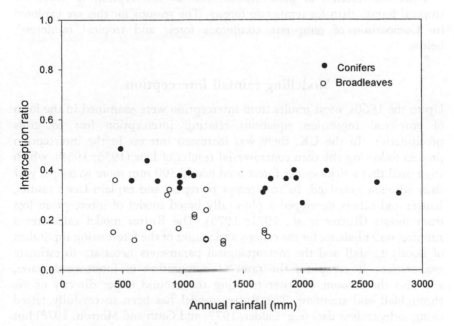

Figure 6.2 Interception ratios plotted against annual rainfall for interception studies involving conifers and broadleaves in the UK.

(~650 mm), Mulder (1985) found that in some years rainfall interception was over 50 percent of the annual gross rainfall. One of the climate change predictions for high latitudes is that winters will become wetter and summers drier. In such a climate change scenario it may be that annual rainfall interception of broadleaf forests will fall, because most rain will occur when trees are not in leaf.

Throughfall below tropical forest canopies appears to be very variable spatially. However, many values of rainfall interception loss must be regarded with some scepticism because of inadequate sampling. Bruijnzeel (1990) has critically reviewed the published rainfall interception data from tropical lowland and montane forests. In the case of tropical lowland forest he chose thirteen studies with an adequate length of record to incorporate seasonal variations and a rain gauge relocation technique or sufficient sampling. The range of interception losses was 7 to 23 percent, with a mean of 15 percent. Since Bruijnzeel's paper, there has been another study in two more lowland tropical rain forests in Brazil (Ubarana, 1996), for which the interception losses were 11.6 and 12.9 percent. For six montane forests thought by Bruijnzeel (1990) to have adequate sampling, the range of interception loss was from 14 to 25 percent, with a mean of 19 percent. The higher values for the montane tropical forests are thought to be due to lower rainfall intensities prevailing on tropical mountains.

Thus the fraction of gross rainfall lost as interception is lower for tropical forests than for temperate forests. The reasons for this are explored in 'Comparison of temperate coniferous forest and tropical rainforest', below.

Modelling rainfall interception

Up to the 1960s, most results from interception were examined in the form of empirical regression equations relating interception loss to gross precipitation. In the UK, there was increased interest in the interception process following the then controversial results of Law (1956; 1958), which suggested that a sitka spruce forest used nearly 300 mm more water per year than adjacent grassland. In an attempt to repeat and explain Law's results, Rutter and others developed a physically based model of interception loss from forests (Rutter *et al.*, 1971; 1975). The Rutter model calculates a running water balance for the canopy and trunks of the forest using input data of hourly rainfall and the meteorological parameters necessary to estimate evaporation. It computes the rates of evaporation of intercepted water, and also the amount of water reaching the ground either directly or via throughfall and stemflow. The Rutter model has been successfully tested using independent data (e.g. Calder, 1977, and Gash and Morton, 1978) but has the disadvantage that it requires hourly rainfall and weather data to be used in forecasting mode.

Gash (1979) described an analytical model that linked Rutter's approach with earlier regression approaches. The Gash model is a storm-based approach that requires a knowledge of the capacity for vegetation to store water and the average rates of evaporation and precipitation during rainfall. It can be used to predict rainfall interception given basic daily rainfall data available from most standard meteorological stations (Lloyd and Marques Filho, 1988). The model has been used successfully in a number of forests such as coniferous forests in the UK (Gash et al., 1980), evergreen mixed forest in New Zealand (Pearce and Rowe, 1981), oak forest in the Netherlands (Dolman, 1987), tropical forest plantations (Bruijnzeel and Wiersum, 1987), and rain forest in Amazonia (Lloyd et al., 1988) and West Africa (Hutjes et al., 1990).

An alternative approach to modelling rainfall interception from forest canopies has been the use of the so-called stochastic model (Calder, 1986). In this model, the wetting of the canopy is related to probabilities of raindrops striking the canopy elements. The model also predicts that, for the same total rainfall amounts, maximum water storage on the canopy will be reached more slowly with larger raindrops. Calder's model has been used successfully for tropical rain forest in Java (Calder et al., 1986) and tropical rain forest and plantations in Sri Lanka and India (Calder, 1996; Calder et al., 1996; Hall et al., 1996).

Increasingly, agroforestry systems involving trees and crops growing in combination are being used to enhance productivity on marginal lands in developing countries. It is important to estimate rainfall interception as this is an important water loss from the agroforestry system. The Gash model was unsuccessful in modelling sparse forest interception loss (Teklehaimanot et al., 1991). Gash et al. (1995) reformulated the original Gash model using evaporation per unit of canopy rather than per unit of ground area. The new model performed well in predicting rainfall interception for Les Landes forest, France (ibid.) and eucalyptus plantations in Portugal (Valente et al., 1997). These two examples are open forest plantations, but it is likely that the model will probably be very valuable for predicting interception losses from native vegetation where tree cover is sparse, e.g. woody savannahs and agroforestry systems.

TRANSPIRATION

Measurement of transpiration

Because there are a wide range of questions relating to transpiration from forests, a wide range of techniques have been used to measure the process. Which technique should be used depends on the temporal and spatial scale over which estimates are needed.

Soil water depletion

Studies of rates of soil water depletion are usually used to estimate transpiration over timescales of at least a few days, and they cannot distinguish water taken up by different species when they are growing in close proximity. Sufficient measurements of soil water content need to be made to account for spatial variation in water storage. In addition, the amount of drainage needs to be measured, or it must be insignificant. Provided drainage from and recharge to the soil is quantified or insignificant, changes in soil water storage allow evaporation to be calculated when rainfall is absent or measured separately. Soil moisture depletion techniques require repeated *in situ* measurements of soil water content, which may be with a neutron probe (Bell, 1987; Greacen, 1981), capacitance probe (Bell *et al.*, 1987; Dean *et al.*, 1987), the impedance 'ThetaProbe' technique (Gaskin and Miller, 1996), or by using time domain reflectometry (Topp *et al.*, 1980; 1984).

Micrometeorological methods

A more revealing approach to measuring evaporation, in which it is possible to discriminate between evaporation from wet canopies and transpiration from dry canopies, exploits some of the micrometeorological methods that are available.

Bowen Ratio method

A common way by which evaporation and transpiration of forests has been determined is using the energy budget method of Bowen (1926). The energy balance equation (equation (6.2)) and the ratio of sensible and latent heat fluxes (the Bowen ratio – equation (6.3)) are solved simultaneously, neglecting the heat flux associated with plant metabolism and assuming that the eddy diffusivities for sensible heat and water vapour are the same (see also Chapter 2). The energy balance of a forest between a reference level and the soil is

$$R_N + G + H + \lambda E + M = 0 \tag{6.2}$$

where R_N is net radiation, G is soil, biomass and trunk space heat flux, H and λE are turbulent fluxes of sensible and latent heat, respectively, and M is the metabolic heat flux. All fluxes are taken as positive downwards. The Bowen ratio (β) is

$$\beta = \frac{H}{\lambda E} = \gamma \frac{K_H \, \Delta T}{K_e \, \Delta e} \tag{6.3}$$

where ΔT and Δe are vertical differences in temperature and humidity, respectively, over the same height interval above the canopy. K_H and K_e are eddy diffusivities for heat and water vapour, respectively, and γ is the psychrometric constant.

Neglecting M and assuming $K_H = K_e$

$$\lambda E = \frac{R_N + G}{1 + \beta} \qquad (6.4)$$

Although evaporation calculated in this way accounts for all water vapour reaching the reference height above the forest, it usually equals the amount of water evaporating from the tree crowns with a fully closed canopy, because soil evaporation is negligible.

The bulk surface conductance (g_c) can then be calculated by inverting the Penman–Monteith equation (Monteith, 1965)

$$g_c = \frac{\lambda E \gamma g_a}{e'(R_N - G) + \rho_a C_p D g_a - \lambda E \,(e' + \gamma)} \qquad (6.5)$$

where e' is the slope of the curve relating saturated vapour pressure to temperature, ρ_a is the density of dry air, C_p is the specific heat of air, D is the saturation deficit of the air above the canopy, and g_a is the aerodynamic conductance between the canopy level and the reference height.

The aerodynamic conductance for the water vapour flux between the evaporating surface and the reference height, g_a, is usually calculated with an equation similar to the one below, with z_0 and d being assumed proportional to the stand height, h, and arbitrarily chosen as $d = 0.8h$ and $z_0 = 0.1h$

$$g_a = \frac{k^2 \, u(z)}{\ln^2 \left[(z - d) \,/\, z_0 \right]} \qquad (6.6)$$

where z is the reference height of the energy balance measurements, z_0 is the surface roughness length, d is the zero plane displacement, k is von Karman's constant and $u(z)$ is the windspeed at height z.

There are some practical difficulties associated with measuring forest evaporation with the energy budget – Bowen ratio – method. There is, of course, a need for a tower to mount instruments some metres above the tallest surrounding trees. The main difficulty, though, is the requirement to measure gradients of temperature and humidity (ΔT and Δe) in the air above the forest. Because of the particularly good mixing of the air above forests, these gradients are often only a few tenths of a °C, substantially smaller than those observed above short vegetation. The measurements of dry- and wet-bulb temperatures to resolve ΔT and Δe required to calculate the Bowen ratio, β, are usually made with thermometers calibrated to one-hundreth of a degree, but systematic differences between sensors may be large enough to partly obscure gradients. An approach that has been adopted to resolve gradients that are close to the instrumental errors has been to devise electronically controlled, mechanical devices (usually trackways or rotating booms) that systematically interchange thermometers between upper and lower levels several times in an hour. An example of such a device has been described by McNeil and Shuttleworth (1975).

Eddy correlation method

Within the turbulent boundary layer generated near the ground by the interaction of wind with surface irregularities, the transfer of water vapour and other atmospheric entities occurs by the process of turbulent diffusion. Although the mean wind speed at each level near the surface is parallel to the ground, the instantaneous air movement at any point can assume any direction, and, in general, there will be a component of wind moving towards and a component moving away from the forest surface, with a mean value of zero but with a finite standard deviation. It is this fluctuating component of wind that transports heat and water vapour away from or to the surface. Evaporation can be measured by making simultaneous measurements of both fluctuations in the wind component normal to the forest surface and fluctuations in the humidity of the air at the same point. The algebraic product of these variables is the instantaneous vapour flux, which can be either positive or negative at any instant (depending on the statistical nature of the turbulent transfer process), but which has a finite mean value, the net evaporation loss from the surface, found by integrating the instantaneous flux. A comprehensive account of the eddy correlation approach can be found in Shuttleworth (1992).

The sensors to measure wind speed and humidity have very rapid responses and must be located close to each other to measure the same parcel of air. Eddy correlation is now the favoured means of measuring evaporation directly from vegetation. This preference has been further enhanced in the last few years because approaches have evolved that have enabled measurements of evaporation and CO_2 exchange to be combined in a single piece of equipment. The CO_2 measurement is made by ducting small air samples, from where wind speed and humidity are measured, into a fast-response infrared gas analyser (Moncrieff et al., 1997).

Until recently, a perceived limitation of micrometeorological techniques was their inability to measure fluxes from components of the forest such as the litter layer or the understorey. A number of recent studies have attempted to measure water vapour fluxes within forests using an eddy correlation device additional to one operating above the forest (e.g. Black et al., 1996) and increasingly, as with above-canopy measurements, measurements of water vapour fluxes are being made in conjunction with those of carbon dioxide. There is a specific interest in CO_2 flux measurements, from which net carbon exchange of forests or their components can be estimated, but additionally, with information about vapour fluxes from transpiration, water use efficiency estimates can be made.

Lysimeters

A lysimeter is a device in which a volume of soil with associated vegetation is isolated hydrologically from the surrounding soil. Drainage is measured or is zero and, in the case of lysimeters that are weighed, changes in water storage are determined by weight difference. The application of weighing lysimeters to studies in mature or semi-mature forests has been very limited. A single large Douglas fir tree was installed in a weighing lysimeter by Fritschen *et al.* (1973), and Reyenga *et al.* (1988) installed a lysimeter in a regenerating eucalyptus forest in New South Wales, Australia. Large potted trees capable of being weighed (up to 500 kg) have proved useful for calibrating other techniques such as isotope tracers (e.g. Dugas *et al.*, 1993). Another form of lysimeter is the drainage lysimeter, which has been used very commonly in short crops but rarely in forests. Calder (1976) described a drainage lysimeter constructed in a Norway spruce plantation in mid-Wales, UK. The base of the lysimeter was sealed naturally by impermeable clay. Data from the lysimeter associated with nearby net rainfall measurements enabled calculation of three separate years of transpiration loss (Calder, 1977).

The cut-tree technique

Excising the bases of large trees underwater and measuring the water uptake was used by Ladefoged (1963) to estimate transpiration. However, Roberts (1977; 1978) showed that removal of soil and root resistances can improve the leaf water status of cut trees compared with controls, leading to differences in stomatal conductance and transpiration between normal trees and those with excised roots. Nevertheless, the tree-cutting technique has proved useful in examining the water relations of mature trees (Roberts, 1977) and the amount of water stored in trees (Roberts, 1976a), and it has proved particularly valuable as a means of calibrating other techniques such as isotopic tracers (Waring and Roberts, 1979).

Sap-flow techniques

Sap-flow techniques provide a means of continuously monitoring rates of sap flow. Information about sapwood cross-sectional area of sampled trees or the leaf area of the sample tree in relation to the leaf area in the forest enables transpiration to be estimated on a land area basis. The range of techniques for measuring sap flow and the limitations of different approaches have been reviewed recently by Swanson (1994) and Smith and Allen (1996).

Heat-pulse velocity

The heat-pulse velocity (HPV) method determines rates of sap flow by determining the velocity of a short pulse of heat carried by the moving sap stream. The technique is only really useful on woody stems, but the depth of sapwood must not be so deep that the sensor probe cannot sample it adequately.

Each set of heat-pulse probes consists of one heater probe and two sensor probes, in which there are miniature thermistors. Typically, four sets of probes are used, and these are distributed equally around the circumference of the stem. The heat-pulse technique is based on a compensation principle; the velocity of sap ascending the stem is determined by correcting the measured velocity of a heat pulse for the dissipation of heat by conduction through the wood matrix. In practice, this is achieved by installing the sensor probes at unequal distances upstream and downstream of the heater probe. The upstream sensor is usually nearer the heater than the downstream one. Heat-pulse velocity (v_h) is calculated from

$$v_h = \frac{X_d - X_u}{2t_o}$$ (6.7)

where X_d and X_u are the distances between the heater and the upstream and downstream sensors, respectively, and t_o is the time taken after the heat pulse for the temperature of the two sensors to become equal again. It is important to realise that sap-flux velocity and heat-pulse velocity are not necessarily equivalent, because of inhomogeneity in wood structure and thermal properties, the influence of wounds and the thermal properties of the materials used to construct the heater and sensing probes. Swanson (1994) and Smith and Allen (1996) discuss corrections to heat-pulse velocity data that take account of some of the important influences on heat conduction.

Stem heat balance

The stem heat balance (SHB) method can be used to measure sap flow in both woody and herbaceous stems, which can be very small in diameter. The approach has been used on branches, small trees and even roots (Smith *et al.*, 1997). The devices are available commercially, and different models are available to fit stems ranging in diameter from 2 to 125 mm. A full description of a SHB gauge is given by Smith and Allen (1996). Heat is applied to the outside of the stem enclosed by the heater and the sap flow derived from the fluxes of heat into and out of the heated section. Sap flow is related to the different heat losses from the stem section (Swanson, 1994) by

$$F = \frac{Q_h - Q_r - Q_v - Q_s}{C_s \Delta T} \text{ g s}^{-1} \tag{6.8}$$

where Q_h is the heater power, Q_v is vertical heat loss, Q_r is radial heat conduction, Q_s is heat storage, C_s is the heat capacity of the sap and ΔT is the temperature difference between the top and bottom of the heated section. The trunk sector heat balance method is similar to the SHB method and is particularly appropriate where stem diameters exceed the largest size of whole SHB units of the largest capacity, i.e. around 120 mm diameter. The stem sector heat balance method, as described by Cermák et al. (1984), is invasive in that five stainless steel electrode plates are inserted into the wood. When there might be substantial variation in sap flow rates around large tree trunks, installations should be made at more than one point.

The thermal dissipation method

Another variant on the sap flow technique was proposed by Granier (1985; 1987). Each probe consists of a pair of needles, which are inserted into the sapwood. The upper needle contains a heating probe and a thermocouple, which is referenced to the second needle inserted into the sapwood lower down the stem. Continuous heating of the upper needle sets up a temperature difference (ΔT) between the two needles. ΔT is at a maximum when the sap flow is at a minimum and decreases as the sap flow increases. Granier (1985) found that for two conifer species and oak volumetric sap flux density (u_v, $\text{m}^3 \text{ m}^{-2} \text{ s}^{-1}$) is related to ΔT by the following relationship:

$$u_v = 0.000119Z \tag{6.9}$$

where

$$Z = \frac{\Delta T_0 - \Delta T}{\Delta T} \tag{6.10}$$

when ΔT_0 is the value of ΔT when there is no sap flow. The mass sap flow rate is then

$$F_m = \rho_s u_v A_{sw} \tag{6.11}$$

where ρ_s is the sap density and A_{sw} is the sapwood cross-sectional area. Granier et al. (1990) suggest that the parameters in equation (6.9) are not dependent on wood properties or tree species and that the technique may possibly be used without calibration. However, this possibility needs testing for a wider range of species than has been achieved so far. This is important, because Granier-type gauges are available commercially at reasonable cost and are likely to become widely used in the future because they are simple to install, the calculations to derive sap flow are straightforward, and the gauges have relatively simple data-logging requirements.

Porometers and infrared gas analysers

A range of porometers and portable infrared gas analysers (IRGAs) with leaf chambers are commercially available. Porometers enable measurements of stomatal conductance, g_s, of individual leaves to be measured *in situ*. An IRGA can also be used to determine CO_2 exchange from the leaf as well as g_s. Additional useful information that can be acquired or calculated are leaf transpiration rates, leaf temperatures and internal CO_2 concentration of the leaves. g_s determined with porometers and IRGAs gives the smallest temporal and spatial scale of information and provides considerable insight into the environmental and internal controls of g_s and hence transpiration.

g_s measured with porometers and IRGAs has also been used to estimate transpiration from plant canopies. This involves multiplying g_s by the leaf area index to produce a surface or canopy conductance (g_c). g_c is used with a canopy boundary layer conductance, g_a (Grace, 1983), to estimate transpiration using the Penman–Monteith equation (Monteith, 1965). It is possible to follow this procedure in very complex canopies. An extreme case would be the tropical rain forest studied by Roberts *et al.* (1993). However, such an enterprise requires canopy access, and considerable effort has to be dedicated to sampling the canopy. Additionally, information about the leaf area index of the vegetation must be available, and in a complex canopy the vertical distribution of the leaf area may need to be known. Nevertheless, particularly with IRGAs, the available technology has become very sophisticated, so that environmental conditions such as light, vapour pressure deficit, temperature and CO_2 concentration can be manipulated in the IRGA and the response of the leaf determined.

Radioactive and stable isotope tracers

A number of tracers have been used to measure transpiration from branches and individual trees. These values can then be scaled up to give stand transpiration. Waring and Roberts (1979) used P^{32} and tritium to measure the transpiration of Scots pine trees. The isotopes were injected at the base of the tree and measured over the next few days as the tracers moved up the tree. Calder *et al.* (1992) described an approach using deuterium oxide (D_2O), which was injected into eucalyptus trees in plantations. Transpired water was collected in polythene bags tied onto selected branches. From the information of the D_2O injected (M) and the concentration in the transpirate produced over a known time interval (Cdt), transpiration (F) can be calculated

$$M = F \int_0^\infty C \mathrm{d}t \qquad\qquad\qquad (6.12)$$

The tracing techniques are simple to apply, and useful results are produced that agree well with other methods. However, there are obvious safety constraints in using radioactive isotopes and, where transpirate needs to be collected, there is still a need for canopy access. The time resolution for tracing techniques is relatively low as well. Transpiration values can only be resolved over as little as a few days. It is, therefore, difficult to use the techniques to understand the influence of short-term environmental fluctuations on transpiration.

Transpiration from forests

Table 6.1 lists a range of vegetation types that have woody plants as a key component of the vegetation. Unfortunately, there are not, as yet, comprehensive hydrological studies for all the classes in the table. For most of the classes there is some physiological information, with details of processes at the leaf level and data relating to biomass and leaf area. However, there are many gaps preventing a fuller understanding of the hydrological processes. It is only recently, largely because of interest in carbon sinks, that intensive studies into the boreal forest have been initiated, and a full output from the BOREAS project (Sellers *et al.*, 1995) should provide considerably more insight than is currently available. In this section, information from a selection of woody vegetation types is examined and attention drawn to the key features of hydrological behaviour. The three types selected arguably represent three quite separate points on the spectrum of annual rainfall supplied to the forest or woodland type.

Temperate forests

These forests comprise both conifers and broadleaves, and, of all forests, there are most data for this type because of the many detailed studies in temperate regions aimed at obtaining information about forest ecology, hydrology and biogeochemistry. Studies have been concentrated in Western Europe and North America.

Table 6.2 lists annual transpiration values from a wide range of studies in Europe involving both conifers and broadleaves. Even though the studies cover a wide range of species and forest ages and were done in different climates, nevertheless, two features emerge that are important in discussions of forest hydrology in temperate regions. First, the annual transpiration values (Table 6.2) and the maximum daily rates of transpiration of between 3 and 4 mm day^{-1} are well below the potential rate of transpiration determined by climatic conditions. Second, there is a close similarity in the annual transpiration values with an average around 335 mm year^{-1}, although a few studies are close to 400 mm year^{-1}.

Table 6.2 Annual transpiration of different vegetation covers.

Species	Country	Transpiration (mm year^{-1})	Forest age (years)	Reference
Broadleaves				
Ash	UK	407	45	Roberts and Rosier (1994)
Ash	UK	294	63	Roberts and Rosier (1998)
Beech	Belgium	344	30–90	Schnock (1971)
Beech	UK	393	64	Roberts and Rosier (1994)
Beech	France	288	–	Chassagneux and Choisnel (1987)
Beech	Germany	283	100	Kiese (1972)
Sweet chestnut (coppice)	France	275	12	Bobay (1990)
Oak (sessile)	Germany	342	18	Brechtel (1976)
		298	54	Brechtel (1976)
		342	165	Brechtel (1976)
Oak	Denmark	293	70	Rasmussen and Rasmussen (1984)
Oak	France	301	32	Bréda *et al.* (1993)
		151	32	Bréda *et al.* (1993)
Oak	France	340	120	Nizinski and Saugier (1989)
		241	120	Nizinski and Saugier (1989)
		284	120	Nizinski and Saugier (1989)
Oak/Beech	Netherlands	267	100	Bouten *et al.* (1992)
		362	100	Bouten *et al.* (1992)
		239	100	Bouten *et al.* (1992)
Conifers				
Norway spruce	Germany	362	70	Tajchman (1971)
Norway spruce	Germany	279	–	Brechtel (1976)
Norway spruce	UK	290	–	Calder (1977)
		340	–	Calder (1977)
		330	–	Calder (1977)
Sitka spruce	UK	340	28	Law (1956)
Scots pine	Germany	327	–	Brechtel (1976)
Scots pine	UK	353	46	Gash and Stewart (1977)
Scots pine	UK	427	–	Rutter (1968)

The principal reason for the low and similar rates of transpiration is a relationship between air humidity deficit and stomatal behaviour that is known to exist in many of the species in Table 6.2. Also, as will be shown later in the chapter for one of the studies listed in the table, both at the leaf (Beadle *et al.*, 1985a, b and c) and the canopy level (Stewart, 1988) a strong negative correlation is observed between air humidity deficit and stomatal and surface conductance. This relationship may be functional, but the mechanisms behind the relationships are not yet clear. In effect, what happens is that on days when evaporative demand is high stomata tend to close, whereas on days when demand is low the stomata are open. The consequences of the humidity deficit–conductance relationship are that daily transpiration rates remain conservative at below 4 mm day^{-1}, while transpiration rates from day to day are quite similar.

It is often the case that a substantial fraction of soil moisture can be lost before there is a reduction in surface conductance. Rutter's (1968) analysis showed the relationship between rates of water use and soil water deficits. Figure 6.3A shows some examples in which initial rates of soil water use are high but fall even when the available soil water has fallen by only 25 percent. Other examples (Figure 6.3B and C) have lower initial transpiration rates, which are sustained over a wide range of soil moisture availability such that transpiration does not decrease until over 60 percent of the available water has been used. Thus, because of the likely modest daily transpiration rates in the studies listed in Table 6.2, it is probable that limiting soil moisture deficits are reached only rarely and therefore play a minor role in generating differences between sites.

Another possible way for species to have similar transpiration rates emerges from studies of the behaviour of stomatal conductance (g_s) in species that have initially different values. Federer (1977) published data from a number of North American species showing the relationship of g_s to air humidity deficit (D). An interesting relationship (Figure 6.4) exists which shows that species with the highest g_s show the greatest decline in g_s with increasing D and that at intermediate values of D the different species have similar g_s, which means that leaf transpiration rates under typical conditions may be more similar than suggested by a consideration of maximum g_s alone. It would be valuable to extend an analysis of this type to mature trees growing in field conditions.

A further consideration in explaining similar transpiration rates of forests, on either a daily or annual basis, is the recognition of the importance of under-storeys and forest litter in contributing to transpiration and evaporation. To some extent, studies of the transpiration of conifer and broadleaf forests have considered only the forest as a whole and tended to neglect the contribution from non-tree components of the forest such as the understorey or litter layers. This could be an important consideration in a hydrological comparison of a dense forest comprising only trees with an open forest with a vigorous understorey. There have been some detailed studies of the contribution of

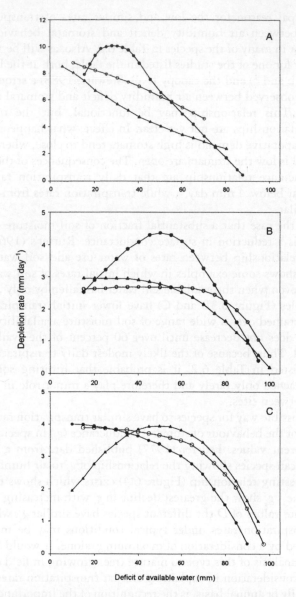

Figure 6.3 The relation between the rate of soil water depletion and the deficit of available water under forests. A: Metz and Douglass (1959) (triangles); Moyle and Zahner (1954) (closed circles); Zahner (1955) (open circles). B: Bass Lake, Rowe and Colman (1951) (closed circles); San Dimas, Rowe and Colman (1951) (triangles); North Fork, Rowe and Colman (1951) (open circles). C: Croft and Monninger (1953) (triangles); Rowe (1948) (closed circles); Zinke (1959) (open circles). Redrawn from Rutter (1968) with permission from Academic Press and A.J. Rutter.

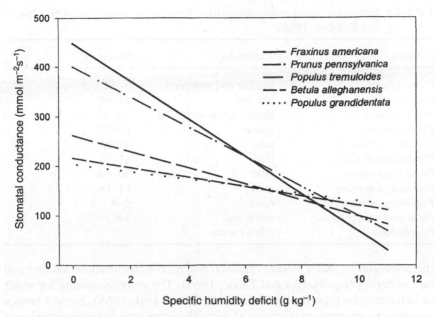

Figure 6.4 Changes in stomatal conductance (g_s) of several species in response to changes in air humidity deficit. From Federer (1977).

the understorey to forest transpiration, which have been reviewed by Black and Kelliher (1989) and are presented in Table 6.3. In some circumstances, transpiration from the understorey can account for over 50 percent of the forest transpiration. The contribution from the understorey is another mechanism by which forests that appear structurally different can have similar transpiration rates. Tan and Black (1976) worked in a Douglas fir forest with a salal (*Gaultheria shallon*) understorey. The contribution from the understorey comprised up to half of the total forest transpiration. Similar contributions from the bracken understorey in a Scots pine forest in eastern England were shown by Roberts *et al.* (1980). These authors showed that particularly on summer days when air humidity deficits were high, tree transpiration was limited, while that from the bracken was not. On such days, transpiration from the bracken could account for up to 60 percent of the forest transpiration. Roberts and Rosier (1994) compared daily and seasonal transpiration rates from a beech forest without an understorey with an ash forest in which there was a vigorous undergrowth of dog's mercury (*Mercurialis perennis* L.). The transpiration of the two forest stands was similar, even though the contribution from the tree components was different.

There is a further component of forest evaporation and that is the water which reaches the forest litter from rainfall and is evaporated. The magnitude of litter interception depends on litter storage capacity (a function of litter

Table 6.3 Percentage stand evapotranspiration from forest understorey (after Black and Kelliher, 1989).

Overstorey	Understorey	Percentage
Pinus sylvestris	heather and cowberry	6–22
Pinus sylvestris	bracken	25–60
Pinus radiata	shrubs, grass and slash	30–50
Pinus ponderosa	sparse	10–27
Pseudotsuga menziesii	salal	40–65
Pseudotsuga menziesii	salal	37–55
Pseudotsuga menziesii	salal	30–42
Pseudotsuga menziesii	none	15–18
Pseudotsuga menziesii	none	3–8
Eucalyptus marginata + E. calophylla	acacia and other shrubs	32–36

thickness and its water storage capacity) and on the frequency of wetting and rate of drying (e.g. Helvey and Patric, 1965). The storage capacity for water in litter may be high (up to 10 mm; Mader and Lull, 1968), but the energy available to promote evaporation is low. Therefore, the litter evaporation is usually small (between 1 and 5 percent of gross rainfall), although it can account for between 10 and 15 percent of the total interception loss (Helvey 1964; 1967; Helvey and Patric, 1965; Rutter, 1966). The litter layer has a particularly important role in facilitating infiltration of rainfall into the soil and the prevention of erosion, particularly on sloping land (Lowdermilk, 1930; Plamondon et al., 1972).

Tropical rain forests

Unfortunately, there are still only a few studies of transpiration of lowland tropical rain forest. Bruijnzeel (1990) refers to nine studies, for which the annual average transpiration is 1045 mm, with a range of 885 to 1285 mm. Even though there will have been periods when the canopies were wetted by rain, these annual totals suggest that daily transpiration rates will have been modest, below 4 mm day^{-1}. Shuttleworth (1988) quoted average daily transpiration rates around 3.6 mm day^{-1}, which are substantially below the potential rate determined by climatic conditions. As explained for temperate forests, the relationship between air humidity deficit and leaf or canopy conductance leads to modest transpiration rates. Case study 2 below examines in more detail the behaviour of the tropical forest at Manaus, Brazil, which was reported on by Shuttleworth (ibid.). This region of the central Amazon has a modest dry season, with only a few months, usually August, September and October, having reduced rainfall and soil water deficits. More recently, studies of the behaviour of the transpiration of the Amazon forest in relation to seasonal soil moisture deficits has been extended into areas where there is a much more strongly developed dry season. Studies of forest transpiration using the eddy correlation technique have been reported for Ji-Paraná (southwest Amazonia) by Wright *et al.* (1996) and from soil moisture measurements at Marabá (eastern Amazonia) by Hodnett *et al.* (1996a). In both these localities, the dry season extends for up to three to four months in which very little rain falls. In both cases, daily transpiration rates were below 4 mm day^{-1}, and there was no evidence that reductions in the available soil water had any influence on transpiration rates. In the study at Marabá, Hodnett *et al.* reported that a limited wet season (November 1992 to April 1993) following the dry season of 1992 failed to fully replenish the soil water store. However, transpiration in the following dry season (May to October 1993) was still unaffected by soil water conditions.

In recent years, an awareness has begun to emerge about deep rooting of trees in the Amazon forest. Nepstad *et al.* (1994) reported that roots occur down to depths of 16 m in parts of the eastern Amazon. They also reported the presence of deep roots in the tropical forest where strong seasonal soil water deficits do not often occur, such as in the central Amazon. Dry years are infrequent in this region, but deep roots might serve as an insurance against the eventuality of such occasional dry years. Nepstad *et al.* regard the need for the forest to remain evergreen to counteract the risk of fire in dry seasons as an important consideration. From a study of twenty-seven years of rainfall record for Manaus in the central Amazon and use of a simple soil water balance model, Hodnett *et al.* (1996b) showed that, on average, once every nine years the tropical forest would need to extract water from below 2 m in the soil to sustain transpiration rates at the average of 3.6 mm day^{-1} (Shuttleworth, 1988).

There are, however, a number of examples for which annual transpiration of seasonally dry tropical forests is substantially less than the figures quoted above (~1000 mm year^{-1}) for lowland tropical rain forests. Studies by Dietrich *et al.* (1982) in Panama and by Gilmour (1977) and Bonell and Gilmour (1978) in Queensland, Australia, indicated values below 500 mm year^{-1}. Bruijnzeel (1990) quotes figures of between 510 and 810 mm year^{-1} for montane tropical forests. Bruijnzeel and Proctor (1995) collected what little information there is on transpiration rates of montane tropical rain forests. Annual and daily rates are very low, presumably because of the low radiation conditions in the predominant cloud cover. There is a need for more comprehensive studies of transpiration as well as other hydrological processes in montane tropical rain forests.

Semi-arid woodlands

In comparison with woody vegetation in other regions, there is a paucity of information about annual, seasonal and daily transpiration from woodlands in the semi-arid tropics. The ecological functioning of such woodlands and fallow savannahs has particular importance in the Sahel, where they provide fuel and fodder for local people, and their animals. Grazing, particularly by sheep and goats, and fuelwood collection prevent the regeneration of natural woodland in areas where bushland has been allowed to regrow as fallow after cropping. A combination of drought, grazing and fuelwood exploitation can impose such pressures on the vegetation that desertification is started, the degraded vegetation having large areas of bare soil, which can be susceptible to erosion. Charney (1975) outlined a mechanism by which the higher albedo of bare soil can lead to atmospheric processes resulting in lower rainfall. The hydrological and surface energy balance processes of semi-arid tropical regions are of considerable importance.

There are at least two problems in interpreting what hydrological data there are for semi-arid woodlands. Usually only total evaporation data are available, with no distinctions being made between evaporation of intercepted water and transpiration. Second, because there are often substantial areas of bare soil, evaporation from the soil is often included in the transpiration term. Nevertheless, some generalisations can be made from the recent studies that have been made. Culf et al. (1993) showed that evaporation of striped woodland 'tiger bush' in Niger, West Africa, accounted for almost all the annual rainfall. During the wet season, evaporation rates were above 4 mm day^{-1} but fell to around 1.5 mm day^{-1} in the dry season. Soil evaporation accounted for about 25 percent of total evaporation. Measurements of evaporation from fallow savannah by Gash et al. (1991) showed similar values of evaporation in the wet and dry season to those reported for the natural vegetation by Culf et al. Similar results were also obtained for thorn scrub (Acacia tortilis and Balanites aegyptiaca) in northern Senegal by Nizinski et al. (1994). Maximum evaporation rates (probably including some interception losses) in the wet season were above 5 mm day^{-1} for Acacia and around 4 mm day^{-1} for Balanites. These rates fell to 0.93 and 0.46 mm day^{-1}, respectively, in the hot dry season. As in the study by Culf et al., Nizinski et al. suggest that very little residual water is left in the profile at the end of the dry season, especially below Acacia. Interception losses were around 12 percent of gross rainfall.

The past decade has seen considerable research activity in the Sahel. One of the key research aims has been to examine the interaction of land-surface processes and climatology. In this context, a number of studies have determined energy and water balances of pristine Sahelian vegetation, fallow savannah and local crops, such as millet. Some of these studies have been referred to above. Unfortunately, other areas of semi-arid woodland of equal relevance in providing fuel and fodder still require investigation. An important example would be the semi-arid caatinga vegetation of northeast Brazil.

Modelling forest transpiration

In most cases, the approach to modelling transpiration has involved modelling of stomatal conductance (g_s) or surface conductance (g_c). There are a wide range of models of g_s and g_c, ranging from largely empirical to highly mechanistic models requiring considerably more information as input variables, to the extent that, in some cases, it may be difficult to obtain the required information in field conditions.

An example of an empirical model is that of surface conductance (g_c) of a pine forest (Gash and Stewart, 1975; see also Case studies, Figure 6.7). In that particular case, g_c varies only as a function of time of day. A similar type of model was used by Shuttleworth (1988) for the Amazonian rain forest (see also Case studies, Figure 6.7). An example of a highly mechanistic model is the type proposed for g_s by Jarvis (1976), where variables such as solar radiation, leaf internal CO_2 concentration, air vapour pressure deficit and leaf water status are used to predict g_s. A drawback of a model of this type is that some of the information about driving variables may be impracticable to obtain in field conditions. A similar approach to that of Jarvis but one for which the information is likely to be more readily available was proposed by Stewart (1988) to model surface conductance (g_c) of pine forest. Stewart's formulation used a multiplicative relationship between functions which relate maximum g_c to each of the driving variables, i.e.

$$g_c = g_{max}.g(l).g(kd).g(\delta q).g(Td).g(\delta\theta) \tag{6.13}$$

where g_c is the surface conductance and g_{max} is the maximum conductance, with each of the moderators relating to leaf area (l), solar radiation (kd), atmospheric humidity deficit (q), leaf temperature (Td) and soil moisture deficit (θ) having a maximum value of unity. The shapes of the functions $g(\ )$ used by Stewart are shown in Figure 6.5. Stewart found that the model was most sensitive to changes in δq and $\delta\theta$.

IMPACTS OF FORESTS ON SOILS

There is general agreement that soils are more variable in forests and woodlands than under other vegetation covers (e.g. Grieve, 1977; Riha et al., 1986; Grigal et al., 1991). Other authors have examined the pattern of variation in soil properties associated with individual trees (Zinke, 1962; Gersper and Holowaychuk, 1970a, b; 1971; Skeffington, 1984; Doettcher and Kaliaz, 1990, Beniamino et al., 1991). The pattern usually observed is a radial change in soil properties away from the tree trunk. Decreases in soil acidity (higher pH) and organic matter content away from the trunk are usually reported, but Crampton (1982) found more acidic soils prevalent under the canopy edge. Wilson et al. (1997) found even greater variability in soil

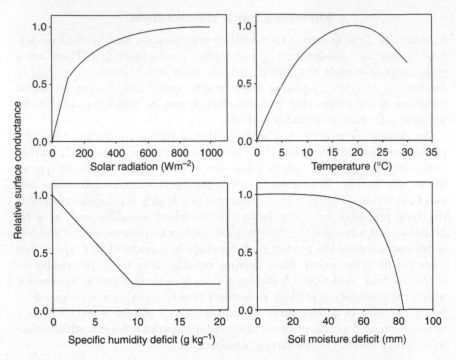

Figure 6.5 The dependence of relative surface conductance on solar radiation, specific humidity deficit, temperature and soil moisture deficit. From Stewart (1988). Reproduced with permission from Elsevier Science.

properties in ancient woodland than in recent woods, but no clear association emerged between the pattern of soil acidity and individual trees. However, in some soils the pattern of carbon distribution was associated with the position of individual trees.

The concentration of carbon content in soils close to the tree is presumably because of the predominance of litter, both leaves and branches, falling close to trees. The explanation for the pattern of pH will be the redistribution of rainfall by the canopies of the trees. In some tree species there is a considerable concentration of throughfall as stemflow. Species with smooth bark surfaces such as beech may transport as much as 12 percent of precipitation as stemflow, whereas pine transfers less than 2 percent (Kittredge, 1948). Examples of high contributions of stemflow to throughfall are also observed in tropical rain forests (Bruijnzeel, 1990). Stemflow may also concentrate water near to the roots of semi-arid species such as *Acacia aneura* (Pressland, 1973). In contrast, other tree species shed considerable throughfall from their canopy edges (Crampton, 1982; Darnhofer *et al.*, 1989).

Where there is a significant spatial heterogeneity of throughfall, there will be consequences for the chemistry and water content of the soil. For example, chemicals deposited previously on the canopy and branches as dry deposition are dissolved by rainwater and added to the soil via localised throughfall or stemflow. As well as the impact that these chemicals have on localised activity in the soil (which depends on chemical species delivered from the canopy), there is also the question of calculating soil nutrient budgets and the need, therefore, for adequate sampling of throughfall when it is delivered in a spatially heterogeneous way. Stemflow may have special significance in delivering potassium, a major plant nutrient, to the area around certain smooth-barked trees, because it is readily leached from the foliage (Gersper and Holowaychuk, 1971).

The heterogeneity in throughfall and associated chemical constituents produced below a woodland canopy consisting of different tree species, tree ages and canopy geometries is likely to have an impact on the species constituting the understorey (Simmons and Buckley, 1992).

The redistribution of rainfall and associated chemicals by forest canopies, the location of roots in the forest floor and the presence of an understorey have practical relevance to the quality of water draining from different forests. Kinniburgh and Trafford (1996) considered that a localised rooting pattern, absence of understorey and wind-throw disturbance were responsible for higher pore water concentrations of NO_3^- below a beech stand compared with an ash stand. The ash stand, on the other hand, was thought to have an extensive root system and had a vigorous understorey (dog's mercury) with a spreading rhizomatous root system. Kinniburgh and Trafford also observed higher NO_3 concentration in the pore water within 50 m of the edge within the beech wood and also found that the the subsoil water content was lower over this region than further inside the wood. Because of the very efficient aerodynamic mixing of the atmosphere with the tree canopies there is enhanced deposition of chemicals at the forest edge. Water use is also likely to be higher, because the aerodynamic conditions will also favour larger interception losses, and the greater concentration of leaf area at the edge will mean higher transpiration. The processes at woodland edges are receiving more attention in a number of European countries because of the increased planting of small areas of woodlands and concerns about chemical deposition. However, the increased water use at the forest edge may also be important in minimising drainage, which otherwise would transport chemicals from the forest to groundwater or streams.

A further consequence of the chemical composition and pH of stemflow was reported by Chappell et al. (1996), who observed a lower hydraulic conductivity in the surface soil horizons closer to trees than some distance away. This effect was thought to be due to acid stemflow water mobilising clay particles, which seal fine pores in the clay close to the tree trunk. These authors also believe that the mass of the tree has a compacting effect, which is greatest near to the tree's axis.

Some early studies showed a decrease in stemflow as a forest becomes older

(Delfs, 1967; Helvey, 1967), but in other studies there was no age effect on stemflow as a percentage of gross rainfall (Brown and Parker, 1970). There are a number of features of older trees that might contribute to a reduction in stemflow. Tree trunks will tend to become less smooth or covered with lichens or mosses, and more distorted with age, and the canopy will comprise a greater percentage of longer, heavier branches that are not angled upwards. It seems reasonable to assume that the young, short and upright branches at the tops of trees direct water to the central trunk, from where it is likely to reach the ground as stemflow. The older, lower branches in tree canopies are larger and more complex. They are more often horizontal or may even curve downwards away from the stem. As the fraction of the canopy having an upright structure is reduced, the proportion of throughfall occurring as stemflow is likely to fall. Using a 9 m tall Douglas fir tree, Hutchinson and Roberts (1981) found that 69 percent of the stemflow came from the upper half of the canopy and explained this distribution in terms of differing branching angle geometry within the tree crown. It might, therefore, be expected that some of the patterns of pH and chemical composition would change as forests became older and stemflow became a smaller percentage of the throughfall.

Despite the observations of reduced hydraulic conductivity of soil close to trees by Chappell et al. (1996), it is generally assumed that when stemflow water reaches the base of the tree it can enter the soil readily along the outside of major roots, which radiate outwards and downwards from the base of the tree. This infiltration may be aided in dry conditions because of shrinkage of the roots to leave air gaps. As well as the infiltration along root channels at the bases of trees, high infiltration rates are usually observed throughout undisturbed forests. The presence of leaf litter at the surface and live and dead root channels deeper down is thought to play a key role in the high infiltration capacity and saturated hydraulic conductivity of undisturbed forest soils (Beasley, 1976; Beven and Germann, 1982). Tracing experiments (Reynolds, 1966) and direct observation (Bonell and Gilmour, 1978) indicate the importance of live and dead root channels for water entry and transmission in forest soils. Gaiser (1952) estimated that there was an average of one major root channel per square metre in an oak forest in Ohio, USA. Live or dead root channels are, of course, not the only pathways for water to be distributed through a forest soil; other conduits include worm and small animal burrows, and termite mounds in tropical soils. Vertical cracks in clay-textured soils would also serve as infiltration channels for water. It is possible that if aerobic conditions allow, tree roots may grow into clay and dry it sufficiently to promote further cracking.

The infiltration characteristics and hydraulic conductivity of forest soils play important roles in determining the hydrological regimes of catchments. The prevailing view is that even in tropical rain forests, as long as they are largely undisturbed, the surface structure of the soil and the saturated hydraulic conductivity of the subsoil are such that even the precipitation input from the most intense storms will infiltrate readily and no ponding or surface runoff will occur. Generally, this perception is sustained by numerous measurements showing the high infiltration and saturated hydraulic conductivity of undisturbed rain forest soils (e.g. Bruinzeel, 1990). In some cases, such as that described by Herwitz (1986) for tropical rain forest in Queensland, Australia, the high concentration of throughfall occurring as stemflow can lead to quantities of water reaching the forest floor in excess of the infiltration capacity of the soil. Ponded conditions are not common in rain forests and are not necessarily associated with stemflow. Typical circumstances are where impeding subsurface layers lead to saturated soils above them (Bonell and Balek, 1993), leading to saturated conditions at the soil surface.

Bruijnzeel (1990) emphasises that the main hydrological response to rainfall after clearing of tropical forest is a partial shift from subsurface flow to some form of overland flow. The stormflow response of a particular catchment can be modified considerably by changes in vegetation cover and depends particularly on the methods used to clear the forest (Lal, 1983; Dias and Nortcliff, 1985a and b). Minimal mechanical clearing, for example by 'slash and burn', produces little extra surface runoff. On the other hand, mechanical clearing with bulldozers produces considerable surface runoff. Mechanical clearing tends to seal the soil surface and compress the upper soil horizons, and provides surface flow routes along the tracks of bulldozers and tractors.

Differences in saturated hydraulic conductivity of soils of forests and short vegetation are observed in comparative studies, even when the non-forest soil has not been recently disturbed. Burch et al. (1987) described the difference in runoff between an 80-year-old grassland and a remnant eucalypt forest. The grassland catchment produced high peak stormflows and large discharge volumes irrespective of antecedent soil moisture status. The forest yielded little runoff provided that soil water content was below 60 percent of the available storage capacity. The difference in runoff behaviour of the two catchments was that the subsurface hydraulic conductivity of the grassland soils was much lower than that of the undisturbed forest. In a study by Ragab and Cooper (1993), higher saturated and unsaturated hydraulic conductivities were measured in woodland than in permanent pasture. Both woodland and permanent pasture had higher hydraulic conductivities than arable soil.

There will be variation in soil moisture conditions determined by the spatial distribution of throughfall and infiltration characteristics. Clearly the live and dead root channels and other fissures described above contribute to the high hydraulic conductivity of forest soils and allow water to be delivered to deep horizons. This may provide an advantage to trees, particularly in semi-arid regions, as some deeply placed water will be out of reach of shallow-rooted plants, allowing the trees competitive advantage.

In addition to variations in the lateral and vertical distribution of soil water caused by heterogeneity of throughfall and infiltration, there will be patterns of soil moisture distribution caused by variations in root uptake. The traditional view of tree root distribution in forests is that it substantially exceeds short vegetation rooting depth and lateral extent from individual plants. Recently, there have been comprehensive reviews of maximum rooting depths (Canadell et al., 1996) and vertical root distributions (Jackson et al., 1996) of global vegetation types. From these two papers it can be concluded that trees of temperate broadleaf and coniferous forests, tropical rain forests and shrubs, especially those in arid or semi-arid areas, have the deepest roots. Additionally, most woody vegetation has about 50 percent of roots in the upper 30 cm of soil. An exception is boreal forest, which along with tundra and grassland has between 80 and 90 percent of its roots in the upper 30 cm of soil. The lateral spread of roots from woody plants clearly varies with the climatic zone. In humid forests, the lateral spread of shallow roots is limited; roots are confined within the canopy area (Kellman, 1979). In semi-arid regions, the area exploited by lateral roots extends beyond the edge of the canopy, sometimes by a considerable distance, as indicated by studies in Australia of *Acacia pendula* (Story, 1967) and in California on chaparral shrubs (Kummerow et al., 1977) (for more information on dryland plants, see Chapter 4).

Where trees and shorter vegetation grow together it is usually the case that both will have roots in the upper soil layers, while only the trees will have roots in the deeper layers (Breman and Kessler, 1995). This type of situation is found in a wide range of woody vegetation, e.g. semi-arid savannah woodlands and temperate broadleaf woodlands with an understorey of shrubs and herbs. When the root distribution is of this form, competition between the trees and short vegetation for the soil water resources in the upper soil layers is likely to be severe. Nevertheless, despite this competition for soil moisture from the upper soil horizons, there is evidence that, in semi-arid woodlands at least, growth of short herbaceous vegetation is greater under trees than in the open (Scholes and Archer, 1997). There are a number of plausible explanations why this should be so. One explanation is that below trees there is a greater availability of water because the soil has a lower bulk density and higher porosity. Additionally, the tree cover reduces radiation load on the vegetation below and, therefore, reduces evaporative demand (Joffre and Rambal, 1988). Under the trees the leaf temperatures of grasses and herbs are probably 2–3 °C lower than in the open, a difference which may mean that physiological conditions are closer to optimal. Another factor that may contribute to improved growth of short vegetation under trees may be additional water provided to the surface soil horizons by tree roots accessing soil water deep in the soil profile. This water is transported upwards and passes out of the plant into dry surface soil horizons when the stomata are closed at night. Evidence for this process, termed 'hydraulic lift' (see Chapter 2) was found by Richards and Caldwell (1987) working with a sagebrush (*Artemisia tridentata*) community in the semi-arid regions of the USA. They observed a regular pattern of soil drying at 35–80 cm depth during the daytime followed by wetting at night. This wetting could not be explained by transfer through the soil alone. Although the amounts of water that were moved to the upper soil layers were small (about 20–30 mm in a dry season), they may be important for plant survival and preventing the drying out of surface roots.

In normal conditions, differences in transpiration both between broadleaf and coniferous forests and within the two groups are small (Table 6.2), but in some circumstances differences between species may occur that will create differences in soil moisture content. These situations will probably occur when conditions are less suitable for one species while another species is largely unaffected or can adapt to the prevailing situation. An example that has practical significance in forestry is the response of different species on waterlogged, poorly aerated peaty soils, where the presence of trees may have substantial impacts on the soil. The potential for drying of peat in the lower soil levels is likely to be greater for trees with deeper root systems. King *et al.* (1986) reported that peaty soils were better aerated and had lower water tables in summer under lodgepole pine compared with sitka spruce. It is known that the roots of lodgepole pine can survive waterlogging, while those of spruce are intolerant of such conditions (Coutts and Phillipson, 1978).

CASE STUDIES

Case study 1: Temperate coniferous forest, Thetford Chase, UK

The site

The Scots pine (*Pinus sylvestris* L.) stand was in Thetford Forest (52°25′N; 0°39′E) and was planted in 1931. In 1976, there were 600 trees ha^{-1}, with the average height being about 16 m. The soil was sand, approximately 1 m deep, overlying 2–3 m of glacial till over pure chalk. Tree roots were found throughout this depth and some reached the chalk (Roberts, 1976b). There was a bracken (*Pteridium aquilinum* (L.) Kuhn.) understorey, the rhizomes of which penetrated to a depth of 0.5 m but were mainly in the mineral soil below the forest litter to a depth of 0.2 m.

The simple structure of the plantation at Thetford had a total leaf area index of 3.67 (2.77 tree and 0.9 bracken) (Beadle *et al.*, 1982; Roberts *et al.*, 1980). Figure 6.6A shows the cumulative downward leaf area index; the vertical pattern of leaf area density (Figure 6.6B) shows the clear distinction between the tree and bracken layers.

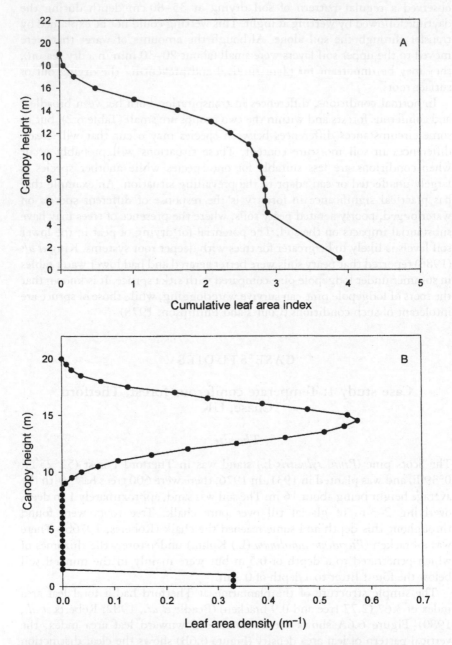

Figure 6.6 A: Downward cumulative leaf area index at Thetford Forest, UK.
B: Vertical variation in leaf area density at Thetford Forest. Data from Beadle *et al.*
(1982). Reproduced with permission from Oxford University Press and C.L.
Beadle.

Micrometeorological measurements

Micrometeorological instrumentation was mounted on a 20 m tower. This instrumentation consisted of a profile of dry- and wet-bulb thermometers from inside the canopy to several metres above the forest. From the gradients of temperature and humidity measured with these instruments it was possible to calculate the Bowen ratio and surface resistance (Gash and Stewart, 1975). An automatic weather station (AWS) was placed at the top of the tower and ran continuously. Gash and Stewart (1975) calculated an average diurnal surface conductance curve from fifty-eight individual days' data. This average function is shown in Figure 6.7 and was used by Gash and Stewart (1977) with AWS data to calculate a transpiration figure of 353 mm for 1975 (Table 6.4).

Rainfall interception

Interception studies at Thetford (Gash and Stewart, 1977) were made by comparing gross rainfall with throughfall collected in twenty-four randomly arranged gauges in a 30 × 15 m plot. Stemflow was measured on five trees in the same plot. Stemflow was less than 2 percent of the gross rainfall.

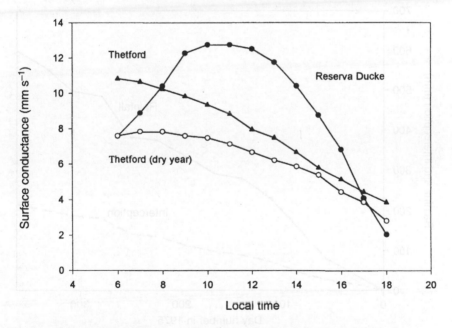

Figure 6.7 Average diurnal curves of canopy conductance for the forest at the Reserva Ducke, Manaus, and Thetford Forest in a normal and a dry year. Redrawn from Shuttleworth (1989). Reproduced with permission from the Royal Society of London and W.J. Shuttleworth.

Table 6.4 Water balance components (mm year⁻¹) for Thetford Forest, UK (1975) and Reserva Ducke, Brazil (1984).

	Rainfall	Interception	Transpiration	Evaporation	Drainage
Thetford	595	213	352	565	30
Ducke	2593	363	1030	1393	1200

Figure 6.8 shows the cumulative total interception loss from Thetford Forest in 1975, which was 37 percent of the gross rainfall, representing an important part of the water balance of this type of vegetation.

Plant physiology

Diurnal measurements of g_s were made on sample days from March to October 1976 in the upper, middle and lower parts of the Scots pine canopy by Beadle *et al.* (1985a, b and c). Over the same period, on selected sample days, diurnal

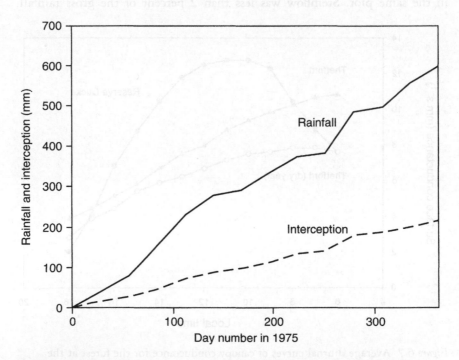

Figure 6.8 Cumulative values of gross rainfall and rainfall interception loss at Thetford Forest (1975). Drawn using data from Gash and Stewart (1977) with permission from Elsevier Science and J.H.C. Gash.

measurements of g_s were also made on the bracken understorey (Roberts *et al.*, 1980). Maximum values of g_s (200–280 mol m^{-2} s^{-1}) were found in the upper part of the tree canopy, and values declined towards the lower levels of the canopy. The maximum values occurred around mid-morning and declined thereafter. Except when temperatures were below 10 °C or light levels were low, at dawn and dusk, changes in g_s were related mainly to changes in air humidity deficit; low stomatal conductances were associated with high air humidity deficits (Figure 6.9). Despite the measurements being made in the drought year of 1976, g_s showed no relationship with variation in leaf water potential (ψ_l).

Maximum values of g_s in the bracken canopy were around 140 mmol m^{-2} s^{-1}. Even in the dry summer of 1976, g_s of the bracken was not strongly influenced by low soil water content, and the bracken was much less sensitive to air humidity deficit than were the pine trees. Transpiration from the bracken canopy as a percentage of the total forest transpiration was typically between 20 and 30 percent. However, in periods when high air humidity deficits prevailed, which reduced g_s of the trees more than that of the bracken,

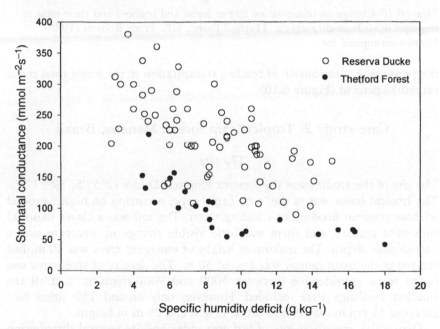

Figure 6.9 The relationship between leaf stomatal conductance and air specific humidity deficit for *Piptadenia suaveolens*, an upper canopy tree species at the Reserva Florestal Ducke, Manaus (unpublished data), and the upper canopy of Scots pine (*Pinus sylvestris* L.) at Thetford Forest, UK. Thetford Forest data redrawn from Beadle *et al.* (1985a) with permission from Blackwell Science and C.L. Beadle.

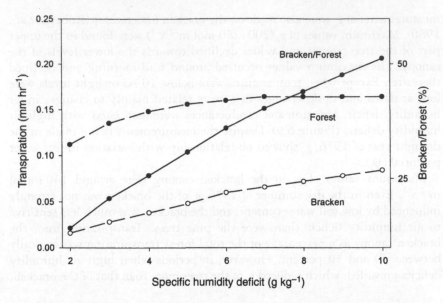

Figure 6.10 Change in transpiration rate of forest and bracken and their ratio in response to air humidity deficit, Thetford Forest, UK. From Roberts (1986). Permission applied for.

the percentage contribution of bracken transpiration to the forest total could exceed 50 percent (Figure 6.10).

Case study 2: Tropical rain forest, Manaus, Brazil

The site

The site of the studies was the Reserva Florestal Ducke (2°57′S; 59°57′W). The tropical forest was of the 'terra firme' type, occurring on higher ground without seasonal flooding or standing water. The soil was a clayey oxisol of consistent texture, and there was little visible change in structure over a considerable depth. The maximum height of emergent trees was 40 m, but the top of the main canopy was around 30 m. The density of tree stems was found to be considerable (between 5000 and 9000 stems ha⁻¹), if all the smallest seedlings were included. However, only around 150 stems ha⁻¹ exceeded 15 cm in diameter at breast height and 25 m in height.

Destructive measurements of leaf area index and its vertical distribution were made on a 20 × 20 plot by McWilliam *et al.* (1993). Figure 6.11A shows the downward vertical accumulation of leaf area index, while the vertical distribution of leaf area density is shown in Figure 6.11B. Leaf area index is around 6 and foliage is distributed throughout the canopy, with some apparent peaks. The foliage is particularly dense around 16 m and close to the

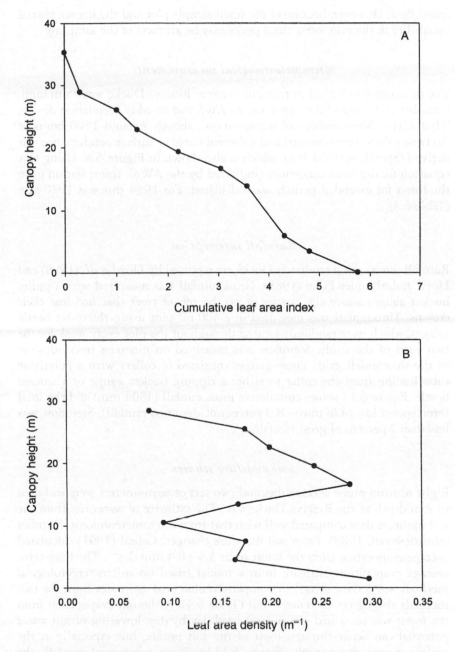

Figure 6.11 A: Downward cumulative leaf area index of the tropical rain forest, Reserva Florestal Ducke, Manaus. B: Vertical variation in leaf area density, Reserva Florestal Ducke, Manaus. Redrawn from McWilliam *et al.* (1993), with permission from Blackwell Science and A.–L.C. Skinner (née McWilliam).

forest floor. However, because of the small sample plot and the known spatial variability in the rain forest these peaks may be artifacts of the sampling.

Micrometeorological measurements

The micrometeorological instrument tower at Reserva Ducke was 45 m high. Installed at the top of the tower was an AWS and an eddy correlation device, 'Hydra' (see 'Measurement of transpiration', above). Around 1500 hours of daytime values were obtained and a diurnal curve of surface conductance was derived (Shuttleworth, 1988), which is also shown in Figure 6.8. Using this equation for dry hour conditions (indicated by the AWS), transpiration from the forest for extended periods was calculated. For 1984 this was 1030 mm (Table 6.4).

Rainfall interception

Rainfall interception studies at Ducke are reported by Lloyd *et al.* (1988) and Lloyd and Marques Filho (1988). Gross rainfall was measured with tipping bucket gauges above the canopy or on the top of trees that had lost their crowns. Throughfall was measured in a 400 m² plot using thirty-six bottle gauges, which were randomly located throughout the plot every week for the two years of the study. Stemflow was measured on nineteen trees adjacent to the throughfall grid. These gauges consisted of collars with a polythene tube leading from the collar to either a tipping bucket gauge or a storage bottle. Figure 6.12 shows cumulative gross rainfall (4804 mm) and the total interception loss (428 mm – 8.9 percent of the gross rainfall). Stemflow was less than 2 percent of gross rainfall.

Soil moisture studies

Eight neutron probe access tubes and two sets of tensiometers were installed to 2 m depth at the Reserva Ducke site. The estimate of water use from the soil moisture data compared well with that from micrometeorological studies (Shuttleworth, 1988). From soil moisture changes, Cabral (1991) calculated average evaporation from the forest to be 3.4 ±0.4 mm day^{-1}. The long-term average evaporation estimate from a model based on micrometeorological methods was 3.6 mm day^{-1}. A comparison of annual estimates from the two methods also agrees extremely well (Table 6.4). Although evaporation from the forest was sustained at a modest level day by day, lowering of soil water potential can occur through most of the soil profile, but especially at the surface during dry periods (Figure 6.13A). With substantial rainfall, the increase in soil water potential occurred first in the middle of the profile, then at the surface and finally in the lower level of the profile (Figure 6.13B). Although there was substantial lowering of soil water potential, the response

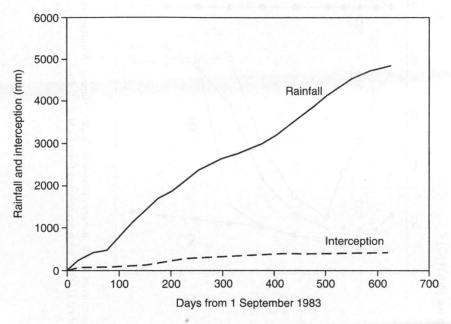

Figure 6.12 Cumulative values of gross rainfall and rainfall interception loss in the Reserva Florestal Ducke, Manaus. Redrawn from Lloyd *et al.* (1988), with permission from Elsevier Science and C.R. Lloyd.

of surface conductance was largely insensitive to changes in lowering of soil water potential.

Plant physiology

Diurnal measurements of g_s were made throughout the rain forest canopy on sample days in dry, wet and intermediate seasons. The highest values of g_s ($\sim$400 mmol m^{-2} s^{-1}) occurred at the top of the canopy. These occurred at mid-morning and showed a decline throughout the day (Figure 6.14). Using data from a number of days, it can be shown that there is a decline in g_s associated with an increase in air humidity deficit (see Figure 6.10). At the bottom of the forest, g_s was only about 20 percent of these values and showed much less of a decline during the day. g_s was measured in three different seasons, and small effects that might relate to lower soil moisture content were seen, but these responses might equally have been an effect of leaf age.

A multi-layer canopy model has been used (Roberts *et al.*, 1993) to examine the contribution of different layers of the forest canopy to total forest transpiration. Figure 6.15 shows that, although there is a substantial fraction of the total forest leaf area at lower levels in the forest, these levels contribute

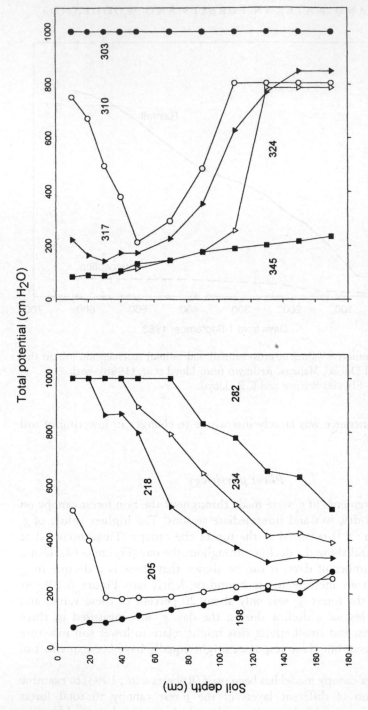

Figure 6.13 Total water potential as a function of soil depth. Measurements at the Reserva Florestal Ducke, Manaus. Numbers on the diagram refer to Julian day number in 1985. Redrawn from Cabral (1991), with permission from O.M.R. Cabral.

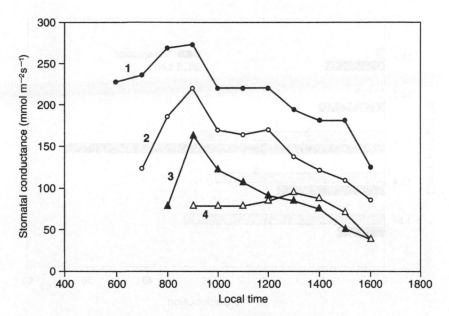

Figure 6.14 Diurnal variation in stomatal conductance of some of the species around the sampling tower at Reserva Ducke, Manaus. 1) *Piptadenia suaveolens* at 33 m; 2) *Licania micrantha* at 25.6 m; 3) *Naucleopsis glabra* at 17 m; 4) *Scheelea* spp. at 3 m. From Roberts *et al.* (1990), with permission from Blackwell Science.

little to total forest transpiration. This is because g_s of this foliage is low, leaf boundary layer conductances are low, and compared with upper parts of the canopy the microclimate is different with a much lower demand for evaporation.

Comparison of temperate coniferous forest and tropical rain forest

A major hydrological difference between these types of forest is in rainfall interception as a percentage of gross rainfall. In Thetford, rainfall interception is around 37 percent of gross rainfall, whereas in the tropical rain forest less than 9 percent of incident rainfall is lost as interception. This difference is due to the very different nature of rainfall in temperate and tropical regions. In temperate regions, particularly when rainfall is frontal, rainstorms are frequent, light in intensity and often of long duration; canopies are very effectively wetted by small raindrops. Evaporation from the wetted canopy can take place during the rainstorms if the atmosphere is not fully saturated. Under these circumstances, interception losses can constitute a considerable percentage of incident rainfall: a fact of both practical and ecological importance when forests, particularly evergreen conifers, are planted.

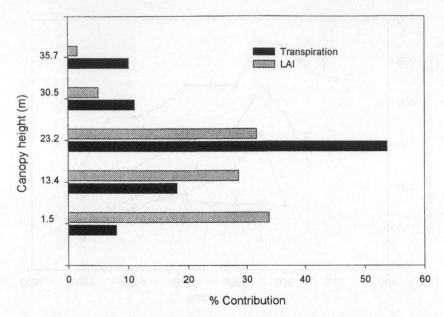

Figure 6.15 Leaf area in each of five canopy layers as a percentage of total leaf area (LAI), and transpiration from each layer as a percentage of total transpiration, Reserva Florestal Ducke, Manaus. From Roberts *et al.* (1996), with permission from the Institute of Hydrology.

The situation in tropical rain forests is quite different; rainstorms are intense and of short duration, and raindrops tend to be large. Under these circumstances, canopies wet slowly and there is also less opportunity for evaporation during the short storm. The high kinetic energy of raindrops in tropical conditions is an important factor in rain splash erosion. In undisturbed rain forest canopies, there is not a serious problem because of the dissipation of kinetic energy by the many layers of vegetation. In tropical forest plantations with no significant undergrowth, however, there is no barrier to the raindrops as they fall from the canopy, and rain splash erosion is likely. The leaves of some tropical trees modify the raindrop spectrum, and the kinetic energy of raindrops falling from leaves such as teak can be substantially greater than that of direct rainfall (Hall and Calder, 1993).

Figure 6.9 shows that g_s of the tropical tree, *Piptadenia suaveolens* (an emergent in the canopy at the Reserva Ducke) is substantially higher than that of *Pinus sylvestris* at Thetford. However, this does not mean that g_s of tropical species is systematically greater than temperate woody species. The values observed for *Piptadenia* are quite moderate, and there are many reported values of temperate broadleaf and conifers equal to these values. A common feature of *Piptadenia* and *Pinus* is the decline in g_s in association with increasing air humidity deficit. The consequences of this relationship will be the same for both tropical and temperate species. The reduction in g_s as D rises has the effect of limiting the levels to which transpiration can rise and makes transpiration levels from day to day rather similar. Soil water will be conserved and critical conditions such as fracture of water columns in the trees (cavitation – see Chapter 2) will be avoided.

There are now many examples of both temperate and tropical trees that show closure of stomata (i.e. a reduction in g_s) with an increase in D. However, this situation is not universal and exceptions can be found among both temperate and tropical species, with important hydrological and ecological consequences. Hall and Allen (1997) describe results of physiological behaviour and water use of poplar short rotation coppice (SRC). Unlike the tree species examined here, SRC does not show a reduction of g_s with D, and high values of g_s can be maintained throughout the day if solar radiation is adequate and soil water is not limiting. Under these circumstances, transpiration rates as high as 10 mm day^{-1} have been observed. Findings such as these may influence decisions on whether SRC is a suitable option where water resources might be compromised by the high water use of SRC.

Results from studies of eucalyptus in India have also shown that the species that are typically planted, *E. camadulensis* and *E. tereticornis*, exhibit high g_s, which is not reduced by the dryness of the atmosphere. When unrestricted by soil water deficits, daily water use can be up to 6 mm day^{-1}. However, g_s can become limited by low soil water content in dry seasons if rooting depth is limited (Roberts *et al.*, 1992), and transpiration rates can fall to around 1 mm day^{-1}. A very important finding from studies in India has been the indication that at some sites rooting is not restricted and eucalypts can exploit deeper and deeper soil water to sustain their transpiration. Calder *et al.* (1997) have shown that in some locations in south India annual water use substantially exceeds the rainfall, causing serious concern for the sustainability of large tracts of eucalypt plantations. An acceptable alternative seems to be a rotational patchwork distribution of eucalypt in the landscape alternating with agricultural crops, which will allow deep soil water to be replenished.

CONCLUDING REMARKS

This chapter has tended to concentrate on evaporation processes in a selection of forest types. Detailed processes in soils have not been addressed, not because they are unimportant but because arguably the differences in the hydrology of different forests are more related to processes in the vegetation itself. For past generations of process hydrologists investigating forests, a key problem was the lack of techniques and equipment to provide useful data on the evaporation process. Now this seems less of a problem; we can measure leaf gas exchange, whole tree transpiration or stand gas exchange almost routinely with commercially available equipment. We also seem to have a good understanding of the interception process and, even though measurement techniques have not advanced, a better idea of the amount of replication that is required. Interception modelling even for sparse canopies now seems to be based on a robust, physical understanding.

Notwithstanding the brief discussion of forest soils, there are a number of important impacts that trees have on soils. Deep litter layers and permeable soils in forests ensure rapid and deep infiltration of water. In contrast to areas with short, shallow-rooted vegetation, the hydrological responses of forested lands will show little or no surface runoff and, because of reserves of water in deep soil layers, sustained streamflow in dry periods. Another key impact of trees on soil is their influence on chemical composition and the variability of chemical composition. Compared with aerodynamically smoother, short vegetation there is enhanced deposition, in dry conditions, of chemicals onto the crowns of trees from the atmosphere. Additional chemicals arrive in the rain, and these are further concentrated on the canopy by the evaporation of intercepted water. The chemical composition of throughfall water can influence physical soil characteristics, such as hydraulic conductivity, by altering the integrity of clay particles. In addition, variation in throughfall chemistry can influence the patterns of woodland flora by altering the distribution of specific site requirements. There is some evidence that root uptake, coupled with high root densities in surface soils of forests, can prevent chemicals leaching from the ecosystem, but more detailed study of rooting patterns and the influence of these on chemistry of drainage waters is required.

Some important remaining problems relate to the scale at which process data are collected (see Chapter 3; Stewart et al., 1996; van Gardingen et al., 1997). Of particular interest is the effect of forests on climate, yet only in passing has this chapter referred to this research issue. The pressing need is for more measurement and modelling studies of the influence of forests on hydrological and climatological processes at the scale of a few to tens of kilometres.

Attention has been drawn to high interception loss, particularly in temperate conifer forests. On the other hand, in tropical forests the interception losses are perhaps surprisingly low as a percentage of rainfall, but we have plausible explanations of why this is. Temperate and tropical forests differ markedly in interception losses as a percentage of gross rainfall but, in sharp contrast, there are similarities in the control of transpiration. The important common feature is the reduction of stomatal conductance with increasing air humidity deficit. The mechanisms behind this behaviour are still debated, but the net effect is a conservative transpiration rate and a similarity in transpiration from day to day. The response of stomata to humidity deficit and conservative transpiration probably delays the onset of stomatal closure due to deficits of soil moisture. Clearly, the mechanisms behind such a widely occurring phenomenon deserve more study. The low transpiration rates of tropical forests seem, at first, surprising but we are becoming more aware of large year to year variation in rainfall in tropical forest regions. Thus, low transpiration rates may be a strategy for conserving water. Only in recent years has the existence of deep roots in tropical forests become known. However, at the moment the purpose of these deep roots is not clear, unless they provide an insurance in the driest years.

The daily transpiration rates of tropical forests are modest and hardly exceed those of temperate forests. As well as feedbacks between the environment and stomatal conductance, the structure of the tropical forest itself contributes to limited transpiration. Although the leaf area of a tropical forest is nearly double that of many temperate forests, substantial amounts of this foliage are in parts of the canopy where conditions for promoting transpiration are not optimal. Stomata are only partially open, wind speeds are low and humidity is close to saturation. Perhaps as little as fifteen years ago we knew very little about the hydrology, micrometeorology and physiology of tropical forests. Through a sustained research effort by a number of groups we now know a lot more, but clearly there is more to learn. Nevertheless, it is perhaps time for attention to be turned to forest types about which we know very little. These are savannahs, boreal forests and semi-arid and mediterranean woodlands.

REFERENCES

Beadle, C.L., Talbot, H. and Jarvis, P.G. (1982) Canopy structure and leaf area index in a mature Scots pine forest, *Forestry* 55: 105–123.

Beadle, C.L., Neilson, R.E., Talbot, H. and Jarvis, P.G. (1985a) Stomatal conductance and photosynthesis in a mature Scots pine forest. I. Diurnal, seasonal and spatial variation in shoots, *Journal of Applied Ecology* 22: 557–571.

Beadle, C.L., Jarvis, P.G., Talbot, H. and Neilson, R.E. (1985b) Stomatal conductance and photosynthesis in a mature Scots pine forest. II. Dependence on environmental variables of single shoots, *Journal of Applied Ecology* 22: 573–586.

Beadle, C.L., Talbot, H., Neilson, R.E. and Jarvis, P.G. (1985c) Stomatal conductance and photosynthesis in a mature Scots pine forest. III. Variation in canopy conductance and canopy photosynthesis, *Journal of Applied Ecology* 22: 587–595.

Beasley, R.S. (1976) Contribution of subsurface flow from the upper slopes of forested watersheds to channel flow, *Soil Science Society of America Journal* 40: 955–957.

Bell, J.P. (1987) Neutron Probe Practice, IH Report No. 19, Institute of Hydrology, Wallingford, UK.

Bell, J.P., Dean, T.J. and Hodnett, M.G. (1987) Soil moisture measurement by an improved capacitance technique. Part II: Field techniques, evaluation and calibration, *Journal of Hydrology* 93: 79–90.

Beniamino, F., Ponge, J.F. and Arpin, P. (1991) Soil acidification under the crown of oak trees. I. Spatial distribution, *Forest Ecology and Management* 40: 221–232.

Beven, K.J. and Germann, P.F. (1982) Macropores and water flow in soils, *Water Resources Research* 18: 1311–1325.

Black, T.A., Den Hartog, G., Neumann, H.H., Blanken, P.D., Yang, P.C., Russell, Nesic, Z., Lee, X., Chen, S.G., Staebler, R. and Novak, M.D. (1996) Annual cycles of water vapour and carbon dioxide fluxes in and above a boreal aspen forest, *Global Change Biology* 2: 219–229.

Black, T.A. and Kelliher, F.M. (1989) Processes controlling understorey evapotranspiration, *Philosophical Transactions of the Royal Society, London* B324: 207–231.

Bobay, V. (1990) Influence d'une éclaircie sur le flux de sève et la transpiration de taillis de châtagnier, these de docteur en sciences, Université de Paris-Sud, Orsay. 218pp.

Boettcher, S.E. and Kalisz, P.J. (1990) Single-tree influence on soil properties in the mountains of eastern Kentucky, *Ecology* 71: 1365–1372.

Bonell, M. and Balek, J. (1993) Recent scientific developments and research needs in hydrological processes of the humid tropics, in M. Bonell, M.M. Hufschmidt and J.S. Gladwell (eds) *Hydrology and Water Management in the Tropics*, Cambridge University Press, Cambridge, UK, pp. 167–260.

Bonell, M. and Gilmour, D.A. (1978) The development of overland flow in a tropical rain forest catchment, *Journal of Hydrology* 39: 365–382.

Bouten, W., Schaap, M.G., Bakker, D.J. and Verstraten, J.M. (1992) Modelling soil water dynamics in a forested ecosystem. I. A site specific evaluation, *Hydrological Processes* 6: 435–444.

Bowen, I.S. (1926) The ratio of heat losses by conduction and by evaporation from any water surface, *Physical Review* 27: 779–787.

Brechtel, H.M. (1976) Influence of species and age of forest stands on evapotranspiration and ground water recharge in the Rhine–Main Valley, *Proceedings of XVI IUFRO World Congress*, Oslo, Norway.

Bréda, N., Cochard, H., Dreyer, E. and Granier, A. (1993) Water transfer in a mature oak stand (*Quercus petraea*): seasonal evolution and effects of a severe drought, *Canadian Journal of Forest Research* 23: 1136–1143.

Breman, H. and Kessler, J.-J. (1995) *Woody Plants in Agro-ecosystems of Semi-arid Regions*, Springer-Verlag, Berlin. 340 pp.

Brown, J.H., Jr and Barker, A.C., Jr (1970) An analysis of throughfall and stemflow in mixed oak stands, *Water Resources Research* 6: 316–323.

Bruijnzeel, L.A. (1990) *Hydrology of Moist Tropical Forests and Effects of Conversion: A State of Knowledge Review*, UNESCO, Paris, and Free University, Amsterdam. 224 pp.

Bruijnzeel, L.A. and Proctor, J. (1995) Hydrology and biogeochemistry of tropical montane cloud forests: what do we really know? in L.S. Hamilton, J.O. Juvik and F.N. Scatena (eds) *Tropical Montane Cloud Forests* Springer-Verlag, Berlin, pp. 38–78.

Bruijnzeel, L.A. and Wiersum, K.F. (1987) Rainfall interception by a young *Acacia auriculiformis* A. Cunn. plantation forest in West Java, Indonesia: application of Gash's analytical model, *Hydrological Processes* 1: 309–319.

Burch, G.J., Bath, R.K., Moore, I.D. and O'Loughlin, E.M. (1987) Comparative hydrological behaviour of forested and cleared catchments in south-eastern Australia, *Journal of Hydrology* 90: 19–42.

Cabral, O.M.R. (1991) Armazenagem da áqua num solo com floresta de terra firme e com seringal implantado, master's thesis, Instituto Nacional de Pesquisas Espaciais, Brazil.

Calder, I.R. (1976) The measurement of water losses from a forested area using a 'natural' lysimeter, *Journal of Hydrology* 30: 311–325.

Calder. I.R. (1977) A model of transpiration and interception loss from a spruce forest in Plynlimon, central Wales, *Journal of Hydrology* 33: 247–275.

Calder, I.R. (1986) A stochastic model of rainfall interception, *Journal of Hydrology* 89: 65–71.

Calder, I.R. (1996) Dependence of rainfall interception on drop size: 1. Development of the two-layer stochastic model, *Journal of Hydrology* 185: 363–378.

Calder, I.R. and Rosier, P.T.W. (1976) The design of large plastic sheet net rainfall gauges, *Journal of Hydrology* 30: 403–405.

Calder, I.R., Hall, R.L., Rosier, P.T.W., Bastable, H.G. and Prasanna, K.T. (1996) Dependence of rainfall interception on drop size: 2. Experimental determination of the wetting functions and two-layer stochastic model parameters for five tropical species, *Journal of Hydrology* 185: 379–388.

Calder, I.R., Kariyappa, G.S., Srinivasalu, N.V. and Srinivasa Murthy, K.V. (1992) Deuterium tracing for the estimation of transpiration from trees. 1. Field calibration, *Journal of Hydrology* 130: 17–25.

Calder, I.R., Rosier, P.T.W., Prasanna, K.T. and Parameswrappa, S. (1997) Eucalyptus water use greater than rainfall input – a possible explanation from southern India, *Hydrology and Earth System Sciences* 1: 249–256.

Calder, I.R., Wright, I.R. and Murdiyarso, D. (1986) A study of evaporation from tropical rain forest – West Java, *Journal of Hydrology* 89: 13–31.

Canadell, J., Jackson, R.B., Ehleringer, J.R., Mooney, H.A., Sala, O.E. and Schulze, E.-D. (1996) Maximum rooting depth of vegetation types at the global scale, *Oecologia* 108: 583–595.

Cermák, J., Jeník, J., Kucera, J. and Zídek, V. (1984) Xylem water flow in a crack willow tree (*Salix fragilis* L.) in relation to diurnal changes of environment, *Oecologia* 64: 145–151.

Chappell, N., Stobbs, A., Ternan, L. and Williams, A. (1996) Localised impact of sitka spruce (*Picea sitchensis* (Bong.) Carr.) on soil permeability, *Plant and Soil* 182: 157–169.

Charney, J.G. (1975) Dynamics of deserts and drought in the Sahel, *Quarterly Journal of the Royal Meteorological Society* 101: 193–202.

Chassagneux, P. and Choisnel, E. (1987) Modélisation de l'évaporation globale d'un couvert forestier. II. Calibrages et résultats du modèle, *Annales des Sciences Forestières* 44: 171–188.

Coutts, M.P. and Phillipson, J.J. (1978) Tolerance of tree roots to waterlogging. I. Survival of sitka spruce and lodgepole pine, *New Phytologist* 80: 63–69.

Crampton, C.B. (1982) Podzolization of soils under individual tree canopies in southwestern British Colombia, Canada, *Geoderma* 28: 57–61.

Croft, A.R. and Monninger, L.V. (1953) Evapotranspiration and other water losses on some aspen forest types in relation to water available for stream flow, *Transactions of the American Geophysical Union* 34: 563–574.

Culf, A.D., Allen, S.J., Gash, J.H.C., Lloyd, C.R. and Wallace, J.S. (1993) Energy and water budgets of an area of patterned woodland in the Sahel, *Agricultural and Forest Meteorology* 66: 65–80.

Czarnowski, M.S. and Olszewski, J.L. (1970) Number and spacing of rainfall gauges in deciduous forest stand, *Oikos* 21: 48–51.

Darnhofer, T., Gatama, D., Huxley P.A. and Akunda, E. (1989) The rainfall distribution at a tree/crop interface, W.S. Reifsnyder and T.O. Darnhofer (eds) in *Meteorology and Agroforestry*, ICRAF, Nairobi, pp. 371–382.

Dean, T.J., Bell, J.P. and Baty, A.J.B. (1987) Soil moisture measurement by an improved capacitance technique. Part I: Sensor design and performance, *Journal of Hydrology* 93: 67–78.

Delfs, J. (1967) Interception and stemflow in stands of Norway spruce and beech in West Germany, in W.E. Sopper and H.W. Lull (eds) *International Symposium of Forest Hydrology*, Pergamon Press, New York, pp. 179–185.

Dias, A.C.C.P. and Nortcliff, S. (1985a) Effects of tractor passes on the physical properties of an oxisol in the Brazilian Amazon, *Tropical Agriculture (Trinidad)* 62: 137–141.

Dias, A.C.C.P. and Nortcliff, S. (1985b) Effects of two land clearing methods on the physical properties of an oxisol in the Brazilian Amazon, *Tropical Agriculture (Trinidad)* 62: 202–212.

Dietrich, W.E., Windsor, D.M. and Dunne, T. (1982) Geology, climate and hydrology of Barro Colorado Island, in E. Leigh, A.S. Rand and D.M. Windsor (eds) *The Ecology of a Tropical Forest*, Smithsonian Institution, Washington DC, pp. 21–46.

Dolman, A.J. (1987) Summer and winter rainfall interception in an oak forest. Predictions with an analytical and a numerical simulation model, *Journal of Hydrology* 90: 1–9.

Dugas, W.A., Wallace, J.S., Allen, S.J. and Roberts, J.M. (1993) Heat balance, porometer and deuterium measurements of transpiration from *Eucalyptus* and *Prunus* trees, *Agricultural and Forest Meteorology* 64: 47–62.

Federer, C.A. (1977) Leaf resistance and xylem potential differ among broadleaf species, *Forest Science* 23: 411–419.

Fritschen, L.J., Cox, L. and Kinerson, R.S. (1973) A 28 meter Douglas fir tree in a weighing lysimeter, *Forest Science* 19: 256–261.

Gaiser, R.N. (1952) Root channels and roots in forest soils, *Soil Science Society of America Proceedings* 16: 62–65.

Gash, J.H.C. (1979) An analytical model of rainfall interception in forests, *Quarterly Journal of the Royal Meteorological Society* 105: 43–55.

Gash, J.H.C., Lloyd, C.R. and Lachaud, G. (1995) Estimating sparse forest rainfall interception with an analytical model, *Journal of Hydrology* 170: 79–86.

Gash, J.H.C. and Morton, A.J. (1978) An application of the Rutter model to the estimation of the interception loss from Thetford Forest, *Journal of Hydrology* 38: 49–58.

Gash, J.H.C. and Stewart, J.B. (1975) The average surface resistance of a pine forest derived from Bowen ratio measurements, *Boundary-Layer Meteorology* 8: 453–464.

Gash, J.H.C. and Stewart, J.B. (1977) The evaporation from Thetford forest during 1975, *Journal of Hydrology* 35: 385–396.

Gash, J.H.C., Wallace, J.S., Lloyd, C.R., Dolman, A.J., Sivakumar, M.V.K. and Renard, C. (1991) Measurements of evaporation from fallow Sahelian savannah at the start of the dry season, *Quarterly Journal of the Royal Meteorological Society* 117: 749–760.

Gash, J.H.C., Wright, I.R. and Lloyd, C.R. (1980) Comparative estimates of interception loss from three coniferous forests in Great Britain, *Journal of Hydrology* 48: 89–105.

Gaskin, G.J. and Miller, J.D. (1996) Measurement of soil water content using a simplified impedance measuring technique, *Journal of Agricultural Engineering Research* 63: 153–160.

Gersper, P.L. and Holowaychuk, N. (1970a) Effects of stemflow water on a Miami soil under a beech tree. I. Morphological and physical properties, *Soil Science Society of America Proceedings* 34: 779–786.

Gersper, P.L. and Holowaychuk, N. (1970b) Effects of stemflow water on a Miami soil under a beech tree. II. Chemical properties, *Soil Science Society of America Proceedings* 34: 787–794.

Gersper, P.L. and Holowaychuk, N. (1971) Some effects of stemflow from forest canopy trees on chemical properties of soils, *Ecology* 52: 691–702.

Gilmour, D.A. (1977) Logging and the environment, with particular reference to soil and stream protection in tropical rain forest situations. FAO Conservation Guide No. 1, FAO, Rome, pp. 223–235.

Grace, J. (1983) *Plant–Atmosphere Relationships*, Chapman & Hall, UK, 92 pp.

Granier, A. (1985) Une nouvelle méthode pour le mesure de flux de sève brute dans le tronc des arbres, *Annales Sciences Forestières* 42: 193–200.

Granier, A. (1987) Evaluation of transpiration in a Douglas fir stand by means of sap flow measurements, *Tree Physiology* 3: 309–320.

Granier, A., Bobay, V., Gash, J.H.C., Gelpe, J., Saugier, B. and Shuttleworth, W.J. (1990) Vapour flux density and transpiration rate comparisons in a stand of maritime pine (*Pinus pinaster* Ait.) in Les Landes forest, *Agricultural and Forest Meteorology* 51: 309–319.

Greacen, E.L. (ed.) (1981) *Soil Water Assessment by the Neutron Method*, CSIRO, Australia.

Grieve, I.C. (1977) Some relationships between vegetation patterns and soil variability in the Forest of Dean, *Journal of Biogeography* 4: 193–199.

Grigal, D.F., McRoberts, R.E. and Ohmann, L.F. (1991) Spatial variation in chemical properties of forest floor and surface mineral soil in the north central United States, *Soil Science* 141: 282–290.

Hall, R.L. and Allen, S.J. (1997) Water use by poplar clones grown as short-rotation coppice at two sites in the United Kingdom, *Aspects of Applied Biology* 49: 163–172.

Hall, R.L. and Calder, I.R. (1993) Drop size modification by forest canopies: measurements using a disdrometer, *Journal of Geophysical Research* 98: D10, 18465–18470.

Hall, R.L., Calder, I.R., Gunawardena, E.R.N. and Rosier, P.T.W. (1996) Dependence of rainfall interception on drop size: 3. Implementation and comparative performance of the stochastic model using data from a tropical site in Sri Lanka, *Journal of Hydrology* 185: 389–407.

Helvey, J.D. (1964) Rainfall interception by hardwood forest litter in the southern Appalachians, US Department of Agriculture, Forest Service, Southeastern Forest Experimental Station, Research Paper SE-8.

Helvey, J.D. (1967) Interception by eastern white pine, *Water Resources Research* 3: 723–729.

Helvey, J.D. and Patric, J.H. (1965) Canopy and litter interception of rainfall by hardwoods of eastern United States, *Water Resources Research* 1: 193–206.

Herwitz, S.R. (1986) Infiltration-excess caused by stemflow in a cyclone-prone tropical rainforest, *Earth Surface Processes and Landforms* 11: 401–412.

Hodnett, M.G., Oyama, M.D., Tomasella, J. and Marques Filho, A. de O. (1996a) Comparisons of long-term soil water storage behaviour under pasture and forest in three areas of Amazonia, in J.H.C. Gash, C.A. Nobre, J.M. Roberts and R.L. Victoria (eds) *Amazonian Deforestation and Climate*, John Wiley, Chichester, pp. 57–77.

Hodnett, M.G., Tomasella, J., Marques Filho, A. de O. and Oyama, M.D. (1996b) Deep soil water uptake by forest and pasture in central Amazonia: predictions from long-term daily rainfall data using a simple water balance model, in J.H.C. Gash, C.A. Nobre, J.M. Roberts and R.L. Victoria (eds) *Amazonian Deforestation and Climate*, John Wiley, Chichester, pp. 79–99.

Hutchinson, I. and Roberts, M.C. (1981) Vertical variation in stemflow generation, *Journal of Applied Ecology* 18: 521–527.

Hutjes, R.W.A., Wierda, A. and Veen, A.W.L. (1990) Rainfall interception in the Tai Forest, Ivory Coast: Application of two simulation models to a humid tropical system, *Journal of Hydrology* 114: 259–275.

Jackson, R.B., Canadell, J., Ehleringer, J.R., Mooney, H.A. Sala, O.E. and Schulze, E.-D. (1996). A global analysis of root distributions for terrestrial biomes, *Oecologia* 108: 389–411.

Jarvis, P.G. (1976) The interpretation of the variations in leaf water potential and stomatal conductance found in canopies in the field, *Philosophical Transactions of the Royal Society, London* B273: 593–610.

Jarvis, P.G. (1993) Water losses from crowns, canopies and communities, in J.A.C. Smith and H. Griffiths (eds) *Water Deficits: Plant Responses from Cell to Community*, Bios Scientific Publications, Oxford, 285–315.

Joffre, R. and Rambal, S. (1988) Soil water improvement by trees in the rangelands of southern Spain, *Acta Oecologia (Oecolia Plantarum)* 9: 405–422.

Kellman, M. (1979) Soil enrichment by neotropical savanna trees, *Journal of Ecology* 67: 565–577.

Kiese, O. (1972) Bestandmeteorologische untersuchungen sur bestimmung des wärmehaushalts eines buchenwaldes, *Berichte des Instituts fur Meteorologie und Klimatologie der Technischen Universitat, Hannover*, No. 6. 132pp.

Kimmins, J.P. (1973) Some statistical aspects of sampling throughfall precipitation in nutrient cycling studies in British Columbian coastal forests, *Ecology* 54: 1008–1019.

King, J.A., Smith, K.A. and Pyatt, D.G. (1986) Water and oxygen regimes under conifer plantations and native vegetation on upland peaty gley soil and deep peat soils, *Journal of Soil Science* 37: 485–497.

Kinniburgh, D.G. and Trafford, J.M. (1996) Unsaturated zone pore water chemistry and the edge effect in a beech forest in southern England, *Water, Soil, and Air Pollution* 92: 421–450.

Kittredge, J. (1948) *Forest Influences*, McGraw-Hill, New York. 394 pp.

Kovner, J.L. and Evans, T.C. (1954) A method for determining the minimum duration of watershed experiments, *Transactions of the American Geophysical Union* 35: 608–612.

Kummerow, J., Krause, D. and Jow, W. (1977) Root systems of chaparral shrubs, *Oecologia* 29: 163–177.

Ladefoged, K. (1963) Transpiration of trees in closed stands, *Physiologia Plantarum* 16: 378–414.

Lal, R. (1983) Soil erosion in the humid tropics with particular reference to agricultural land development and soil management, *International Association of Hydrological Sciences Publication* 140: 221–239.

Law, F. (1956) The effect of afforestation upon the yield of water catchment areas, *Journal of the British Waterworks Association* 38: 344–354.

Law, F. (1958) Measurement of rainfall, interception and evaporation losses in a plantation of sitka spruce trees, *International Association of Hydrological Sciences* 2: 397–411.

Lloyd, C.R., Gash, J.H.C., Shuttleworth, W.J. and Marques Filho, A. de O. (1988) The measurement and modelling of rainfall interception by Amazonian rainforest, *Agricultural and Forest Meteorology* 43: 277–294.

Lloyd, C.R and Marques Filho, A. de O. (1988) Spatial variability of throughfall measurements in Amazonian rainforest, *Agricultural and Forest Meteorology* 42: 63–73.

Lowdermilk, W.C. (1930) Influence of forest litter on run-off percolation and erosion, *Journal of Forestry* 42: 63–73.

Mader, D.L. and Lull, H.W. (1968) Depth, weight and water storage of the forest floor. US Department of Agriculture, Forest Service, Northeast Forest Experimental Station, Paper 109.

McNeil, D.D. and Shuttleworth, W.J. (1975) Comparative measurements of energy fluxes over a pine forest, *Boundary Layer Meteorology* 9: 297–313.

McWilliam, A.-L.C., Roberts, J.M., Cabral, O.M.R., Leitão, M.V.B.R., da Costa, A.C.L., Maitelli, G.T. and Zamparoni, C.A.G.P. (1993) Leaf area index and above-ground biomass of terra firme rainforest and adjacent clearings in Amazonia, *Functional Ecology* 7: 310–317.

Metz, L.J. and Douglass, J.E. (1959) Soil moisture depletion under several Piedmont cover types. US Department of Agriculture, Technical Bulletin 1207, 23 pp.

Moncrieff, J.B., Massheder, J.M., de Bruin, H.A.R., Elbers, J., Friborg, T., Heusinkveld, B.G., Kabat, P., Scott, S., Søgaard, H. and Verhoef, A. (1997) A system to measure fluxes of momentum, sensible heat, water vapour and carbon dioxide, *Journal of Hydrology* 188–189: 589–611.

Monteith, J.L. (1965) Evaporation and Environment. *Symposium of the Society for Experimental Biology*, 19: 205–234.

Moyle, R.C. and Zahner, R. (1954) Soil moisture as affected by stand conditions. US Department of Agriculture, Forest Service, Southern Forest Experimental Station, Occasional Paper 137, 14 pp.

Mulder, J.P.M. (1985) Simulating interception loss using standard meteorological data, in B.A. Hutchison and B.B. Hicks (eds) *The Forest–Atmosphere Interaction*, Dordrecht, the Netherlands: D. Reidel, 177–196.

Nepstad, D.C., de Carvalho, C.R., Davidson, E.A., Jipp, P.H., Lefebvre, P.A., Negreiros, G.H., da Silva, E.D., Stone, T.A., Trumbore, S.E. and Vieira, S. (1994) The role of deep roots in the hydrological and carbon cycle of Amazonian forests and pastures, *Nature* 372: 666–669.

Nizinski, J., Morand, D., and Fournier, C. (1994) Actual evapotranspiration of a thorn scrub with *Acacia tortilis* and *Balanites aegyptiaca* (north Senegal), *Agricultural and Forest Meteorology* 72: 93–111.

Nizinski, J. and Saugier, B. (1989) A model of transpiration and soil-water balance for a mature oak forest, *Agricultural and Forest Meteorology* 47: 1–17.

Pearce, A.J. and Rowe, L.K. (1981) Rainfall interception in a multi-storied, ever-green mixed forest: estimates using Gash's analytical model, *Journal of Hydrology* 49: 341–353.

Plamondon, A.P., Black, T.A. and Goodell, B.C. (1972) The role of hydrologic properties in the forest floor in watershed hydrology, in Csallany, S.C., McLaughlin, T.G. and Striffler, W. (eds) *National Symposium on Watersheds in Transition, Proceedings*, Series 14, 341–348. AWRA, Urbana, Illinois.

Pressland, A.J. (1973) Rainfall partitioning by an arid woodland (*Acacia aneura* F. Muell.) in south-western Queensland, *Australian Journal of Botany* 21: 235–245.

Ragab, R. and Cooper, J.D. (1993). Variability of unsaturated zone water transport parameters: implications for hydrological modelling. 1. In situ measurements, *Journal of Hydrology* 148: 109–131.

Rasmussen, K.R. and Rasmussen, S. (1984) The summer water balance in a Danish Oak stand, *Nordic Hydrology* 15: 213–222.

Reyenga, W., Dunin, F.X., Bautovich, B.C., Rath, C.R. and Hulse, L.B. (1988) A weighing lysimeter in a regenerating eucalypt forest: Design, construction and performance, *Hydrological Processes* 2: 301–314.

Reynolds, E.R.C. (1966) The percolation of water through soil demonstrated by fluorescent dyes, *Journal of Soil Science* 17: 127–132.

Richards, J.H. and Caldwell, M.M. (1987) Hydraulic lift: substantial nocturnal water transport between soil layers by *Atremisia tridentata* roots, *Oecologia* 73: 486–489.

Riha, S.J., Senesac, G. and Pallant, E. (1986) Effects of forest vegetation on spatial variability of surface mineral soil pH, soluble aluminium and carbon, *Water, Air, and Soil Pollution* 31: 929–940.

Roberts, J.M. (1976a) An examination of the quantity of water stored in mature *Pinus sylvestris* L. trees, *Journal of Experimental Botany* 27: 473–479.

Roberts, J.M. (1976b) A study of root distribution and growth in a *Pinus sylvestris* L. (Scots pine) plantation in East Anglia, *Plant and Soil* 44: 607–621.

Roberts, J.M. (1977) The use of tree cutting techniques in the study of the water relations of mature *Pinus sylvestris* L. I. The technique and survey of the results, *Journal of Experimental Botany* 28: 751–767.

Roberts, J.M. (1978) The use of the tree cutting technique in the study of the water relations of Norway spruce *Picea abies* (L.) Karst, *Journal of Experimental Botany* 29: 465–471.

Roberts, J.M. (1986) Stomatal conductance and transpiration from a bracken understorey in a pine plantation, in *Proceedings of Conference Bracken 1985*, Leeds, R.T Smith and J.A. Taylor (eds) Parthenon Press, Carnforth, UK.

Roberts, J.M., Cabral, O.M.R., and de Aguiar, L. de F. (1990) Stomatal and boundary layer conductances measured in a terra firme rainforest, *Journal of Applied Ecology* 27: 336–353.

Roberts, J.M., Cabral, O.M.R., Fisch, G., Molion, L.C.B., Moore, C.J. and Shuttleworth, W.J. (1993) Transpiration from an Amazonian rainforest calculated from stomatal conductance measurements, *Agricultural and Forest Meteorology* 65: 175–196.

Roberts, J.M., Cabral, O.M.R., McWilliam, A.-L.C., da Costa, J. de P. and Sá, T.D. de A. (1996) An overview of the leaf area index and physiological measurements during ABRACOS, in J.H.C. Gash, C.A. Nobre, J.M. Roberts and R.L. Victoria (eds) *Amazonian Deforestation and Climate*, Chichester: Wiley, pp. 287–306.

Roberts, J.M., Pymar, C.F., Wallace, J.S. and Pitman, R.M. (1980) Seasonal changes in leaf area, stomatal and canopy conductances and transpiration from bracken below a forest canopy, *Journal of Applied Ecology* 19: 409–422.

Roberts, J.M., Rosier, P.T.W. and Srinavasa Murthy, K.V. (1992) Physiological studies in young eucalyptus stands in southern India and their use in estimating forest transpiration, in I.R. Calder, R.L. Hall and P.G. Adlard (eds) *Growth and Water Use of Forest Plantations*, Chichester: Wiley, pp. 226–243.

Roberts, J.M. and Rosier, P.T.W. (1994) Comparative estimates of transpiration of ash and beech forest at a chalk site in southern Britain, *Journal of Hydrology* 162: 229–245.

Rowe, P.B. (1948) Influence of woodland chaparral on water and soil in Central

California, California Department of Natural Resources, Division of Forestry, 70 pp.

Rowe, P.B. and Coleman, E.A. (1951) Disposition of rainfall in two mountain areas of California, US Department of Agriculture, Technical Bulletin 1048.

Rutter, A.J. (1966) Studies on the water relations of *Pinus sylvestris* in plantation conditions. IV. Direct observations on the rates of transpiration, evaporation of intercepted water and evaporation from the soil surface, *Journal of Applied Ecology* 3: 393–405.

Rutter, A.J. (1968) Water consumption by forests, in T.T. Koslowskii (ed.) *Water Deficits and Plant Growth*, Volume II, London: Academic Press, 23–84.

Rutter, A.J., Kershaw, K.A., Robins, P.C. and Morton, A.J. (1971) A predictive model of rainfall interception in forests. I. Derivation of the model from observations in a plantation of Corsican pine, *Agricultural Meteorology* 9: 367–389.

Rutter, A.J., Morton, A.J. and Robins, P.C. (1975) A predictive model of rainfall interception in forests. II. Generalization of the model and comparison with observations in some coniferous and hardwood stands, *Journal of Applied Ecology* 12: 367–380.

Schnock, G. (1971) Le bilan d'eau dans l'écosystème d'une forêt – application à une chênaie mélangée de haute Belgique, Conf. UNESCO, Paris, 1969, Productivité des écosystèmes forestieres. Actes Colloq., Bruxelles, 41–47.

Scholes, R.J. and Archer, S.R. (1997) Tree–grass interactions in savannas, *Annual Review of Ecology and Systematics* 28: 517–544.

Sellers, P.J., Hall, F.G., Margolis, H., Kelly, R., Baldocchi, D.D., den Hartog, G., Cihlar, J., Ryan, M.G., Goodison, B., Crill, P., Ranson, K.J., Lettenmaier, D. and Wickland, D.E. (1995) The Boreal Ecosystem–Atmosphere Study (BOREAS): An overview and early results from the 1994 field year, *Bulletin of the American Meteorological Society* 76: 1549–1577.

Shuttleworth, W.J. (1988) Evaporation from Amazonian rainforest, *Proceedings of the Royal Society, London* B233: 321–346.

Shuttleworth, W.J. (1989) Micrometeorology of temperate and tropical forest, *Philosophical Transactions of the Royal Society, London* B324: 321–346.

Shuttleworth, W.J. (1992) Evaporation, in D.R. Maidment (eds) *Handbook of Hydrology*, New York: McGraw-Hill, pp. 4.1–5.53.

Simmons, E.A. and Buckley, G.P. (1992) Ground vegetation under planted mixtures of trees, in M.G.R. Cannell., D.C. Malcolm and P.A. Robertson (eds) *The Ecology of Mixed-species Stands of Trees*, Oxford: Blackwell Scientific Publications, pp. 211–231.

Skeffington, R.A. (1984) Effects of acid deposition and natural soil acidification processes on soil: some studies in the Tillingbourne catchment, Leith Hill, Surrey, *SEESOIL* 2: 18–36.

Smith, D.M. and Allen, S.J. (1996) Measurement of sap flow in plant stems, *Journal of Experimental Botany* 47: 1833–1844.

Smith, D.M., Jackson, N.A. and Roberts, J.M. (1997) A new direction in hydraulic lift: can tree roots siphon water downwards? *Agroforestry Forum* 8, 1: 23–26.

Stewart, J.B. (1977) Evaporation from the wet canopy of a pine forest, *Water Resources Research* 13: 915–921.

Stewart, J.B. (1988) Modelling surface conductance of pine forest, *Agricultural and Forest Meteorology* 43: 19–35.

Stewart, J.B., Engman, E.T., Feddes, R.A. and Kerr, Y. (eds) (1996) *Scaling up in Hydrology Using Remote Sensing*, Chichester: Wiley, 255 pp.

Story, R. (1967) Pasture patterns and associated soil water in partially cleared woodland, *Australian Journal of Botany* 15: 175–187.

Swanson, R.H. (1994) Significant historical developments in thermal methods for measuring sap flow in trees, *Agricultural and Forest Meteorology* 72: 113–132.

Tajchman, S. J. (1971) Evapotranspiration and energy balance of forest and field, *Water Resources Research* 7: 511–523.

Tan, C.S. and Black, T.A. (1976) Factors affecting the canopy resistance of a Douglas fir forest, *Boundary Layer Meteorology* 10: 475–488.

Teklehaimanot, Z., Jarvis, P.G. and Ledger, D.C. (1991) Rainfall interception and boundary layer conductance in relation to tree spacing, *Journal of Hydrology* 123: 261–278.

Topp, G.C., Davis, J.L. and Annan, A.P. (1980) Electromagnetic determination of soil water content; measurement in coaxial transmission lines, *Water Resources Research* 16: 574–582.

Topp, G.C., Davis, J.L., Bailey, W.G. and Zebchuk, W.D. (1984) The measurement of soil water content using a portable TDR probe, *Canadian Journal of Soil Science* 64: 313–321.

Ubarana, V.N. (1996) Observations and modelling of rainfall interception at two experimental sites in Amazonia, in J.H.C. Gash, C.A. Nobre, Roberts, J.M. and R.L. Victoria (eds) *Amazonian Deforestation and Climate*, Chichester: Wiley, pp. 151–162.

Valente, F., David, J.S. and Gash, J.H.C. (1997) Modelling interception loss for two sparse eucalypt and pine forests in central Portugal using reformulated Rutter and Gash analytical models, *Journal of Hydrology* 190: 141–162.

van Gardingen, P.R., Foody, G.M. and Curran, P.J. (eds) (1997) *Scaling up: From Cell to Landscape*, Society for Experimental Biology, Seminar Series 63, Cambridge University Press, Cambridge, UK, 386pp.

Waring, R.H. and Roberts, J.M. (1979) Estimating water flux through stems of Scots pine with tritiated water and phosphorus-32, *Journal of Experimental Botany* 30: 459–471.

Wilson, B.R., Moffat, A.J. and Nortcliff, S. (1997) The nature of three ancient woodland soils in southern England, *Journal of Biogeography* 24: 633–646.

Woodward, F.I and Steffen, W.L. (1996) Natural disturbances and human land use in dynamic global vegetation models, IGBP Global Change, Report No. 38, Stockholm.

Wright, I.R., Gash, J.H.C., da Rocha, H.R. and Roberts, J.M. (1996) Modelling surface conductance for Amazonian pasture and forest, in J.H.C. Gash, C.A.

Nobre, J.M. Roberts and R.L. Victoria (eds) *Amazonian Deforestation and Climate*, Chichester: Wiley, pp. 437–457.

Zahner, R. (1955) Soil water depletion by pine and hardwood stands during a dry season, *Forest Science* 1: 258–264.

Zinke, P.J. (1959) The influence of a stand of *Pinus coultori* on the soil moisture regime of a large San Dimas lysimeter in southern California, *International Association of Scientific Hydrology Publication* 49: 126–138.

Zinke, P.J. (1962) The pattern and influence of individual trees on soil properties, *Ecology* 43: 130–133.

7

PLANTS AND WATER IN STREAMS AND RIVERS

Andrew R.G. Large and Karel Prach

INTRODUCTION

It has been estimated that between 1 and 2 percent of higher plants in the world are aquatic (Cook, 1996). This represents approximately 4000 species, most of which are adapted to survival and reproduction in standing waters. While only a relatively small number of species can grow successfully in flowing water, their occurrence in streams and rivers is generally expected to have great hydrological, ecological and even economic importance (e.g. Haslam, 1987; Wiegleb, 1988). Many engineering texts consider the effects of plants on the roughness coefficient of streamflow equations such as Manning's empirical equation and Chezy's semi-empirical equation. However, while numerous studies have been made of flow over grass surfaces (e.g. Hydraulics Research, 1992), aquatic plants that live permanently in water have received little attention, despite the fact that, hydrologically, aquatic plants can cause a variety of problems relating to water quality and quantity. Ecologists, on the other hand, are more interested in how plants are affected by flow regime because, in streams, species composition and abundance of aquatic macrophytes are regulated by physico-chemical stream conditions (as well as management regimes), which are themselves in turn modified by the macrophytes (Sand-Jensen *et al.*, 1989). This chapter explores both themes and considers how flow micro-environments, water quality and channel morphology are affected by, and in turn affect, plant growth and survival.

Although diverse adaptive strategies allow plants to colonise a wide range of different habitats in a river landscape (Large *et al.*, 1996), the fact remains that plants *per se* are sensitive environmental indicators. Lotic (running water) systems are unique in that they display many of the attributes of both terrestrial and aquatic environments, attributes that often intergrade over extremely short spatial and temporal scales. Recent years have seen an increased interest in the ecology and hydrology of these systems as we have come to realise that these systems play an important role in regulating the flow of energy and material across the landscape (see, for example, Holland, 1988; Naiman *et al.*, 1988; Naiman and Décamps, 1990; Prach *et al.*, 1996). However, from the source of the river to its mouth, the in-stream habitat changes, and thus so do the plant communities and the individual species of which they are comprised. In the downstream direction, system parameters such as channel size (width, depth and stream order) generally increase, slope generally decreases, the dominant sediment type changes, system trophic status alters from oligotrophic towards eutrophic, and the amount of light reaching the channel bed decreases due to the increased depth and turbidity of the water. A comprehensive work summarising the occurrence of water macrophytes in relation to a variety of river characteristics has been published by Haslam (1987). Despite a large number of publications appearing since that time, this work still remains a basic source of information and provides an in-depth overview of European rivers.

The situation for plants and water in streams and rivers is complicated by the fact that fluvial hydrosystems form a nested system comprised not only of the wetted perimeter of the river *channel* itself but also incorporating the *river corridor* (including the riparian or bankside zone) and the *floodplain*. The *valley floor*, within which these features are found, encompasses all the landforms dominated by processes of deposition and erosion (Newson, 1992), including the legacies of glacial activity in higher latitudes. River systems are interactive along three spatial dimensions (Ward, 1989). These are the lateral (river–floodplain), vertical (channel–aquifer) and longitudinal (channel–channel) and are shown in Figure 7.1.

In terms of the effect of flow patterns on the plants inhabiting river systems, the principal dimension under consideration in this chapter is the longitudinal one. However, when we come to consider the influence of water quality on plants (and *vice versa*), as well as the influence of drought and flooding on the plants of streams and rivers, we can see that the lateral dimension assumes equal importance. The fringing floodplain with its diverse range of flowing and standing water bodies and mosaic of different vegetation communities is an integral part of the river system (Welcomme, 1979; Junk *et al.*, 1989; Pinay *et al.*, 1990; Prach *et al.*, 1996), and this lateral dimension becomes particularly important in braided reaches (Ward and Stanford, 1995).

Plants inhabiting watercourses must, by definition, tolerate or even prefer having water around all or just some of their components (Haslam, 1978). While they can, in the simplest sense, be classified as described in Table 7.1, difficulties can arise due to the plasticity of form in amphibious habitats (Jeffries and Mills, 1994). A special growth form is represented by annual plants growing on emergent substrates, which often possess long-lived seeds. Not included in this classification are plants of riparian and riverine wetland habitats, a number of which, as mentioned later in this chapter, have distinct influences on the quality of running waters. For the channel itself, Haslam

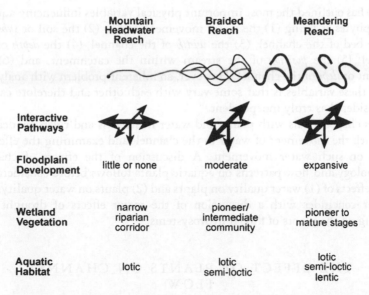

Figure 7.1 Three-reach model system. The arrows show interaction between the river channel and the surrounding corridor. Relative importance of the interactions is shown by the thickness of the arrows. After Ward and Stanford (1995). Reproduced with permission of John Wiley & Sons Ltd.

Table 7.1 Basic four-fold classification of macrophytes of streams and rivers. Species belonging to the first three categories do not necessarily need to be attached to the substrate.

Form	Habit
Submerged macrophytes	Grow primarily under water, although some can resist and respond to exposure. Reproductive organs can be submerged, emergent or floating.
Floating leafed macrophytes	Possess at least some leaves floating at the surface attached by stems to the substrate within the channel. Many also have submerged leaves.
Emergent macrophytes	Include plants whose aerial structures are produced in a similar fashion to terrestrial plants. Many species grow well on exposed substrate as well as surviving completely submerged.
Free-floating macrophytes	Grow primarily unattached to the substrate. Some float on the surface with much of the emergent structure growing clear of the water, while others lie under the water surface.

(1978) has outlined the most important physical variables influencing aquatic macrophytes as being (1) the water movement or *flow* (2) the soil or *substrate* on the bed of the channel, (3) the *width* of the channel, (4) the *depth* of the channel, (5) the *position* of the stream within the catchment, and (6) the gradient or *slope* of the channel. However, an inherent problem with analysing any of these variables is that some vary with each other and therefore cannot be considered as truly independent.

This chapter deals with plants and water in streams and rivers by dealing first with the movement of water in the channel, and examining the effect of plants on such water movement. A discussion of the effects of channel morphology and flow patterns on aquatic plants follows before consideration of the effects of (1) water quality on plants and (2) plants on water quality. The chapter concludes with a discussion of the major effects of drought and flooding on the plants of fluvial hydrosystems.

THE EFFECT OF PLANTS ON CHANNEL FLOW

In rivers, flow characteristics such as average and maximum depths, mean velocities and secondary circulation patterns are governed by channel properties including cross-sectional shape, slope of the reach in question, and bank and bed material properties, as well as the presence and form of in-stream and riparian vegetation (Bathurst, 1997). The water within a channel is subject to two main forces: that of gravity, which causes flow downslope through a catchment, and frictional forces leading to resistance against this downslope movement. In its simplest form, a smooth and steeply sloping channel where frictional forces are low and gravitational forces are high will have high flow velocities, while low velocities will be found in a low-gradient channel with high boundary roughness.

Mean water velocity in an open channel can be calculated using Manning's equation:

$$V = \frac{R^{2/3}S^{1/2}}{n} \qquad (7.1)$$

where V is mean water velocity (m s^{-1}), R is the hydraulic radius (ratio of wetted cross-sectional area to wetted perimeter), S is the channel slope (dimensionless), and n is a measure of channel roughness (dimensionless in this form of the equation – see Dingman, 1984: p. 113). Petts and Foster (1985) provide some examples of Manning's n but warn that estimation of empirical roughness coefficients is problematic because roughness is controlled not only by vegetation type and extent of cover but also by a number of other factors, which include bed material size and bed form. In

240

addition, these factors do not operate individually but rather they interact in an often extremely complex manner. Wilson (1974) draws attention to the implications for estimating discharge, noting that, because n for natural streams is often about 0.035, an error in n of 0.001 gives an error in discharge of 3 percent. Smith *et al.* (1990) point out that difficulties in using the Manning equation arise in attempting to define the hydraulic radius of channels containing a significant volume of submerged vegetation, while Barnes (1967) attempted to produce a roughness classification based on photographs of channels of known roughness. Figure 7.2 shows the general relationship between Manning's n and VR (the velocity–hydraulic radius product). In the upper roughness regime, the vegetation controls the water flow, but as the VR value increases, Manning's n declines due to drowning out of the vegetation and flattening towards the bed of the channel. As the degree to which vegetation bends influences boundary resistance to flow (Kouwen and Li, 1980), it can be surmised that denser stands of submerged vegetation will offer greater resistance to flow (Watson, 1987). At the lower roughness regime shown in Figure 7.2, the vegetation effects have largely been drowned out and water flow becomes the principal factor determining boundary roughness. Plant growth is not the only control of roughness in vegetated sites. If discharge rises and is accompanied by a velocity increase during a plant growth period, the value of Manning's n can decrease despite the increased plant cover at the time. However, as plant biomass shows a general inverse relationship to velocity, vigorous plant growth is best viewed as exaggerating that relationship (*ibid.*).

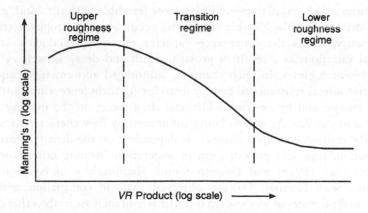

Figure 7.2 General relationship between Manning's n and the velocity and hydraulic radius product (VR) at constant biomass. After Watson (1987). Reproduced with permission of John Wiley & Sons Ltd.

Types of flow resistance

As Bathurst (1985) notes, there are several types of flow resistance in natural open channels. These all contribute to the total resistance. Bathurst defines these as 'skin' resistance (the interaction between the flow and the channel boundary), channel form resistance, free surface resistance and vegetation resistance. Most equations describing the interaction between the flow and the channel boundary are based on turbulent boundary layer theory. Here the boundary layer is defined as the zone where velocity changes with depth due to the friction of the boundary surface, i.e. the channel bed itself. In natural channels, however, the boundary layer extends to the surface (Richards, 1982). In addition, where the flow depth is of a similar scale to the size of the bed material, the velocity profile will be disrupted (Broadhurst *et al.*, in press). In a channel free from the influence of vegetation, the channel form resistance can be significant, and the resistance to flow is then seen to comprise two main elements: channel topography and river plan form. In relation to topography, irregular bed features such as riffles and pools introduce additional components of resistance, and the resultant changes in velocity profile causes increased turbulence, encouraging flow separation from the boundary and introducing flow reversal to maintain continuity of discharge. As Broadhurst *et al.* point out, while attempts have been made to model such flow resistance effects in gravel-bed rivers (Bathurst, 1981; Miller and Wenzel, 1985; Hey, 1988), few attempts have been made to model these effects in channels with beds consisting of bedrock.

Resistance to flow is greatly affected by the nature of the channel boundary. Boundaries may be rigid (non-erodible), loose (erodible with silt, sand, gravel or boulders) or flexible. Flexible boundaries occur where macrophyte growth significantly affects the conveyance capacity of rivers, producing strong seasonal variations as a result of growth, death and decay as well as from management regimes. In open channels, submerged and emergent aquatic vegetation affects resistance to flow by introducing turbulence, thus causing a loss of energy, and by exerting additional drag forces on the moving fluid (Bakry *et al.*, 1992). As well as being influenced by flow characteristics, the hydraulic resistance in open channels is dependent on the density, maturity, distribution, type and growth form of vegetation. Among other workers, Brooker *et al.* (1978) and Dawson found Manning's *n* to be positively correlated with biomass. Dawson observed that, in comparison with its decline as dead material was washed out, the rate of increase in the value of the coefficient for the system slowed with increasing biomass as the growing season progressed. The resultant hysteretic effect (Figure 7.3) was accentuated by higher velocities and discharges in the rising phase of the growing season, whereas, later in the season when biomass is at a maximum, discharges are comparatively low, a common feature of temperate rivers. Brookes (1995), in pointing out that the relative importance of vegetation on *n* is a function of

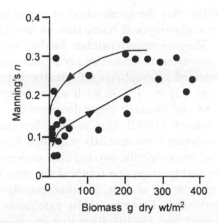

Figure 7.3 Relationship between the biomass of *Ranunculus penicillatus* var. *calcareus* and Manning's *n* (after Dawson, 1978). Reproduced with permission of the European Weed Research Society.

the depth of flow and the density, distribution and type of vegetation, highlighted the fact that aquatic macrophytes react differently to flow. Some wash out during a flood flow, while others that bend with flow may have a lesser effect on *n*. Masterman and Thorne (1992) have concluded that seasonal vegetation growth can have a significant effect in reducing channel capacity where the width–depth ratio of the channel is less than 16.

Patchy macrophyte distribution creates a diverse in-stream habitat with low flow velocities and fine-grained sediments within macrophyte patches, and high flow velocities and coarser sediments in the deeper and often narrow channels between patches (Sand-Jensen and Madsen, 1992). According to Sand-Jensen and Madsen, macrophytes form a mutually protecting structure against high flow velocities in the channel, which tend to be concentrated at the upstream and lateral margins of the patches themselves. This protection has two main benefits for the plant communities: (1) reduction of physical stress and the possibility of uprooting; and (2) enhanced sedimentation with a consequent increase in nutrient supply. Although patches enhance particle sedimentation and provide a habitat for algae, invertebrates and fish (Bijl *et al.*, 1989), shading by other plants is a negative consequence that can reduce growth through reduction of photosynthetic carbon gain.

As flow and substrate vary, their effects on plants present within and around the channel also varies. In any given river channel, there may be a wide variety of micro-habitats, each with specific flow and substrate conditions, and therefore with different dominant vegetation types (Stephan and Wychera, 1996). As soon as vegetation colonises a section of river, it alters the flow velocity within the channel as well as the substrate, thus encouraging the further spread of vegetation across the river cross-section. On the other hand, it is often possible that the acceleration of the deflected flow around patches may generate a self-imposed restriction on lateral patch expansion within the channel. The vegetation patches further increase biodiversity in the stream habitat as they are colonised by periphyton and invertebrates. In lotic conditions, submerged macrophytes are usually distributed in distinct patches of single species (Hynes, 1970), with a more extensive distribution in slow-flowing streams and channels (Sand-Jensen et al., 1989). According to Sand-Jensen and Madsen (1992), this patchy distribution results from vegetative growth emanating from spatially separated sources. Macrophytes therefore develop dense mono-specific patches that undergo dynamic change over time in both size and location as a result of the flow regime, incoming solar radiation altering over the seasons, intrinsic population dynamics and potential effects of herbivores. However, while patchiness is common in all aquatic habitats, it is clear from the literature that its causes (e.g. sources and rates of colonisation, establishment and expansion, recession and mortality) and relation to driving forces (physical disturbance, environmental gradients, biotic influences) have only been studied occasionally. Mono-specific patches are often typical of extreme site conditions (Grime, 1979).

While the distribution of aquatic macrophytes is aligned along the two principal environmental gradients of water depth and substrate type (Day et al., 1988), competition is a third factor that partly controls macrophyte distributions (Wilson and Keddy 1986; 1991; Grace, 1987). The first two gradients determine the extent of a species' occurrence, which in turn affects the outcome of competition (Gopal and Goel, 1993). A pattern of increasing flow rate (the result of shear stress at the surface and the viscous nature of the water in the stream itself) can be seen as one moves away from the submerged vegetation surface and into what Losee and Wetzel (1993) have termed the hydrodynamic boundary layer.

The relationships between flow rates, the structure and species dominance of the submerged plant community are complex. In-stream flow rates are affected by the interaction of the flowing water with the entire plant community. The effects for an individual plant as a member of a submerged community include not only reduced flow rates but also protection from shear under rapid flow rates. Rapid fluctuations occur in littoral flow rates, and this means that the submerged plant boundary layer may be unstable (ibid.). Plants bend in the current and deflect flow over and around the vegetation. As a consequence of the pressure drag of the vegetation, flow is accelerated around the macrophyte patches (Sand-Jensen and Mebus, 1996), and the resultant energy is then dissipated in the pronounced vortices downstream of the patches. Results obtained by Newall (1995) show that the presence of aquatic plants provides very different environments from those found in open flow situations. While downstream flows are likely to be affected by in-stream stands of macrophytes, how far the effects extend downstream depends on the structure and species content of the macrophyte stand and has direct implications for the habitat diversity of these micro-flow environments. The potential for habitat partitioning in aquatic macrophyte communities is great as these environments are often heterogeneous with respect to flow velocity, depth and sediment texture (French and Chambers, 1996). Supporting this is the fact that greater numbers of macroinvertebrates are found in macrophyte patches than anywhere else in the aquatic environment (Marshall and Westlake, 1978). However, while habitat partitioning has been extensively studied in terrestrial plant communities, less attention has been paid to the aquatic environment.

Macrophyte stands shelter other macrophytes from scouring currents, thereby facilitating recruitment and increasing survival rates. Figure 7.4 summarises the main roles that rooted macrophytes play in regulating the physical conditions and biological structure of lowland streams. As different plant species have different effects on flow turbulence (Watts and Watts, 1990), it can be seen how stands of aquatic macrophytes provide refugia from the main body of flow at varying stages in the life-cycle for biota such as invertebrates and fish. While current and light are the major controlling factors of the 'micro-environment' (Dodds, 1991) in filamentous algae communities, plant chemistry also has an impact on the in-stream environment. Secretions by macrophytes partly determine their epiphytic algal cover (Sand-Jensen et al., 1989; Gopal and Goel, 1993; Jeffries and Mills, 1994) and there is evidence that, while they support invertebrate communities, macrophytes can also inhibit invertebrate colonisation. As an example, Dhillon et al. (1982) observed that profuse growth of Myriophyllum spicatum in stagnant water habitats regulates the population density of chironomid midges and mosquito larvae. This is primarily due to the effects of plant growth on

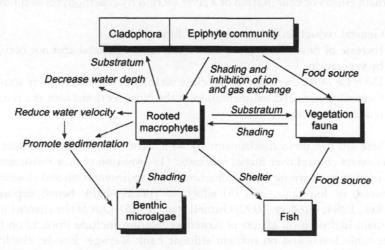

Figure 7.4 The roles played by rooted macrophytes in lowland streams (after Sand-Jensen *et al.*, 1989). Reproduced with permission of Blackwell Science Ltd.

oxygen availability, but plant alleles and other secretions may also play a role here (see Gopal and Goel (1993) for a comprehensive review of allelopathy in aquatic plant communities).

High velocity and turbulence intensity can have the effect of enhancing the flux of dissolved CO_2, oxygen and nutrients by reducing the thickness of the diffusion boundary layer. Within this layer, the rate of gas diffusive and dissolved substance transport across the macrophyte–periphyton interface (Losee and Wetzel, 1993) with the water column can limit biological and chemical activity within the macrophyte–epiphytic assemblage. Thus the effect of diffusion layer thinning is the enhancement of photosynthesis and growth (Wheeler, 1980, cited in Sand-Jensen and Mebus, 1996). On the other hand, it has been suggested that lower velocities and reduced turbulence lead to oxygen accumulation and carbon depletion during daytime photosynthesis, and, therefore, potentially to limitation of photosynthesis and plant growth. Sand-Jensen and Mebus (1996), in citing a range of research, point out that the extensive plant surfaces and the reduced velocities within patches generate an environment that favours the accumulation of fine-grained, nutrient-rich particles and the development of dense invertebrate communities. In discussing factors governing export of macrophyte detritus from patches, Carpenter and Lodge (1986) also suggested that macrophyte beds act as sieving mechanisms retaining particulate detritus (cf. Figure 7.4). Within streams, the processes of sedimentation and colonisation by vegetation typically occurs at low flows, followed by wash-out as discharge rises (Dawson, 1976).

Submerged macrophytes therefore form permanent biotic communities in shallow temperate lowland streams. Stephan and Wychera (1996) summarise the main effects of colonisation of a river section by macrophytes as follows:

- General reduction of cross-sectional flow velocity.
- Increase of flow velocity in parts of the cross-sectional area not occupied by vegetation.
- Decrease of flow velocity in the close vicinity of the macrophyte stand.
- Generally increased sedimentation in the cross-sectional area as a result of reduced flow velocity.

There are five main mechanisms by which the vegetation of streams and rivers exerts control over fluvial processes: (1) provision of flow resistance, (2) provision of bank strength, (3) influence on bar sedimentation and erosion, (4) formation of log jams, and (5) effect on concave-bank bench deposition (Hicken, 1984; Gregory, 1992; Gurnell et al., 1995). Out of the channel itself, the main hydrological effects of rooted vegetation include impacts on bank water table levels and on influent/effluent bank seepage. Besides the living biomass produced by aquatic macrophytes, dead biomass of varying origin, transported by the stream's waters, is also of great ecological importance, being a substrate for many heterotrophs. Moreover, larger woody debris, especially tree trunks falling into the channel, provides habitat for colonisation by aquatic algae, bryophytes and higher plants, as well as invertebrates and fungi, thus increasing the overall biodiversity of the channel.

Perhaps the most dramatic way in which plants directly affect the flow of water in streams and rivers is in the form of woody debris dams. These impede low to medium flows, producing a step-pool sequence in smaller rivers, and a variety of plunge-, marginal- and backwater-pool habitats in medium-sized rivers (Triska and Cromack, 1980; Bisson *et al.*, 1987). This is advantageous for the biota through the maintenance of pool areas, as well as causing local variations in bank hydrology (Gurnell and Gregory, 1987). The attenuation of low to medium flows and the enhanced interaction with the alluvial aquifer may have very important impacts on water quality as well as on the hydrology of the river system (Gurnell, 1997). Such dams also dissipate the river's energy at specific locations along the long profile, resulting in the development of plunge pools and local sorting of bed material in small streams and the development of marginal scour zones in larger rivers.

THE EFFECTS OF CHANNEL MORPHOLOGY AND FLOW PATTERNS ON AQUATIC PLANTS

While plants alter the flow pattern of a stream, they are also themselves significantly affected by alterations in flow regime and flow pattern. Indeed, water movement is perhaps the most important factor affecting plants and their distribution in streams and rivers. While Gilvear and Bravard (1996) state that a knowledge of river flow hydraulics is essential for an understanding of river geomorphology, a knowledge of flow hydraulics is also important for an understanding of plant growth and distribution, as hydraulic processes operating within the channel determine the nature of the aquatic biotopes (the physical environment of the biotic community) found therein. As well as affecting plants directly, flow also has an indirect effect by virtue of its influence on the bed of the channel (Haslam, 1978). Haslam divided the flow conditions affecting river plants into five main categories – *negligible flow*; *slow flow*, where trailing plants hardly move; *moderate flow*, where trailing plants clearly move and the surface is disturbed; *fast flow*, where the vegetation moves vigorously and the surface is markedly disturbed; and, finally, *spate flow*, The spate flow typically causes the most damage to aquatic plant communities. A combination of changing channel morphology and flow patterns is well expressed in the downstream direction influencing plant distribution. Distribution histograms produced by Haslam (Figure 7.5) show species to have ranges of tolerance, including ranges in which they grow well, ranges in which they grow less well or less frequently and ranges where they

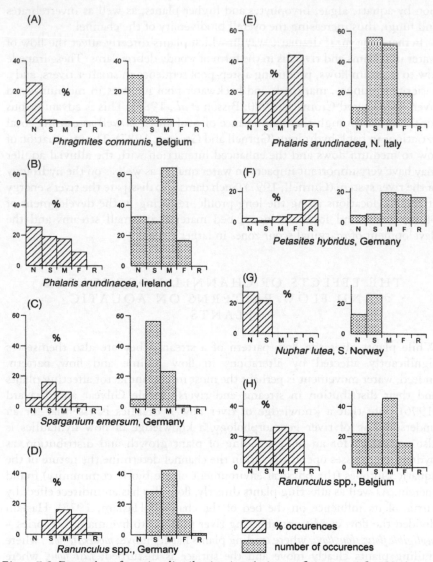

Figure 7.5 Examples of species distribution in relation to flow type (after Haslam, 1978). N = negligible flow, S = slow flow, M = moderate flow, F = fast flow and R = rapid (spate) flow. Both aquatic macrophytes and species growing along river banks are considered here. Reproduced with permission of Cambridge University Press.

are absent altogether. It should be borne in mind, however, that channels are seldom uniform (except in channelised cases) and that these patterns may be complicated due to habitat variations within the same reach.

Plants are affected by flow at all stages of their life-cycle. Hydrochory, i.e. dispersal by water, is one of the major dispersal mechanisms for plants along river corridors, as both seeds and vegetative propagules are transported and dispersed by water. While the importance of hydrochory in dispersing plant propagules of single species along watercourses is recognised (e.g. Schneider and Sharitz, 1988), its importance in structuring whole plant communities along rivers is not well understood (Nilsson *et al.*, 1991; Johansson and Nilsson, 1993).

Remarks on invasive species

Some undesirable exotic species spread along rivers because of easy movement of their diaspores by water (Malanson, 1993; Pyšek and Prach, 1993; Pyšek, 1994) and frequent occurrence of disturbed micro-habitats suitable for their establishment. In temperate Europe, species such as *Fallopia japonica*, *Impatiens glandulifera* (Figure 7.6) and *Heracleum mantegazzianum* are among the most aggressive colonisers of riverside habitats, having distinct

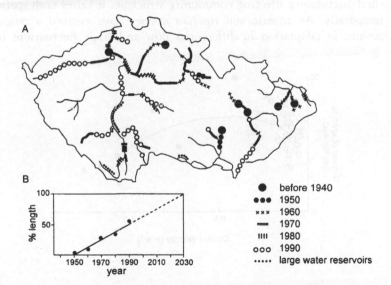

Figure 7.6 Colonisation of rivers and streams with a minimum average discharge of 5 cumecs in the Czech Republic by *Impatiens glandulifera* over the 50-year period from 1940 to 1990. The accompanying graph shows the increasing proportion of river length occupied by *I. glandulifera*, with the extrapolated line predicting full occupation by *c*. 2030. After Pyšek and Prach (1995). Reproduced with permission of Elsevier Science.

competitive superiority over ecologically similar species, which allows them
to become a permanent part of the local flora (Pyšek and Prach, 1993).

Many aquatic plants share the properties of being perennial, producing
clones and relying on vegetative units for reproduction and dispersal
(Sculthorpe, 1967). In summarising the literature, Johansson and Nilsson
(1993) point out that a wide range of vegetative units are dispersed by water,
ranging from whole plants (e.g. *Eleocharis* spp.) to rhizomes (e.g. *Nymphaea*
spp., *Limosella aquatica*), tubers (e.g. *Potamogeton pectinatus, Sagittaria sagittifo-
lia*), turions (e.g. *Myriophyllum* spp.) and fragments of stems (*Fallopia* spp.).
While these vegetative propagules often have a greater potential than seeds
for successful establishment, they are larger and move differently in water and
tend to be shorter-lived than seeds. Vegetative propagules often have a higher
chance of establishment, as the requirements for seed germination of a species
control and often restrict the occurrence of that species (Farmer and Spence,
1987).

Natural physical disturbance plays a well-documented and important
role in shaping general community structure along streams (e.g. Vannote *et
al.*, 1980, Nilsson, 1987). A number of studies show a gradual change in
the community structure of aquatic macrophytes along a gradient of flow
velocity (Figure 7.7). However, while flow velocity may serve as an estimate
of natural disturbance affecting community structure, it varies both spatially
and temporally. As aquatic and riparian species have evolved a range of
mechanisms of adaptation to differential flow and stage fluctuations (e.g.

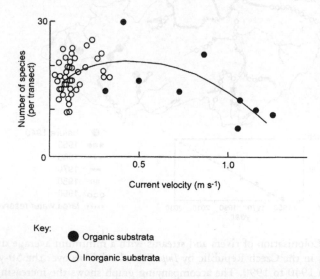

Figure 7.7 Relationship between number of species of water macrophyte and
current velocity for fifty-eight transects across the Sävarån River, northern Sweden
(after Nilsson, 1987). Reproduced with permission of Blackwell Science Ltd.

Haslam, 1978), it is possible that such responses will also vary at the community level. Nilsson (1987) illustrated the fact that species richness in aquatic and riparian habitats is affected by current velocity, with few species being found in faster-flowing stretches. This is principally due to very few plant species being well adapted to moving water (Hynes, 1970). In addition to flow-governed restrictions on distribution, groundwater movement has long been known to be an important factor influencing both the distribution of fauna and patterns of macrophyte growth in river corridors. There are problems, however, in defining the spatial extent of the true hyporheic zone (*sensu* Orghidan, 1955), and thus the term 'interstitial' habitat (Angelier, 1953) is perhaps best used to describe these heterogeneous environments. Dole-Olivier and Marmonier (1992) have also highlighted the patchiness of the interstitial habitat in terms of its hydrology, physico-chemistry and biology, concluding that there are considerable changes in these attributes over space and through time.

Flow moves sediment and therefore acts on plants directly, as plant tissue is abraded by the sediment load (Brookes, 1995). The degree of hydraulic resistance that a plant presents to the flow varies from species to species as well as with the size and shape of the individual plant. Trailing, submerged plants (e.g. *Ranunculus penicillatus*) are able to grow under greater flow forces than are floating leafed plants such as *Nuphar lutea* (Newall, 1995). The pull exerted by the water movement will also affect the plant's attachment to the substrate, which depends on anchor strength. Different plant species have different anchor strengths. In addition, the substrate affects attachment ability – plants usually have shallow roots in coarse sediment, while plants in slower, deeper channels root deeply into the fine deposits (Figure 7.8). In periodically flooded gravel-bed rivers, some plants possess deep, thick anchor roots, enabling them to withstand high flows and substrate movement.

While amphibious plant species can form continuous populations extending well above and below the waterline in the absence of severe physical perturbation, the large number of species that occupy the transition zone or the 'moving littoral' (*sensu* Junk et al., 1989) suggest that the land–water interface does not present as difficult a barrier to adaptation to permanent or temporary submergence as has been suggested in much of the literature. The amphibious features evolved by a number of terrestrial species (e.g. *Berula erecta*, *Myosotis palustris*, *Rorippa amphibia*) enhance their capability to establish themselves in these habitats characterised by less intense competition for space. Despite this, many plants do strive against oxygen deprivation through flooding, and some possess distinct biochemical, anatomical or morphological adaptations (Jackson et al., 1991; Brändle et al., 1996). Sand-Jensen et al. (1992) point out that the evolution of aquatic angiosperms from terrestrial ancestors is ongoing, and that many submerged species retain terrestrial traits such as stomata, leaf cuticles and aerial flowers. Submergence means that water availability is not a limiting factor, and thus tissue for protection

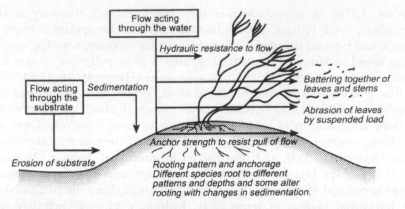

Figure 7.8 How flow affects macrophyte growth and survival (after Haslam, 1978, in Jeffries and Mills, 1994). Reproduced with permission of John Wiley & Sons Ltd.

against water loss, as well as for structure and solute transport, may be saved. In addition, both leaves and roots can take up nutrients (Moeslund *et al.*, 1990), and the fact that seasonal fluctuations in temperature are reduced in running waters allows growth for longer periods throughout the year. Nilsson (1987) concluded that, along the stream edge, increasing current velocity is related to species richness, with both riparian and aquatic vegetation showing unimodal responses to this current velocity gradient. This pattern also corresponds to the theory relating to an increase in diversity at certain levels of disturbance (Grime, 1979). Ham *et al.* (1982) found high discharges increased the growth of *Ranunculus* species in two ways: by directly increasing the rate of uptake of carbon and by preventing the growth of epiphytic algae on the leaf surfaces, which would otherwise depress the rate of photosynthesis.

A rather specific effect of channel water on plant species distribution is due to waves caused by river traffic. The effect of such disturbance is usually to reduce species diversity, as the majority of water plants are sensitive to this type of disturbance, with only a few species actually flourishing under these disturbance regimes. An example of such species loss is shown in Figure 7.9, where the abundance of *Hydrocharis morsus-ranae* (in common with other species) in the Elbe River in the Czech Republic was seen to decline markedly over the twenty-year period 1976–1996, coinciding with a substantial increase in boat traffic on the river (Rydlo, 1996).

A variety of works summarise qualitative or semi-quantitative information on the occurrence of European water macrophytes in relation to various channel morphology and flow characteristics. Foremost of these is the work of Haslam (1978; 1987), while, for Britain, Holmes (1983) describes how rivers can be classified according to their flora.

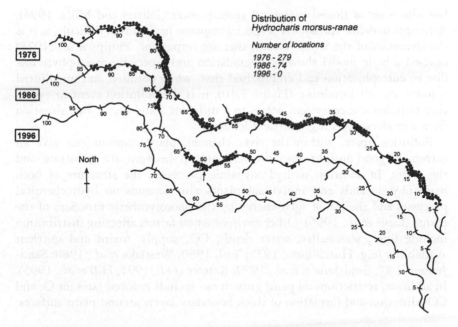

Figure 7.9 Distribution of *Hydrocharis morsus-ranae* along a 100 km section of the River Elbe in the Czech Republic as surveyed over the period 1976–1996, showing the effect of boat traffic on plant abundance (after Rydlo, 1996). Reproduced with permission.

WATER QUALITY

With respect to water quality, the most basic ecological fact is that the more polluted the site, the fewer the number of species of water plant that will be found there (one possible exception being if a strongly oligotrophic stream is subjected to slight enrichment). Haslam (1990) summarises the main effects of pollution on water plants, highlighting the fact that pollution can act twice on the plant. First, when exposed to pollution a plant may grow poorly, particularly if its roots are affected. Second, if its roots are affected, the potential for washout (often of the entire plant) by higher discharge increases. This is particularly important in the case of shorter-rooted plants such as *Callitriche* spp. and *Elodea canadensis*, where even a small decrease in root length will dramatically decrease anchorage. Haslam (*ibid.*) illustrates this with respect to *Myosotis scorpioides* and *Berula erecta*, with results obtained showing a significant reduction in relative amounts of root growth with increasing sewage effluent concentrations. In general, however, the picture is very complex, as toxicity varies with a number of factors, including pollutant speciation, organic matter content, oxygen concentration and temperature. In summary, what characterises eutrophication is not just the supply of nutrients

253

but also a set of linked biological consequences (Jeffries and Mills, 1994). Attempts to define thresholds of species response have proved difficult, as it is the dynamics of the whole system that are important. Philips *et al.* (1978) devised a basic model showing degradation and especially macrophyte loss due to eutrophication and emphasised that, while polluted and unpolluted systems are self-sustaining (Figure 7.10), it is the pollution event or events that switches the ecosystem across to a different state, i.e. in this situation from a moderate to a high nutrient loading.

Pollution apart, within the river channel, plant communities give an extremely good insight into the relationship between the substrate and the water. In addition to hydrodynamic processes, the structure of both macrophyte stands and individual plants also depends on hydrochemical processes and the use of solar energy by the photosynthetic processes of the plants (Large *et al.*, 1996). Other environmental factors affecting distribution include light, seasonality, water depth, CO_2 supply, toxins and nutrient availability (e.g. Hutchinson, 1975; Teal, 1980; Westlake *et al.*, 1980; Sand-Jensen 1987; Sand-Jensen *et al.*, 1989; Rattrey *et al.*, 1991; Hill *et al.*, 1995). In addition, restrictions on plant growth can include reduced rates for O_2 and CO_2 diffusion and formation of thick boundary layers around plant surfaces.

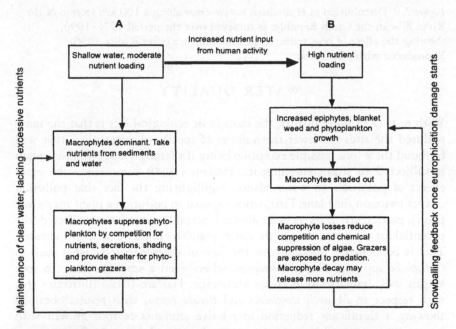

Figure 7.10 Model showing influence of eutrophication on water plants. The systems A and B are self-sustaining, and the pollution perturbation switches the system across to a new state (after Philips *et al.*, 1978). Reproduced with permission of Elsevier Science.

The potential for light to limit rates of photosynthesis is particularly influential on aquatic communities because of reflection at the surface and attenuation within both the water column and macrophyte stands. The attenuation of light arising from the presence of submerged macrophytes varies about four-fold between species (Carpenter and Lodge, 1986), with obvious implications for relationships between neighbouring species. In addition, the vertical temperature gradients within macrophyte stands can be as great as 10 °C m^{-1}, compared with gradients of less than 0.2 °C m^{-1} in neighbouring unvegetated areas (Dale and Gillespie, 1977). To offset the environmental stresses associated with submergence under lotic waters, a number of species have evolved the ability to utilise bicarbonate (HCO_3^-) as an additional source of inorganic carbon (Hutchinson, 1975; Lucas and Berry, 1985; Madsen and Maberly, 1991). Examples of these plants include *Ceratophyllum demersum*, *Elodea canadensis*, *Myriophyllum spicatum*, *Potamogeton crispus* and *P. lucens* (Large *et al.*, 1996).

In nutrient-rich temperate rivers, the seasonal maximum biomass attained by submerged plant communities is about 400–700 g dry wt. m^{-2} (Westlake, 1975, in Wright *et al.*, 1982). However, the continuous effect of disturbance under lotic conditions, coupled with both the influence of shading (Ham *et al.*, 1982) and the fact that photosynthesis can be restricted by reduced diffusion rates of O_2 and CO_2 (Sand-Jensen and Madsen, 1992), means that the channel bed is rarely completely covered with macrophytic vegetation, and average biomass is usually much lower. Organic debris dams can also play a role in affecting the quality of the riverine environment as they retain organic matter within the stream (Bilby and Likens, 1980), thereby allowing it to be processed into smaller fractions further upstream than would otherwise be the case, and where it can more directly affect the community structure.

Self-purification

All the hydrochemical processes mentioned above may be substantially altered by changes in water quality, including thermal pollution. On the other hand, plants, by their activities, influence water quality and may also contribute to its purification (e.g. Osborne and Kovacic, 1993). There is no doubt that, in the presence of macrophytes, more pollutants are fixed or degraded more quickly than is the case where macrophytes are absent. In general, plants purify sediment and water as they break down toxic products through uptake, alteration, production of root exudates and the influence of microphytic micro-organisms. In addition, Haslam (1987) mentions that the micro-organisms *Escherichia coli*, *Salmonella* and *Enterococcus* are eliminated to different degrees by plant species such as *Mentha aquatica*, *Alisma plantago-aquatica*, *Iris pseudacorus*, *Juncus effusus*, *Scirpus lacustris* and *Phragmites australis*, principally through allelopathic mechanisms. This self-purification ability is particularly evident in riparian areas and wetlands contained within the river corridor, and it has important implications in relation to water quality management and improvement. Numerous publications have addressed this in recent years. Riparian zones have been described as the 'kidneys' of the water landscape (Fennessy, 1993) and, via the action of the vegetation, can act as nutrient sinks, thereby protecting surface and subsurface waters from enrichment by agricultural nutrients and other sources of pollution. This is in addition to a number of other functions that they fulfil in the riparian landscape (Large and Petts, 1994). The ability of plants to remove nitrogen is primarily a function of plant uptake and denitrification (e.g. Lowrance *et al.*, 1985; Haycock and Pinay, 1993; Muscutt *et al.*, 1993; Vought *et al.*, 1994). Of the two, denitrification is the most dominant and longer-lasting removal process (Fennessy, 1993). Plant communities therefore carry out a number of important functions: they take up nutrients directly, binding them in biomass and freeing exchange sites in the soils, while their litter provides habitats for microbe colonisation and, along with root exudates, forms the primary source of organic carbon used in denitrification. In addition, it has been stated that the presence of vegetation significantly increases the amount of oxygen in the soil through the formation of oxidised rhizospheres (*ibid.*), although other workers question the importance of this process (see also Chapter 5). Questions still to be answered effectively include what constitutes the optimal width for a vegetated buffer strip, and what species are most efficient in taking up nutrients (Large and Petts, 1994). Decomposition processes and nutrient release in freshwater systems have a strong seasonality (Polunin, 1982; 1984), raising concerns as to what are the seasonal variations in effectiveness. Furthermore, nutrient release may occur as a result of senescence (Large and Petts 1994) and decomposition of the vegetation *in situ*. Thus, problems may arise when the buffer strip becomes saturated with nutrients (Large and Petts, 1992). In summary, the plants of riparian habitats can be viewed as bidirectional buffers – while they buffer the effects of surface and subsurface runoff on river waters, they also buffer the effects of river flooding on adjacent areas within the river corridor. These effects vary on a temporal basis due to seasonal variation in water flows in rivers themselves.

DROUGHT AND FLOOD

While individual river plants are influenced by both 'normal' and extreme flows, extreme events are often critical in determining ecosystem composition and functioning (Haslam 1978). Community composition is determined by the presence or absence of flooding and its duration and frequency. Haslam (*ibid.*) suggests that drought flows can be important if they substantially alter turbulence or silting, but they generally have a more localised effect than flooding. He may, however, have underestimated the effects of drought. For example, depleted groundwater levels in chalk catchments in southern England during the late 1980s and the 1990s dramatically reduced flows in some channels and is likely to have had a profound effect on plant distributions. Intensified research efforts will be necessary to ascertain and quantify these effects. Flooding, on the other hand, constitutes an important environmental factor that influences plant composition and plant growth in many parts of the world. While gradients in stream size, water depth, current velocity, substrate composition and standing crop all produce predictable patterns in riverbank vegetation (Day *et al.*, 1988; Nilsson and Grelsson, 1989; Nilsson *et al.*, 1989) in systems unmodified by humans, disturbance via flooding constitutes a key factor in the distribution of plant species and the intensity of their production. A disturbance here is defined as a significant change from the normal pattern in an ecological system (Forman and Godron, 1986, Pickett and White, 1985, Resh *et al.*, 1988). In fluvial hydrosystems, such disturbance can be direct (e.g. inundation, desiccation, erosion, siltation, bar formation) or indirect (e.g. alterations to the hydrochemical gradient) and can follow on from other changes in the ecosystem (e.g. redirection of succession sequences). Any plant can be damaged by extreme flow events, although different species are likely to be affected in different ways. Plants that grow on soft sediments are particularly vulnerable to scour and damage. Recovery of these species depends on the presence of plant fragments and is enhanced if roots or rhizomes remain *in situ*. Haslam (1978) provides a good summary of the susceptibility of individual species to storm damage and concludes that plant vigour has an additional bearing on resistance to damage from extreme flows.

The flood pulse

As Junk *et al.* (1989) have concluded, the principal driving force for the existence, productivity and interactions of the major biota in river–floodplain systems is the flood pulse. They define the flood pulse as a 'batch' process, as distinct from the continuous processes in flowing water environments considered in the 'River continuum concept' (Vannote *et al.*, 1980), which hypothesises that a continuous gradient of physical conditions exists from headwater to mouth in a stream system. A flood event has two types of effect on the mosaic pattern of the fluvial unit. First, it can, by inundating the unit, affect the communities without altering the biotope. Alternatively, the flood can transform the biotope through erosion and deposition of alluvium, provoking a marked change in the species composition of the plant communities. This has been described as the 'biological resetting' of the system. Besides importing nutrients, inundating floods carry organic debris, which accumulates locally in the form of litter throughout the fluvial environment and principally around and within the main channel. As well as providing valuable habitats for in-stream fauna such as fish and invertebrates, this debris is consumed by the detritivores and mineralised by the decomposers to fuel secondary production within the system. Depending on loading, excess nutrients encourage rapid growth and multiplication of algae and phytoplankton and an increase in the growth of macrophytes (see Figure 7.10). In the riparian zone, the import of nutrients by flooding events favours primary productivity (Brown and Lugo, 1982), but the difference here, compared with the main channel, is that the major part of the biomass remains stocked in the form of wood and the degradation and recycling of the biotic resource operates over a much longer timescale. When this nutrient movement is expressed in terms of a helical flow (not to be confused with helical flow within a meander), this stocking of organic matter manifests itself in the form of tightening of the length of the helix – in other words as a slowing down of the transfer of material in a downstream direction.

In effect, the influence of the flood pulse (Junk *et al.*, 1989) is to form a mosaic of units at different ages and stages of development. It therefore contributes to an increase in the total biodiversity and primary productivity of the fluvial hydrosystem. Overall, this determines the zonation of vegetation communities in river systems (Blom *et al.*, 1990), as plants from different populations or species growing in the same riverine habitats and exposed to environmental stresses such as flooding or drought may possess different life-history strategies to ensure survival. In addition, the severity of the effects of such external environmental factors on the behaviour of plants in streams and rivers will depend on the stage of the life-cycle during which the disturbance event occurs. Frequent disturbances will keep plant communities in an early successional stage dominated by a small number of opportunistic species that have a capacity for rapid colonisation and growth (Grime, 1979), often represented by species with ephemeral root systems such as *Potamogeton pectinatus* or *Elodea canadensis* (Haslam, 1978). It can be seen that the life-cycles of the majority of the biota in fluvial systems (and in particular the plant communities occupying riparian and floodplain areas) are related to the flood pulse in terms of its timing, duration and rate of rise and fall (Figure 7.11). The timing of the flood event is particularly important in temperate rivers, where the seasonality of temperature regimes and light availability also serve to regulate primary and secondary productivity.

CONCLUSIONS

Modern concepts in plant ecology have, in general, been derived from terrestrial situations. Examples of this include the relationships of plants and plant communities to environmental gradients, their life-history strategies, mechanisms of competition, the role of disturbance, succession, biodiversity and the role of herbivores (see, for example, Begon *et al.*, 1990). Moreover, we certainly possess a greater understanding of the ecology of aquatic plants in

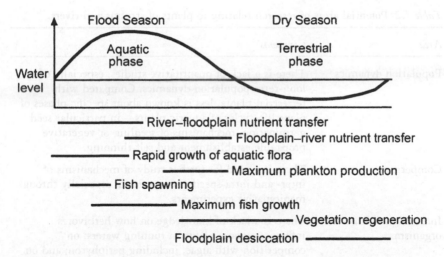

Figure 7.11 The influence of the flood pulse within the river–floodplain complex (after Ward, 1989). Reproduced with permission of the North American Benthological Society.

standing water bodies than we do for flowing-water systems. Studies in running waters have mostly concentrated on water level dynamics, water quality, and invertebrate and fish communities. Calow and Petts (1992; 1994) provide a useful synthesis of the extent of current research in river ecology. In addition, much of our understanding of the relationships between a variety of stream parameters and the plants colonising those systems has been based on qualitative and semi-quantitative field observations instead of on quantitative data. Table 7.2 illustrates some of the more pressing areas for research.

From the discussions above, it can be seen that patchy macrophyte distributions create a diverse in-stream habitat with locally low and high flow velocities and coarse- and fine-grained sediment distribution within the aquatic environment. However, while patchiness is common in all aquatic habitats, it is clear that both its causes (e.g. sources and rates of colonisation, establishment and expansion, recession and mortality) and its relationship to driving forces (physical disturbance, environmental gradients, biotic influences) have only been studied occasionally.

Water movement is perhaps the most important factor affecting plants and their distribution in streams and rivers as, while plants alter the flow pattern of a stream, they are also themselves significantly affected by alterations in flow regime and flow pattern. In general, in-stream macrophyte growth has the effect of reducing mean flow velocity, enhancing sedimentation, and providing habitats for invertebrates, fish and epiphytic algae. However, more research needs to be done on the subject of habitat partitioning in lotic environments, perhaps incorporating the concept of disturbance as this

Table 7.2 Potential areas of research relating to plants of streams and rivers.

Area	Comments
Population dynamics	There is a lack of quantitative studies, especially of long-term population dynamics. Compared with terrestrial plants, less is known about specific phases of plant life-cycles of aquatic plants – in particular seed germination, mechanisms of seedling or vegetative propagule establishment and self-thinning.
Competition	There is a need for further study of mechanisms of inter- and intra-specific competition, especially through manipulative experiments.
Interactions with other organisms	There is a lack of knowledge on how herbivores influence aquatic plants in running waters; on competition with algae, including periphyton; and on interactions with fungi.
Succession and other temporal changes	While something is known regarding dynamics of some aquatic plants in relation to seasonal fluctuations in discharge, less knowledge exists relating to their long-term dynamics on both a population and community level. It is often difficult to separate a long-term succession from fluctuations caused by unpredictable discharge and other changes.
Influence of plants on water dynamics and water quality and *vice versa*	While there is a comparably large amount of information relating to this study area, much of it is non-systematic in nature. There is a need to increase the amount of manipulative experiments, both *in situ* and under laboratory conditions. Questions include: how do changing flow parameters influence plants? and how do changes in plant growth influence flow parameters?

constitutes a key factor in the distribution of plant species and the intensity of their production. Finally, water quality – in terms of both its effects on plants and the effects of plants on water quality – needs to be taken into account when examining plants in streams and rivers. While the most basic ecological interpretation is that the more polluted the site the lower the number of species of water plant that will be found there, the picture in reality is very complex as toxicity varies with a wide range of factors, including pollutant speciation, organic matter content, oxygen concentration and temperature.

While many papers have addressed the subject of plants and water in streams and rivers, a range of subject areas require further study to elucidate more clearly the processes and interactions occurring between the flowing water environment and the plant life found therein (*cf*. Table 7.2). For example, in discussing the effects of current velocity on plant species richness along rivers, Nilsson (1987) has highlighted a number of gaps in the knowledge base relating to aquatic vegetation that could usefully be addressed, pointing out that information is needed on the relationships between edaphic conditions, vegetative dispersal, seed availability and germination success along the current velocity gradient, as well as on gap formation and deposition of propagules. More than a decade later, this would still appear to be the case. Johansson and Nilsson (1993) conclude that modelling of vegetation dynamics should aim at combining within-site processes like competition, stress and disturbance with between-site processes (i.e. hydrochory) as mechanisms structuring plant communities. Sand-Jensen *et al.* (1989) have also pointed out that, despite the marked influence that submerged macrophytes have on lowland stream systems, little is known about their seasonal development, or the factors that result in variation in macrophyte development and abundance both among streams and between years in a particular stream. Future effort should concentrate on these and related areas of study because, as the above discussion shows, while only a relatively small number of species can grow successfully in-stream, their occurrence in streams and rivers has important hydrological and ecological consequences. In many cases too, these consequences have considerable economic implications.

REFERENCES

Angelier, E. (1953) Recherches écologiques et biogéographiques sur la faune de sables submergés, *Archives de Zoologie Expérimentale et Générale* 90: 37–161.

Bakry, M.F., Gates, T.K. and Khattab, A.F. (1992) Field-measured hydraulic resistance characteristics in vegetation-infested canals, *Journal of Irrigation and Drainage Engineering* 118: 256–274.

Barnes, H.H. (1967) Roughness characteristics of natural channels. US Geological Survey Water Supply Paper 1949, US Government Printing Office, Washington DC.

Bathurst, J.C. (1981) Discussion of bar resistance of gravel bed streams, *Journal of the Hydraulics Division, ASCE* 107: 1276–1278.

Bathurst, J.C. (1985) *Literature Review on Some Aspects of Gravel Bed Rivers*, Wallingford: Institute of Hydrology.

Bathurst, J.C. (1997) Environmental river flow hydraulics, in Thorne, C.R., Hey, R.D. and Newson, M.D. (eds) *Applied Fluvial Geomorphology for River Engineering and Management*, Chichester: Wiley, pp. 69–93.

Begon, M., Harper, J.L. and Townsend, C.R. (1990) *Ecology: Individuals, Populations and Communities*, Oxford: Blackwell.

Bilby, R.E. and Likens, G.E. (1980) Importance of organic debris dams in the structure and function of stream ecosystems, *Ecology* 61: 1107–1113.

Bijl, L. van der, Sand-Jensen, K. and Hjermind, A.L. (1989) Photosynthesis and canopy structure of a submerged plant, *Potamogeton pectinatus*, in a Danish lowland stream, *Journal of Ecology* 72: 947–962.

Bisson, P.A., Bilby, R.E., Bryant, M.D., Dolloff, C.A., Grette, G.B., House, R.A., Murphy, M.L., Koski, K.V. and Sedell, J.R. (1987) Large woody debris in forested streams in the Pacific northwest: past, present and future, in Salo, E.O. and Cundy, T.W. (eds) *Streamside Management: Forestry and Fishery Interactions*, Seattle: College of Forest Resources, University of Washington, pp. 143–190.

Blom, C.W.P.M., Bögemann, G.M., Laan, P., van der Sman, A.J.M., van de Steeg, H.M. and Voesenek, L.A.C.J. (1990) Adaptations to flooding in plants from river areas, *Aquatic Botany* 38: 29–47.

Brändle, R., Cizková, H. and Pokorny, J. (1996) *Adaptation Strategies in Wetland Plants: Links between Ecology and Physiology*, Uppsala: Opulus Press.

Broadhurst, L.J., Heritage, G.L., van Niekerk, A.W. and James, C.S. (in press) *Translating Discharge into Local Hydraulic Conditions on the Sabie River: An Assessment of Channel Roughness*, Pretoria: Water Research Commission Report No. 474/2/96.

Brooker, M.P., Morris, D.L. and Wilson, C.J. (1978) Plant–flow relationships in the River Wye catchment, in *Proceedings of the European Weed Research Symposium*, 5th Symposium on Aquatic Weeds, Amsterdam: European Weed Research Society, pp. 63–69.

Brookes, A. (1995) The importance of high flows for riverine environments, in Harper, D.M. and Ferguson, A.J.D. (eds) *The Ecological Basis for River Management*, Chichester: Wiley, pp. 33–49.

Brown S. and Lugo A.E. (1982) A comparison of structural and functional characteristics of saltwater and freshwater forested wetlands, in Gopal B., Turner R.E., Wetzel R.G. and Whigham D.F. (eds) *Wetlands Ecology and Management*, Jaipur: International Scientific Publications.

Calow, P. and Petts, G.E. (eds) (1992) *The Rivers Handbook*, Volume 1. Oxford: Blackwell.

Calow, P. and Petts, G.E. (eds) (1994) *The Rivers Handbook*, Volume 2. Oxford: Blackwell.

Carpenter, S.R. and Lodge D.M. (1986) Effects of submersed macrophytes on ecosystem processes, *Aquatic Botany* 26: 341–370.

Cook, C.D.K. (1996) *Aquatic Plant Book*, Amsterdam: SPB Academic.

Dale, H.M. and Gillespie, T.J. (1977) The influence of submersed aquatic plants on temperature gradients in shallow water bodies, *Canadian Journal of Botany* 55: 2216–2225.

Dawson, F.H. (1976) The annual production of the aquatic macrophyte *Ranunculus penicillatus* var. *calcareous*, *Aquatic Botany* 2: 51–73.

Dawson, F.H. (1978) The seasonal effects of aquatic plant growth on the flow of water in a stream, in *Proceedings of the European Weed Research Symposium*, 5th Symposium on Aquatic Weeds, Amsterdam: European Weed Research Society, pp. 71–78.

Day, R.T, Keddy, P.A., McNeill, J. and Carleton, T. (1988) Fertility and disturbance gradients: a summary model for riverine marsh vegetation, *Ecology* 69: 1044–1054.

Dhillon, M.S., Mukka, M.S. and Hwang, Y.-S. (1982) Allelochemicals produced by the hydrophyte *Myriophyllum spicatum* affecting mosquitoes and midges, *Journal of Chemical Ecology* 8: 517–526.

Dingman, S.L. (1984) *Fluvial Hydrology*, New York: Freeman, 383 pp.

Dodds, W.K. (1991) Micro-environmental characteristics of filamentous algal communities in flowing freshwaters, *Freshwater Biology* 25: 199–209.

Dole-Olivier, M.J. and Marmonier, P. (1992) Patch dynamics of interstitial communities: prevailing factors, *Freshwater Biology* 27: 177–191.

Farmer, A.M. and Spence, D.H.N. (1987) Flowering, germination and zonation of the submerged aquatic plant *Lobelia dortmanna* L., *Journal of Ecology* 75: 1065–1076.

Fennessy, S. (1993) *Riparian Buffer Srips: Their Effectiveness for the Control of Agricultural Pollution*, London: Institute for European Environmental Policy Report 210/89, University College.

Forman, R.T.T. and Godron, M. (1986) *Landscape Ecology*, New York: Wiley.

French, T.D. and Chambers, P.A. (1996) Habitat partitioning in riverine macrophyte communities, *Freshwater Biology* 36: 509–520.

Gilvear, D.J. and Bravard, J.-P. (1996) Geomorphology of temperate rivers, in Petts, G.E. and Amoros, C. (eds) *Fluvial Hydrosystems*, London: Chapman & Hall, pp. 68–97.

Gopal, B. and Goel, U. (1993) Competition and allelopathy in aquatic plant communities, *Botanical Review* 59: 155–210.

Grace, J.B. (1987) The impact of pre-emption on the zonation of two *Typha* species along lakeshores, *Ecological Monographs* 57: 283–303.

Gregory, K.J. (1992) Vegetation and river channel process interactions, in Boon, P.J., Calow, P. and Petts, G.E. (eds) *River Conservation and Management*, Chichester: Wiley, pp. 255–269.

Grime, J.P. (1979) *Plant Strategies and Vegetation Processes*, Chichester: Wiley.

Gurnell, A.M. (1997) Woody debris and river–floodplain interactions, in Large, A.R.G. (ed.), *Floodplain Rivers: Hydrological Processes and Ecological Significance*, Newcastle: British Hydrological Society Occasional Paper No. 8, University of Newcastle upon Tyne, pp. 6–17.

Gurnell, A.M. and Gregory, K.J. (1987) Vegetation characteristics and the prediction of runoff: analysis of an experiment in the New Forest, Hampshire, *Hydrological Processes* 1: 125–142.

Gurnell, A.M., Gregory, K.J. and Petts, G.E. (1995) The role of coarse woody debris in forest aquatic habitats: implications for management, *Aquatic Conservation: Marine and Freshwater Ecosystems* 5: 143–166.

Ham, S.F., Cooling, D.A., Hiley, P.D., McLeish, P.R., Scorgie, H.R.A. and Berrie, A.D. (1982) Growth and recession of aquatic macrophytes on a shaded section of the River Lambourn, England, from 1971 to 1980, *Freshwater Biology* 12: 1–15.

263

Haslam, S.M. (1978) *River Plants: The Aquatic Vegetation of Watercourses*, Cambridge: Cambridge University Press.

Haslam, S.M. (1987) *River Plants of Western Europe: The Macrophytic Vegetation of Watercourses of the European Economic Community*, Cambridge: Cambridge University Press.

Haslam, S.M. (1990) *River Pollution: An Ecological Perspective*, Chichester: Wiley.

Haycock, N.E. and Pinay, G. (1993) Groundwater–nitrate dynamics in grass and poplar vegetated riparian buffer strips during the winter, *Journal of Environmental Quality* 22: 273–278.

Hey, R.D. (1988) Bar form resistance in gravel bed rivers, *Journal of the Hydraulics Division, ASCE* 114: 1498–1508.

Hicken, E.J. (1984) Vegetation and river channel dynamics, *Canadian Geographer* 28: 111–126.

Hill, W.R., Ryon, M.G. and Schilling, E.M. (1995) Light limitation in a stream ecosystem: responses by primary producers and consumers, *Ecology* 76: 1297–1309.

Holland, M.M. (1988) SCOPE/MAB technical consultations on landscape boundaries: report of a SCOPE/MAB workshop on ecotones, *Biology International* Special Issue 17: 47–106.

Holmes, N.T.H. (1983) *Typing British Rivers According to their Flora*, Focus on Nature Conservation 4, Peterborough: Nature Conservancy Council.

Hutchinson, G.E. (1975) *A Treatise on Limnology. III Limnological Botany*, New York: Wiley.

Hydraulics Research (1992) *The Hydraulic Roughness of Vegetated Channels*, HR Wallingford Report SR 305, Wallingford: Hydraulics Research.

Hynes, H.B.N. (1970) *The Ecology of Running Waters*, Liverpool: Liverpool University Press.

Jackson, M.B., Davies, D.D. and Lambers, H. (eds) (1991) *Plant Life Under Oxygen Deprivation: Ecology, Physiology and Biochemistry*, Amsterdam: SPB Academic.

Jeffries, M. and Mills, D. (1994) *Freshwater Ecology: Principles and Applications*, Chichester: Wiley.

Johansson, M.E. and Nilsson, C. (1993) Hydrochory, population dynamics and distribution of the clonal aquatic plant *Ranunculus lingua*, *Journal of Ecology* 81: 81–91.

Junk, W.G., Bayley, P.B. and Sparks, R.E. (1989) The flood pulse concept in river-flooding systems, *Canadian Journal of Fisheries and Aquatic Sciences* 106: 110–127.

Kouwen, N. and Li, R.M. (1980) Biomechanics of vegetative channel linings, *Journal of the Hydraulics Division, ASCE* 106: 1085–1103.

Large, A.R.G. and Petts, G.E. (1992) *Buffer Zones for Riverbank Conservation*, National Rivers Authority Project 340/Y/5, Loughborough: International Centre of Landscape Ecology, Loughborough University, 71 pp.

Large, A.R.G. and Petts, G.E. (1994) Rehabilitation of river margins, in Calow, P. and Petts, G.E. (eds) *The Rivers Handbook*, Volume 2, Oxford: Blackwell Scientific, pp. 401–418.

Large, A.R.G., Pautou, G. and Amoros, C. (1996) Primary production and primary producers, in Petts, G.E. and Amoros, C. (eds) *Fluvial Hydrosystems*, London: Chapman & Hall, pp. 117–136.

Losee, R.F. and Wetzel, R.G. (1993) Littoral flow rates within and around submersed macrophyte communities, *Freshwater Biology* 29: 7–17.

Lowrance, R., Leonard, R. and Sheridan, J. (1985) Managing riparian ecosystems to control non-point pollution, *Journal of Soil and Water Conservation* 40: 87–91.

Lucas, W.J. and Berry, J.A. (1985) Inorganic carbon transport in aquatic photosynthetic organisms, *Physiologia Plantarum* 65: 117–124.

Madsen, T.V. and Maberly, S.C. (1991) Diurnal variation in light and carbon limitation of photosynthesis by two species of submerged freshwater macrophyte with a differential ability to use bicarbonate, *Freshwater Biology* 26: 175–187.

Malanson, G.P. (1993) *Riparian Landscapes*, Cambridge: Cambridge University Press.

Marshall, E.J.P. and Westlake, D.F. (1978) Recent studies on the roles of aquatic macrophytes in their ecosystem, in *Proceedings of the European Weed Research Symposium*, 5th Symposium on Aquatic Weeds, Amsterdam: European Weed Research Society, pp. 183–188.

Masterman, R. and Thorne, C.R. (1992) Predicting influence of bank vegetation on channel capacity, *Journal of Hydraulic Engineering* 118: 1052–1058.

Miller, B.A. and Wenzel, H.G. (1985) Analysis and simulation of low flow hydraulics. *Journal of Hydraulic Engineering* 111: 1429–1446.

Moeslund, B. Løjtnant, B., Mathiesen, H., Mathiesen, L., Pedersen, A., Thyssen, N. and Schou, J.C. (1990) *Danske Vandplanter*, Danish Environmental Protection Agency.

Muscutt, A.D., Harris, G.L., Bailey, S.W. and Davies, D.B. (1993) Buffer zones to improve water quality: a review of their potential use in UK agriculture, *Agriculture, Ecosystems and Environment* 45: 59–77.

Naiman, R.J., Décamps, H., Pastor, J. and Johnston, C.A. (1988) The potential importance of boundaries to fluvial ecosystems, *Journal of the North American Benthological Society* 7: 289–306.

Naiman, R.J. and Décamps, H. (eds) (1990) *The Ecology and Management of Aquatic–Terrestrial Ecotones*, Carnforth: UNESCO Paris and Parthenon Publishing Group.

Newall, A.M. (1995) The micro-flow environment of aquatic plants: an ecological perspective, in Harper, D.M, and Ferguson, A.J.D. (eds) *The Ecological Basis for River Management*, Chichester: Wiley, pp. 79–92.

Newson, M.D. (1992) *Land, Water and Development: River Basin Systems and their Sustainable Management*, London: Routledge.

Nilsson, C. (1987) Distribution of stream-edge vegetation along a gradient of current velocity, *Journal of Ecology* 75: 513–522.

Nilsson, C. and Grelsson, G. (1989) The effects of litter displacement on riverbank vegetation, *Canadian Journal of Botany* 68: 735–741.

Nilsson, C., Gardfjel, M. and Grelsson, G. (1991) Importance of hydrochory in structuring plant communities along rivers, *Canadian Journal of Botany* 69: 2631–2633.

Nilsson, C., Grelsson, G., Johansson, M. and Sperens, U. (1989) Patterns of plant species richness along riverbanks, *Ecology* 70: 77–84.

Orghidan, T. (1955) Un nou domeniu de viata acvatica subterana: 'biotopul hiporeic', *Buletin Stiintific sectia de Biologie si stiinte Agricole si sectia de Geologie si Geografie* VII: 657–676.

Osborne, L.L. and Kovacic, D.A. (1993) Riparian vegetated buffer strips in water-quality restoration and management, *Freshwater Biology* 29: 243–258.

Petts, G.E. (1994) Rivers: dynamic components of catchment ecosystems, in Calow, P. and Petts, G.E. (eds) *The Rivers Handbook*, Volume 2, Oxford: Blackwell, pp. 3–22.

Petts, G.E. and Foster, I.D.L. (1985) *Rivers and Landscape*, London: Edward Arnold.

Philips, G.L., Eminson, D.F. and Moss, B. (1978) A mechanism to account for macrophyte decline in progressively eutrophicated freshwaters, *Aquatic Botany* 4: 103–126.

Pickett, S.T.A. and White, P.S. (1985) Patch dynamics: a synthesis, in Pickett, S.T.A. and White, P.S. (eds) *The Ecology of Natural Disturbances and Patch Dynamics*, New York: Academic Press, pp. 371–384.

Pinay, G., Décamps, H., Chauvet, E. and Fustec, E. (1990) Functions of ecotones in fluvial systems, in Naiman, R.J. and Décamps, H. (eds) *The Ecology and Management of Aquatic–Terrestrial Ecotones*, Carnforth: UNESCO Paris and Parthenon Publishing Group, pp. 141–169.

Polunin, N. (1982) Processes contributing to the decay of reed (*Phragmites australis*) litter in fresh water, *Archiv für Hydrobiologie* 114: 401–414.

Polunin, N. (1984) The decomposition of emergent macrophytes in fresh water, *Advances in Ecological Research* 14: 115–166.

Prach, K., Large, A.R.G., and Jeník, J. (1996) River floodplains as ecological systems, in Prach, K., Jeník, J., and Large A.R.G. (eds) *Floodplain Ecology and Management: The Luznice River in the Trebon Biosphere Reserve, Central Europe*, Amsterdam: SPB Academic, pp. 1–9.

Pyšek, P. (1994) Ecological aspects of invasion by *Heracleum mantegazzianum* in the Czech Republic, in de Waal, L.C., Child, L.E., Wade, P.M. and Brock, J.H. (eds) *Ecology and Management of Invasive Riverside Plants*, Chichester: Wiley, pp. 45–54.

Pyšek, P. and Prach, K. (1993) Plant invasions and the role of riparian habitats – a comparison of four species alien to Central Europe, *Journal of Biogeography* 20: 413–420.

Pyšek, P. and Prach, K. (1995) Invasion dynamics of *Impatiens glandulifera* – a century of spreading reconstructed, *Biological Conservation* 74: 41–48.

Rattrey, M.R., Brown, J.M.A. and Howard-Williams, C. (1991) Sediment and water as sources of nitrogen and phosphorus for submerged rooted aquatic macrophytes, *Aquatic Botany* 40: 225–237.

Resh, V.H., Brown, A.V., Covich, A.P., Gurtz, M.E., Li, H.W., Minshall, G.W., Reice, S.R., Sheldon, A.L., Wallace, J.B. and Wissmar, R.C. (1988) The role of disturbance in stream ecology, *Journal of the North American Benthological Society* 7: 433–455.

Richards, K. (1982) *Rivers: Form and Process in Alluvial Channels*, London: Methuen.

Rydlo, J. (1996) Water macrophytes in the section of the River Elbe in 1976, 1986 and 1996, *Museum a soucasnost (Rydlia)* 11: 87–128.

Sand-Jensen, K. (1987) Experimental control of bicarbonate use among freshwater and marine macrophytes, in Crawford, R.M.M. (ed.) *Plant Life in Aquatic and Amphibious Habitats*, British Ecological Society Special Publication 5, Oxford: Blackwell, pp. 99–112.

Sand-Jensen, K. and Madsen, T.V. (1992) Patch dynamics of the stream macrophyte *Callitriche cophocarpa*, *Freshwater Biology* 27: 277–282.

Sand-Jensen, K. and Mebus, J.R. (1996) Fine-scale patterns of water velocity within macrophyte patches in streams, *Oikos* 76: 169–180.

Sand-Jensen, K., Jeppesen, E., Nielsen, K., van der Bijl, L., Hjermind, L., Nielsen, L.W. and Iversen, T.M. (1989) Growth of macrophytes and ecosystem consequences in a lowland Danish stream, *Freshwater Biology* 22: 15–32.

Sand-Jensen, K. Pedersen, M.F. and Nielsen, S.L. (1992) Photosynthetic use of inorganic carbon among primary and secondary water plants in streams, *Freshwater Biology* 27: 283–293.

Schneider, R.L. and Sharitz, R.R. (1988) Hydrochory and regeneration in a bald-cypress–water Tupelo swamp forest, *Ecology* 69: 1055–1063.

Sculthorpe, C.D. (1967) *The Biology of Aquatic Vascular Plants*, London: Edward Arnold.

Smith, R.J., Hancock, N.H. and Ruffini, J.L. (1990) Flood flow through tall vegetation, *Agricultural Water Management* 18: 317–332.

Stephan, U. and Wychera, U. (1996) Analysis of flow velocity fluctuations in different macrophyte banks in a natural open channel, *Proceedings of Symposium Ecohydraulics 2000*, Quebec, Canada, June 1996, A191–A202.

Teal, J.M. (1980) Primary production of benthic and fringing plant communities, in Barnes, R.S.K. and Mann, K.H. (eds) *Fundamentals of Aquatic Ecosystems*, Oxford: Blackwell, pp. 67–83.

Triska, F.J. and Cromack, K. (1980) The role of wood debris in forests and streams, in Waring, R.H. (ed.) *Forests: Fresh Perspectives from Ecosystem Analysis*, Oregon State University, pp. 171–189.

Vannote, R.L., Minshall, G.W., Cummins, K.W., Sedell, J.R. and Cushing, C.E. (1980) The river continuum concept, *Canadian Journal of Fisheries and Aquatic Sciences* 37: 130–137.

Vought, L.B.-M., Dahl, J., Pedersen, C.L. and Lacoursière, J.O. (1994) Nutrient retention in riparian ecotones, *Ambio* 23: 342–348.

Ward, J.V. (1989) The four dimensional nature of lotic ecosystems, *Journal of the North American Benthological Society* 8: 2–8.

Ward, J.V. and Stanford, J.A. (1995) The serial discontinuity concept: extending the model to floodplain rivers, *Regulated Rivers: Research and Management* 10: 159–168.

Watson, D. (1987) Hydraulic effects of aquatic weeds in UK rivers, *Regulated Rivers: Research and Management* 1: 211–227.

Watts, J.F. and Watts, G.D. (1990) Seasonal change in aquatic vegetation and its effect on channel flow, in Thornes, J.B. (ed.) *Vegetation and Erosion*, Chichester: Wiley.

Welcomme, R.L. (1979) *Fisheries Ecology of Floodplain Rivers*, London: Longman.

Westlake, D.F. (1975) Macrophytes, in Whitton, B.A. (ed.) *River Ecology*, Oxford: Blackwell, pp. 106–128.

Westlake, D.F., Adams, M.S., Bindloss, M.E., Ganf, G.G., Gerloff, G.C., Hammer, U.T., Javornický, P., Koonce, J.F., Marker, A.F.H., McCracker, M.D., Moss, B., Nauwerck, A., Pyrina, I.L., Steel, J.A.P., Tilzer, M. and Walters, C.J. (1980) Primary production, in Le Cren, E.D. and Lowe-McConnell, R.H. (eds) *The Functioning of Freshwater Ecosystems*, International Biological Programme 22, Cambridge: Cambridge University Press, pp. 141–246.

Wheeler, W.H. (1980) Effects of boundary layer transport on the fixation of carbon by the giant kelp *Macrocystis pyrifera*, *Marine Biology* 56: 103–110.

Wiegleb, G. (1988) Analysis of flora and vegetation in rivers: concepts and applications, in Symoens, J.J. (ed.) *Vegetation of Inland Waters*, Dordrecht: Kluwer, pp. 311–340.

Wilson, E.M. (1974) *Engineering Hydrology*, London: Macmillan Press.

Wilson, S.D. and Keddy, P.A. (1986) Measuring diffuse competition along an environmental gradient: results from a shoreline plant community, *American Naturalist* 127: 862–869.

Wilson, S.D. and Keddy, P.A. (1991) Competition, survivorship and growth in macrophyte communities, *Freshwater Biology* 25: 331–337.

Wright, J.F, Cameron, A.C., Hiley, P.D. and Berrie, A.D. (1982) Seasonal changes in biomass of macrophytes on shaded and unshaded sections of the River Lambourn, England, *Freshwater Biology* 12: 271–283.

8

PLANTS AND WATER IN AND ADJACENT TO LAKES

Robert G. Wetzel

INTRODUCTION

The structure, metabolism and biogeochemistry of a lake are strongly coupled to inputs from its drainage basin. A lake reflects the developmental conditions and disturbances within the catchment, its drainage streams, and the vegetation above, at and below the land–water interface.

Those allochthonous inputs are largely transported to the lake in particulate and especially soluble compounds with water. Inputs from atmospheric sources are usually small in comparison with loadings from ground and surface-water sources. These compounds, particularly nutrients, influence the autochthonous development and productivity of micro- and macro-vegetation within the lake. The vegetation in turn can have profound effects on the movements of water within the basin, export of water from the basin via evapotranspiration, and alteration of habitats within lakes. Indeed, the macro-vegetation influences the entire ontogeny (successional development) of lake ecosystems (Wetzel, 1979; 1983; 1990; 1992).

THE LAND–WATER INTERFACE REGION OF AQUATIC ECOSYSTEMS

The land–water interface zone includes a large area of most lake ecosystems that extends from hydrosoils that are at least periodically saturated at the upland boundary through to and including the wetland and littoral areas to depths that have sufficient light to support larger aquatic plants (Wetzel, 1983). This interface zone constitutes an intensely metabolically active region that strongly influences the biotic metabolism of both standing and flowing fresh waters. Much of the nutrient loading, cycling and recycling within many fresh waters is controlled by the metabolism of the wetland and littoral macrophytes and their associated microflora (Wetzel, 1990). Because most

269

aquatic ecosystems occur in terrain of gentle slopes with very small eleva-
tional gradients and are small in area and shallow in depth (Figure 8.1), the
wetland–littoral components usually dominate in productivity and synthesis
of organic matter in freshwater ecosystems (Wetzel, 1990; Meybeck, 1995).

The land–water interface is, for all practical purposes, always the most
productive region per unit area along the gradient from land to open water in
both lake and stream ecosystems. The region of greatest productivity is the
emergent macrophyte zone. The emergent plants have a number of structural
and physiological adaptations that not only tolerate the hostile anaerobic
sediments but also take advantage of the relatively abundant nutrient and
water conditions of this habitat. Nutrients imported with influent water to
the zone of emergent macrophytes are assimilated largely by the microflora,
primarily bacteria, fungi and attached algae, that occur on and among the
sediments and macrophytic detrital particles of organic matter. These
nutrients are recycled within the microflora associated with particulate
organic detritus, largely of aquatic plant origins, and are reassimilated by the
emergent macrophytes from interstitial solutions through their rooting
tissues (Wetzel, 1990; 1993). Export from the emergent plant zones is
dominated by dissolved organic compounds released from decomposition of
the plant detrital material. Nutrients, especially phosphorus and nitrogen, are
tightly conserved among the attached microflora.

Submerged macrophytes are limited physiologically primarily by slow
diffusion rates of nutrients and gases at the boundary layers at the surfaces of
leaves and by reduced underwater light availability. Many physiological and
biochemical adaptations allow improved growth and competitiveness of these

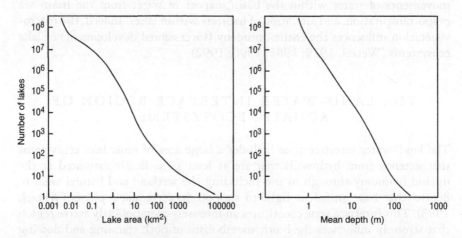

Figure 8.1 Number of permanent lakes in the world in relation to lake area and
approximate mean water depth (after Wetzel, 1990). Printed with permission of
the International Association of Theoretical and Applied Limnology.

plants (Sculthorpe, 1967; Hutchinson, 1975; Wetzel, 1990). Recycling of resources, both gases of metabolism and critical nutrients, within the plants and their immediate micro-environments, are important parts of the abilities of submerged plants to function and grow as well as they do under chronic light and gas resource limitations. The growth and productivity of submerged plants are, however, less than those of emergent and floating macrophytes (Table 8.1).

The autotrophic phytoplankton, consisting of planktonic algae and cyanobacteria, within the water is the least productive along the gradient from land to water, regardless of the amounts of nutrient loading and availability. Limitations are spatial, related first to sparse distribution in a dilute environment where nutrient recycling is limiting. When nutrient recycling and availability are increased, depth limitations of light preclude appreciable development of phytoplankton densities. Photosynthetic productivity of phytoplankton per unit area is among the lowest in the biosphere, even under optimal growth conditions (Table 8.2). Phytoplanktonic productivity can be increased only slightly by extending the growing season, as in the tropics, or by increasing water mixing rates in the trophogenic zones of lakes and thereby increasing the frequency of interception of adequate light.

The second most productive component of the gradient from land to open water, after the emergent hydrophytes, is the attached microflora. In any water with submerged vegetation and particulate detritus, the surface area for

Table 8.1 Averages (and ranges) of rates of primary production of higher aquatic plants of inland waters (summarised from Wetzel, 1983; Carter, *et al.*, in press; and Wetzel and Howe, in press).

Type	Seasonal maximum biomass or above-ground biomass $(g\,dry\,m^{-2})^a$	Rate of production $(g\,dry\,m^{-2}\,year^{-1})^b$
Emergent macrophytes	2500 (<2–9900)	5000 (3000–10,300)
Temperate	–	3800 (3000–4500)
Tropical	–	7500 (6500–10,300)
Floating	(630–1500)	(1500–4400)
Floating-leaved rooted	140 (25–340)	300 (110 – 560)
Submerged	220 (15–500)	1300 (100–2000)
Temperate	–	600 (100–700)
Tropical	–	1700 (1200–2000)

[a] Approximate only; generally but not always above-ground biomass only.
[b] Rates of turnover vary greatly. Most aquatic plants are herbaceous perennials and often exhibit multiple cohorts (1.5 to 3 in temperate to 10 per year in tropical regions) or continuous growth with seasonal variations.

Table 8.2 Annual net primary productivity of aquatic communities on fertile sites in comparison with that of other communities (modified from Westlake, 1963; 1965).

Type of ecosystem	Approximate organic (dry) productivity ($t\ ha^{-1}\ year^{-1}$)*	Range ($t\ ha^{-1}\ year^{-1}$)
Marine phytoplankton	2	1–4.5
Lake phytoplankton	2	1–9
Freshwater submerged macrophytes		
Temperate	6	1–7
Tropical	17	12–20
Marine submersed macrophytes		
Temperate	29	25–35
Tropical	35	30–40
Marine emergent macrophytes (salt marsh)	30	25–85
Freshwater emergent macrophytes		
Temperate	38	30–45
Tropical	75	65–110
Desert, arid	1	0–2
Temperate forest		
Deciduous	12	9–15
Coniferous	28	21–35
Temperate herbs	20	15–25
Temperate annuals	22	19–25
Tropical annuals	30	24–36
Rain forest	50	40–60

* t = tonne. Values ×100 = $g\ m^{-2}\ year^{-1}$ (approximately) and ×50 = $g\ C\ m^{-2}\ year^{-1}$ (approximately).

attached microfloral colonisation and development increases manifold in comparison with waters lacking appreciable submerged vegetation. The physiology and growth of attached microflora are intimately coupled to the physical and physiological dynamics of the living and dead substrata upon which they grow (Wetzel, 1990; 1993). Again, recycling of limiting essential gases and both inorganic and organic dissolved nutrients is the key to the commonly observed high sustained growth of attached microflora. By means of intensive recycling of nutrients and gases to maintain extant community biomass, losses are minimised and imported external nutrients and gases can be directed primarily to net growth.

The land–water interface zone of wetland–littoral communities produces the primary sources of organic matter of most freshwater ecosystems. Most of the particulate organic matter remains within the land–water interface region and undergoes degradation near sites of production. Organic matter is exported from these regions to the open water predominantly as dissolved organic matter of relatively recalcitrant chemical compounds (Wetzel, 1991; 1992). These dissolved organic compounds of wetland–littoral origins function in several important ways in aquatic ecosystems. The dissolved organic compounds, often dominated by organic acids and polyphenolic substances of higher plant origin, have an array of regulatory functions on the availability of metals, inorganic and organic micronutrients, carbonate and clay particle reactivities, and extracellular enzyme activities, among others (Wetzel, 1992). The dissolved organic matter forms a large reservoir of organic matter, usually an order of magnitude or more greater than particulate organic matter. Although decomposed slowly (c. 1 percent per day), the collective microbial metabolism is a major component of the energy dissipation of the ecosystem (Wetzel, 1995).

MOVEMENTS OF WATER TO AND WITHIN LAKES

The water balance of a lake is evaluated from the change in the water volume of the lake per unit time. In some cases, changes in lake volume can be calculated from changes in lake area. The change in storage is equal to the rate of inflow from all sources minus the rate of water loss.

Water income to a lake or reservoir includes several sources:

1 *Precipitation (P) directly on the lake surface.* Most lakes exist in exorheic regions, within which rivers originate and from which they flow to the sea, and receive a relatively small proportion (<10 percent) of their total water income from direct precipitation. This percentage varies greatly, however, and increases in very large lakes. Extreme examples include Lake Victoria of equatorial East Africa, which receives most (>70 percent) of its water from precipitation on its surface (Kilham and Kilham, 1990).

2 *Water from surface influents (R) of the drainage basin.* Many lakes of endorheic regions, within which rivers arise but never reach the sea, receive nearly all of their water from runoff. The rate of runoff from the drainage basin and corresponding changes in lake level are influenced by the nature of the soil and vegetation cover of the drainage basin (*cf.* Wetzel, 1983; Likens, 1985).

3 *Groundwater seepage (G_i) below the surface of the lake.* Seepage of groundwater is commonly a major source of water for lakes in rock basins and lake basins in glacial tills that extend well below the water table. Sub-lacustrine groundwater seepage represents a major source of water flow into, and out of, karst and doline lakes of limestone substrata. In wetlands and littoral areas of lakes, there is often a complex seepage hydrology with groundwater flow oscillating between upwelling and

downwelling. This complex behaviour is mediated by variations in precipitation and evapotranspiration.

4 *Groundwater entering lakes as discrete springs.* Sub-lacustrine springs from groundwater occur frequently in hard water lakes of calcareous drift regions, where the basin is effectively sealed from groundwater seepage by deposits within the basin (e.g. Wetzel, 1965; 1966; Wetzel and Otsuki, 1974). Such groundwater inputs are often closely coupled to rates of precipitation and evapotranspiration. These inputs vary with any modifications, such as climate changes, to rates of precipitation and evaporation. For example, in the Dead Sea contributions from springs and shore seepage decreased as river influents were purloined by partial diversions (Serruya and Pollinger, 1984).

Losses of water occur by *flow from an outlet* (D) in drainage lakes and by *seepage into the groundwater* through the basin walls in seepage lakes. Deposition of clays, silts and carbonates commonly forms a very effective seal in drainage lakes, from which most or all of the outflow leaves by a surface outlet. In seepage lakes, the sediments over much of the deeper portions of the basin also usually form an effective seal; losses to groundwater (G_o) usually occur from the uppermost portions (<2 m depth) of the basin in these lakes.

Further losses of water occur directly by *evaporation* (E), or by *evapotranspiration* from emergent and floating-leaved aquatic macrophytes. The extent and rates of evaporative losses are highly variable according to season and latitude, and are greatest in endorheic regions. Closed lakes of semi-arid regions commonly have no outflow and lose water only by evaporation. Because of the major importance of aquatic macrophytes on the water budgets of lakes, this ecological mediation of water fluxes will be treated separately (see 'Plants as controls on lake hydrological processes', below).

The water balance states that the rate of change of the lake volume (V) with time (t) is the difference between the water input and output rates (L^3T^{-1}) (Hutchinson, 1957; Szesztay, 1974; Mason *et al.*, 1994), expressed as a differential equation:

$$\frac{dV}{dt} = (R + P + G_i) - (D + E + G_o) \tag{8.1}$$

Each of these hydrological components is dynamic and optimally requires continuous measurement over large spatial areas in order to evaluate accurately the water budget of a lake. Uncertainty boundaries and variations between measurements within spatial distributions must be evaluated (Winter, 1981). Some atmospheric components of a water budget such as precipitation are variable spatially at relatively small distances (several kilometres) even though controlled by circulation of the atmosphere. Continuous measurement of other parameters is less reliable. For example, evaporative loss of water vapour from a water body is controlled by wind and vapour fluxes over the water body (Winter, 1995), both of which are very variable.

The most accurate method of determining evaporation over periods of months or more is by the energy budget method, which is expensive in terms of instruments and labour. Gains and losses of heat from all sources of radiation fluxes, latent heat, sensible heat, precipitation, surface water and groundwater must be evaluated (Figure 8.2). The complexities of the energy budget method of evaluating evaporation in lakes is treated in detail by Brutsaert (1982), Hostetler (1995), Imboden and Wüest (1995), and Winter (1995). Surface water fluxes can be measured accurately both directly by river inflows/outflows and indirectly by variations in lake depth and area.

Subsurface inputs and losses of water can be measured directly, although this is often difficult and demanding to accomplish accurately, with *in situ* underwater piezometers (*cf.* Lee, 1977; 1985; Lee and Cherry, 1978; Lee and Harvey, 1996). Seepage measurements give some insights into subsurface loadings and allow limited application of two- and three-dimensional models of groundwater flow dynamics. Advected groundwater and its energy contribute to the mass and energy budgets of closed lakes. For example, advected groundwater accounts for more than 50 percent of the annual water input to a moderately sized (40 ha) closed lake in northern Minnesota, USA (Sturrock *et al.*, 1992). More than 50 percent of the water loss from the lake occurs annually by groundwater flow.* Even though seepage into the lake is large, groundwater contributes little thermal energy input to the lake, because the groundwater flowing into the lake is relatively cold at *c.* 7 °C. Similarly, because the lake water is warm for only brief periods in the summer, seepage from the basin results in only small losses of thermal energy to groundwater.

Although groundwater energy inputs and losses may not affect evaporation in major ways, these subsurface inputs can be of major importance in the loading of inorganic and organic solutes that influence biotic distributions within lakes and their growth and productivity. Temporal and spatial patterns of seepage vary greatly in lake sediments because of heterogeneous sediment composition (e.g. Lee, 1977; Shaw and Prepas, 1990), transpiration from littoral and nearshore wetland vegetation (see 'plants as controls on lake hydrological processes', below), and non-uniform bending of groundwater flow paths where a sloping water table from surrounding terrestrial deposits meets the flat lake surface (Winter, 1995). Because energy storage and losses are related to mass rather than volume, additional measures of density

---

* Whether groundwater flows to or from a lake often depends on transpiration from littoral vegetation and its effect on lakeshore water tables (see 'plants as controls on lake hydrological processes', below).

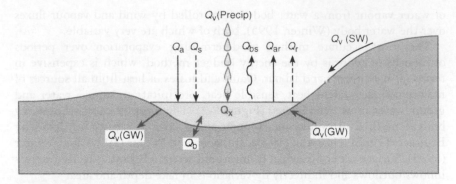

Figure 8.2 Energy sources and sinks associated with the energy budget of lakes. Q_s is incoming short-wave solar radiation, Q_r reflected short-wave radiation, Q_a incoming long-wave terrestrial radiation, Q_{ar} reflected long-wave radiation, Q_{bs} long-wave radiation emitted from the water, Q_b net energy advected to and from the sediments, Q_v net energy advected to and from the lake, and Q_x change in heat stored in the lake (modified from Winter, 1995).

are required for snow and ice. Means of measurement and estimation are summarised in Wetzel and Likens (1991). Once the densities of the snow and ice types have been determined, then latent heat effects on storage and water fluxes, especially for evaporation, condensation and sublimation, can be estimated. Energy budgets must be corrected for latent heat of fusion of ice, particularly when ice cover is considerable (e.g. Adams and Lasenby, 1978; Wetzel, 1983).

Changes in lake volume determined by continuous stage height measurements of the lake surface and detailed bathymetry integrate all fluxes to and from the water body. These changes can be used to verify other measures of water balance.

Measurements of changes in the level (L) or in the area (A_o) have been substituted for determinations of changes in volume, particularly for shallow closed lakes (those without outlet). A_o can increase monotonically as a function of L, from which estimates of rates of change of volume can be made (Street, 1980; Mason *et al.*, 1994). These parameters respond to differences in basin geometry and changes in climate in predictable ways, particularly in closed lakes, and to runoff variations in open lakes. For large (>100 km²) closed lakes, satellite remote sensing of lake levels and areas allows monitoring of variations in aridity and average basin precipitation on a timescale of years to decades.

NUTRIENT SOURCES AND THEIR DISTRIBUTION IN LAKES

Evaluation of the biogeochemistry of a lake ecosystem requires an understanding of the hydrological inputs and losses. Nutrient inputs and losses occur by geological, meteorological and biological vectors (Likens *et al.*, 1977; Wetzel, 1983; Likens, 1984). Nutrients without a significant gaseous phase (e.g. Ca, Mg, K) are loaded primarily in drainage waters and in precipitation, and are lost largely in sediments within the lake and in outflow waters. Nutrients with a gaseous phase (e.g. C, N, H, O, S) enter the lake ecosystem as gases, as well as in solution with surface and groundwater influents and, to a generally small extent, with immigrating biota. Losses are large via gaseous escape to the atmosphere, water outflows (in solution in surface or groundwater), sedimentation and long-term storage in the sediments, and, to a limited extent, by emigrating biota.

Nutrients are constantly exchanged or cycled within the lake ecosystem in many forms: dissolved, adsorbed, absorbed, complexed, ingested, fixed, exchanged, excreted, exuded, leached, precipitated, oxidised, reduced, among other processes (Likens, 1985). The cycling or exchange processes among inorganic nutrients, nutrients incorporated into living and dead organic matter, and minerals are largely regulated by biotic metabolism. Distribution of nutrients in dissolved and particulate forms, and as a result biotic metabolism, is influenced by water movements within the lake basin.

Lakes of modest depth (> *c.* 6 m) commonly stratify as a result of thermal or salinity differences in density, often on a distinct seasonal basis (Figure 8.3). Because diffusion of oxygen from the atmosphere into and within water is a relatively slow process, turbulent mixing of water is required for dissolved oxygen to be distributed in equilibrium with that of the atmosphere. The subsequent distribution of oxygen and dissolved nutrients in the water of thermally stratified lakes is controlled by a combination of solubility conditions, hydrodynamics, inputs from photosynthesis, and losses to chemical and metabolic oxidations (Wetzel, 1983). In unproductive oligotrophic lakes that stratify thermally in the summer, the oxygen content of the upper stratum (epilimnion) decreases as the water temperature increases. The oxygen content of the lower, more dense stratum (hypolimnion) is higher than that of the epilimnion because the oxygen-saturated colder water from the spring period of complete mixing (turnover) is exposed to limited oxidative consumption. The loading of organic matter to the hypolimnion and sediments of productive eutrophic lakes increases the consumption of dissolved oxygen. As a result, the oxygen content of the hypolimnion is reduced progressively during the period of summer stratification. The rate of dissolved oxygen consumption is usually greatest in the deepest portion of the basin, where water strata contain smaller volumes of water and receive a focusing concentration of sedimenting particulate organic matter undergoing decomposition.

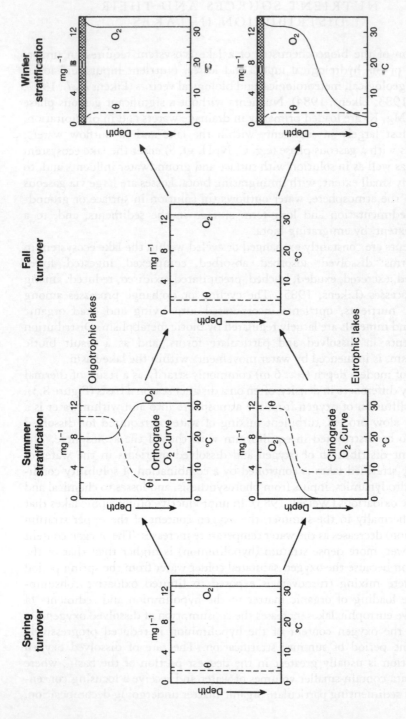

Figure 8.3 Idealised vertical distribution of oxygen concentrations and temperature (θ) during the four main seasonal phases of an oligotrophic (upper) and a eutrophic (lower) dimictic lake that circulates completely twice annually (from Wetzel, 1983). Cross-hatching denotes ice.

Oxygen saturation at existing water temperatures usually returns to the entire water column during autumnal circulation (fall turnover). Under ice cover, the oxygen content and saturation levels are reduced at lower depths in productive lakes, but not to the extent observed during summer stratification, because of prevailing lower water temperatures (greater solubility; reduced respiration).

Incoming river water or groundwater to a lake or reservoir flows into water strata of density similar to its own. Inflow water movements are dependent on density differences due to differences in temperature and differences in concentrations of dissolved materials and particulate matter. If density differences are moderately large, inflows from river waters or groundwaters do not mix readily and flow occurs in complex overflow and interflow patterns. Discharge varies widely seasonally, not only in volume but also in the accompanying load of dissolved and suspended materials. The extent of intrusion and current generation in the receiving lake is also a function of discharge volume of the river in relation to the volume of the lake or reservoir.

The theoretical retention time of a lake, given by total influents minus outflows divided by the total volume of the lake, is realised only approximately in most lakes. Retention time varies with the dimensions and shape of the basin, seasonal rates of inflow, and stratification characteristics (Kajosaari, 1966). At high discharge rates, overflow and interflow of rivers may channel across or through the water mass of the stratified basin, whereas at lower discharge rates, river water may penetrate more into the main water mass and be mixed through more normal circulatory mechanisms. In larger lakes, geostrophic deflection of incoming water, whether as surface flows or interflows, is consistently observed. Subaqueous spring or seepage waters flow similarly along density gradients, often through sediments and in littoral areas, where the nutrient concentrations are often much greater than in the overlying waters. Comparative analyses of the hydrodynamics of river, reservoir and natural lake ecosystems emphasise the transition from rapid, gravity-mediated, unidirectional (downstream) and horizontal flushing in rivers to slow, wind-driven, relatively constant, three-dimensional movements in lakes (Wetzel, in press; see also the discussion of three-dimensional water exchanges in river systems in Chapter 7).

Water movements in lakes are often caused principally by wind, which transfers energy to the water. Rhythmic motions (oscillations) due to wind result both at the surface of the water and internally deep within the basin (Wetzel, 1983; Imberger and Patterson, 1990). These motions and their attendant currents may be in phase or in opposition. The ultimate fate of these movements is to degrade into arrhythmic turbulent motions, which disperse the water as well as chemicals and organisms within it. These wind-mediated water movements are exceedingly important for nutrient cycling and biotic metabolism within lakes. Although properties such as density of planktonic organisms or concentrations of dissolved organic matter could modify thermal conditions and influence evaporation, stratification and related mixing processes, the organisms of the open-water pelagic zone of lakes are of relatively minor consequence to the hydrology of lake basins *per se*.

The stratification and hydrological mixing patterns of course affect the distribution and availability of nutrients, which in turn affect the levels of biotic productivity of the lake ecosystems. Some of the couplings of nutrient distributions are direct, such as importation to the surface waters of the photic zone with alteration of conditions essential for photosynthetic production of organic matter. Other effects are indirect, such as alterations of oxygen concentrations in stratified waters, which can result in marked changes in nutrient availability to biota as redox conditions fluctuate. Space here does not allow development of these relationships; interested readers are referred to detailed syntheses such as Hutchinson (1957) and Wetzel (1983).

Wind-mediated water movements can influence basin morphology and the sediments needed for colonisation of floating-leaved and emergent macrophytes. Lundqvist (1927) separated lakes into 'littoral' versus 'profundal' based on their geomorphology, rates of sediment formation and sites for colonisation by littoral macrophytes. For example, the colonisation and development of macrophytes is markedly greater within protected littoral regions of lakes in the leeward side of prevailing wind-induced wave turbulence in comparison with development along opposing wave-swept regions of lakes (e.g. Gessner, 1959; Rich *et al.*, 1971). Encroaching wetland vegetation, particularly emergent macrophytes, in littoral areas of lakes severely reduces water movement from wave-induced currents. As a result, coarse and fine detrital particles of organic matter accumulate above and below the lake boundary. This rapid accrual of coarse organic detritus results in compacted deposits, anaerobic conditions, and rates of decomposition that are slower than rates of deposition. These deposits allow rapid colonisation and expansion of emergent macrophytes in a lakeward direction in an accelerating manner over periods of years (Wetzel, 1979; Godshalk and Wetzel, 1984).

In littoral areas affected by prevailing wave action, colonisation and establishment of macrophytes is reduced. Detrital particles from dead plants are rendered smaller during relatively rapid decay under conducive conditions in sediments of moderate temperature and high dissolved oxygen (Godshalk and Wetzel, 1978). Surficial materials that are deposited in these regions are frequently resuspended by movements of surface waters, and ultimately transported to the central depression of the lake. Deeper littoral areas (e.g. <1–2 m) are often colonised by submerged vascular plants and are less subject to surface-water currents, which tend to remove residues. Sediment accretion continues until decreasing depths of *c.* 1 m from the surface are approached in most small lakes. At that point, the water movements of wave action can effectively displace and transport sediment deposits to the central basin. In some lakes, submerged macrophytes growing in shallow areas can accelerate sediment accretion. Where basin geomorphology has formed irregular raised lake basin areas within the photic zone that can support submerged macrophytes, sedimentation can accelerate in mounded areas ('lake mounts'). Sediments in exposed areas, however, accrete only to within approximately 1 m of the water surface, after which surface water movements resuspend sediments and disperse them laterally to deeper areas. In a lake in northeast Indiana, USA, for example, no appreciable sediment accumulation occurred on exposed lake-mount littoral areas for over two millennia (Wetzel, 1970). Therefore, water movements in lakes cause various zones of the lake bottom to be subjected to erosional and depositional processes. Water movements and gravitational force act in concert to cause a net movement of fine suspended detrital particles to the deepest portion of the lake. Such heterogeneity can affect the rates of biotically mediated evapotranspiration and water losses (see 'plants as controls on lake hydrological processes', below).

Development of emergent and floating-leaved rooted macrophytes is further enhanced by the presence of nutrients in the sediments. The effect of sediments on nutrient accumulation and availability can be illustrated by comparing two small post-glacial kettle lakes, one a soft water lake with sparse macro-vegetation and the other a well-vegetated hard water lake (Moeller and Wetzel, 1988). The stratigraphy of organic C:P ratios was examined for evidence of progressive P depletion within the sediment. In the soft water lake (Mirror Lake, New Hampshire, USA), sediment accumulated only in the profundal zone, but in a calcareous hard water lake (Lawrence Lake, Michigan, USA), the distinct littoral deposit accounted for 25 percent of the sedimentary P, 38 percent of the organic carbon and 51 percent of the carbonate C. Molar ratios of organic C:P were higher in the littoral sediment (c. 700) than in the profundal sediment (c. 350). The ratio increased linearly with depth into the littoral sediment, which suggested that the littoral sediment includes a surficial store of slowly remineralised P that sustains the rooted macro-vegetation.

These two lakes were compared with respect to estimated phosphorus loading, and productivity of phytoplankton, periphyton and macrophytes (Table 8.3). P loading at Lawrence was two to three times that at Mirror Lake. The four- to five-fold greater productivity in Lawrence Lake reflected mainly littoral metabolism, since phytoplankton was only one to two times more productive than in Mirror Lake. Phosphorus retention as sediment was higher in Mirror (83 percent) than in Lawrence Lake (60 percent); in part this reflected more rapid hydraulic flushing in Lawrence Lake (0.66 year versus 1.06 year, respectively). Total microfloral production in Lawrence was 2.6 times that in Mirror, and proportional to P loading. Because aquatic macrophytes obtain nearly all of their nutrients from the sediments (Wetzel, 1983; Moeller et al., 1988), macrophyte utilisation of sediment P does not compete with its availability to algae. By providing physical and some nutritional support to epiphytes, macrophytes may subsidise periphyton at the expense

Table 8.3 Comparison of the productivity and loading characteristics of Lawrence Lake, Michigan, and Mirror Lake, New Hampshire (modified from Moeller and Wetzel 1988).

		Lawrence Lake		*Mirror Lake*
Limnological parameters[a]				
A_L	Lake area (10^4 m²)	5.0		15.0
A_C	Catchment area (10^4 m²)	>35		103
Z	Maximum depth (m)	12.6		11
V	Volume (10^3 m³)	290		862
O	Outflow (10^3 m³ year⁻¹)	437		813
[P]	P in outflow (mg m⁻³)	12		3
[Ch]	Chlorophyll *a* (mg m⁻³)	1–2		1–2
Primary and bacterial production[b] ($g\ C\ m^{-2}\ year^{-1}$)				
P_P	Phytoplankton	51		37
P_B	Bacterioplankton	68		2
P_A	Attached algae	40		2
P_M	Macrophytes	93		1.5
PE	(Allochthonous)	(25)		(17)
	Total	252	(277)	42 (59)

		Littoral	+ Profundal	= Total	Total
Sediment accumulation ($g\ m^{-2}\ year^{-1}$)					
S_C	Carbon – organic	10.1	16.5	26.6	12.8
S_N	Nitrogen	0.95	1.59	2.54	0.93
S_P	Phosphorus	0.038	0.112	0.15	0.10
S_M	Carbon – CO_3	42	40	82	0
Phosphorus loading ($g\ m^{-2}\ year^{-1}$)					
L_I	Sum of Inputs[c]				0.09
L_S	$S_P + (O[P]/A_L)$			0.25	0.12

[a] Hydrology from Wetzel and Otsuki (1974), Winter (1985); epilimnial chlorophyll.
[b] Lawrence data from Wetzel (1983: p. 699) and Coveney and Wetzel (1995), Mirror data from Jordan *et al.* (1985).
[c] From Likens *et al.* (1985: p. 153), excludes estimated leaching of septic fields.

of the phytoplankton, which compete for the same limiting element, phosphorus (Moeller *et al.*, 1998). The relatively high production of macrophytes in Lawrence represents an independent allocation of limiting P that has no negative effects on total algal production but can favour periphytic microbial production at the cost of the phytoplankton.

As rain and snowmelt water moves over and through the rock and soils of drainage basins, the chemical composition of these dilute solutions is modified as they acquire substances by dissolution or as they lose substances (e.g. through adsorption). Inputs contain nutrients from the surrounding terrestrial–forest–stream ecosystems in response to differences (depending on site geology, season, diurnal cycles, etc.) in processes such as weathering, chemical exchange rates, organic biomass accumulation, and evapotranspiration within the ecosystem.

Several examples exist of the influence of input sources on nutrient budgets in streams, and these are summarised in Table 8.4. From the table it can be

Table 8.4 Nitrogen mass balances of five streams in percentage of total in each category.[a]

	Watershed 10, Oregon	Beaver Creek Riffle, Quebec	Sycamore Creek, Arizona	Bear Brook, New Hampshire	Mare's Egg Spring, Oregon
Inputs (% of total)					
Dissolved inorganic N	3	15	16–58	73	(24)
Dissolved organic N	69	67	37–69	11	(24)
Particulate organic N	0	18	9–15	2	43.8
Precipitation and throughfall	2	0.02	0	3	4.2
Litter	19	0.12	–	11	0.2
N₂ fixation	5	0.004	?	?	3.8
Pools retained N (% of total)					
Fine particulate inorganic nitrogen	40	19	0	–	–
Large particulate organic nitrogen	59	80	0	–	–
Producers	0.6	?	86–93	–	–
Consumers	0.2	?	6–14	–	–
Outputs (% of total)					
Dissolved inorganic N	4	15	18–58	84	(21)
Dissolved organic N	74	67	37–72	12	(21)
Particulate organic N	23	18	7–22	3.7	57[b]
Coarse PON	8	0.1	0.1–6	3	
Fine PON	15	18	7–16	0.7	
Emergence	0.2	0.1	0.1–1	?	0.001
Denitrification	–	–	–	–	0.906

[a] Data extracted from Meyer et al. (1981), Naiman and Melillo (1984), Triska et al. (1984), Dodds and Castenholz (1988), and Grimm (1987).
[b] Includes particulates (16%) and sedimented burial of PON (41%).

seen that nitrogen storage was largely associated with woody debris (59–80 percent) in forested streams (Watershed 10, Oregon and Beaver Creek, Quebec), whereas in an open desert stream (Sycamore Creek, Arizona) as much as 93 percent was in algae and autochthonous detritus. Some nutrients are stored in consumer animals, although turnover rates in these organisms are slow. Dissolved organic nitrogen exports are less than inputs and indicate biological utilisation associated with mineralisation, and potentially some sorption and flocculation processes (Triska *et al.*, 1984). Most particulate organic inputs increased in nitrogen concentrations prior to export to lakes.

The nitrogen budgets of lakes vary greatly. For example, in Lake Mendota, Wisconsin, USA, roughly equivalent contributions from runoff, groundwater and precipitation are balanced by major losses of nitrogen via sedimentation, denitrification and outflow (Keeney, 1972). Loss by seepage out of the basin is probably small, since most lake basins are well sealed (*cf.* Wetzel and Otsuki, 1974). Mirror Lake, a kettle lake located in a crystalline granitic bedrock region of the White Mountains of central New Hampshire, receives appreciable (half) nitrogen loading from atmospheric sources (Table 8.5). Primary outputs were via permanent sedimentation and river outflow in approximately equal quantities of inorganic and organic forms.

Phosphorus loadings to lakes also vary greatly with patterns of land use, geology and morphology of the drainage basin, soil productivity, and in relation to contamination of surface and groundwaters from human activities. In a large, undisturbed arctic lake, loadings of phosphorus are nearly entirely from stream inputs, largely as dissolved P (Table 8.6). Over two-thirds of the phosphorus leaves the lake in surface outflows, also largely as dissolved phosphorus. Numerous mass balance models have been developed to predict, on the basis of phosphorus loading and retention times, the anticipated responses of algal biomass and productivity. The models give reasonably accurate estimations of permissible phosphorus loading needed to achieve a certain level of productivity, if loading is lowered. In certain shallow lakes with greater than average turbulence, large littoral areas and small anaerobic hypolimnia, reduced productivity does not always occur as rapidly in response to decreased loadings as predicted from the models (Wetzel 1983). In these lakes, phosphorus release from sediment stores ('internal loading') is much greater than values (10 to 30 percent of total loading) common to deeper, more stratified lakes.

BASIN MORPHOLOGY AND AQUATIC PLANT COMMUNITIES

Variation in basin morphology is great because of a diversity of geological and geomorphological origins (Hutchinson, 1957; Håkanson, 1981; Wetzel, 1983). Most lake basins, however, originated from geological processes, and

Table 8.5 Average annual nitrogen budget for Mirror Lake, New Hampshire, 1970–1975.[a]

	kg year^{-1} ±SE
Inputs	
Precipitation	
Inorganic	112 ± 10.7
Organic	7
Litter	13.6
Dry deposition	~15
Fluvial	
Dissolved	
Inorganic	35 ± 7.2
Organic	46
Particulate, organic	9
Total	238
Outputs	
Gaseous flux	?
Fluvial	
Dissolved	
Inorganic	61 ± 8.8
Organic	54
Particulate, organic	11
Net insect emergence	13
Permanent sedimentation	127–139
Total	266–278
Change in lake storage	–6
(decrease in NO_3^- storage)	

[a] Modified from Likens (1985).

particularly from glacial processes (see below). These processes have resulted in predominantly shallow lake basins with a morphology that approximates a semi-ellipse (Neumann, 1959; Håkanson and Jansson, 1983; Wetzel, 1983). The mean depth of an average lake is slightly less than one-half (0.46) of its maximum depth. The result is that, with modest depth, most lake basins would have a large proportion of their basin in shallow waters within which light can reach the sediments and allow colonisation by photosynthetic biota.

As noted above, many lake basins have been formed by glacial activity. These glacial processes have resulted in predominantly shallow lake basins (Hutchinson, 1957). Consequently, most lakes and other standing freshwater bodies are quite shallow, with mean depths much less than 10 m (see Figure 8.1). In immense areas of the Northern Hemisphere, literally millions of lakes of considerable area exist with mean depths of less than 3 m (Wetzel, 1990;

Table 8.6 Phosphorus budget for a deep, arctic tundra lake, Toolik Lake, Alaska.[a]

| | $mmol\ m^{-2}\ year^{-1}$ | | |
	Fractional P	Total P	% of total
Inputs:			
Stream inflows			
Dissolved P	3.28		70.7
Particulate P	1.12	4.40	24.1
			94.8
Direct precipitation			
Dissolved P	0.09		1.9
Particulate P	0.15	0.24	3.3
			5.2
Total inputs			
Dissolved P	3.37		72.6
Particulate P	1.27	4.64	27.4
			100.0
Exports:			
Sedimentation from water column	1.4–1.7	1.4–1.7	30.1–36.6
Stream outflow			
Dissolved P	2.02		42.1
Particulate P	1.23	3.25	25.6
			67.7
Total outputs			
Dissolved P	2.02		42.1
Particulate P	2.63–2.93	4.65–4.95	56.6–59.2
			100.0

[a] Numbers rounded; from data cited in Whalen and Cornwell (1985).

Meybeck, 1995). As one progresses northwards, the mean depth decreases progressively to the point where in tundra regions mean depths are frequently less than 1 m (Hobbie, 1980).

Most lake basins are destined to fill by sedimentation because decomposition of organic matter and exports as gases by escape to the atmosphere or as soluble compounds with outflows are nearly always less than allochthonous loadings of organic matter and clastic (minerogenic) sediments and autochthonous loadings of organic matter. The rates of these loadings and sedimentation within the basin are dependent upon (1) the nutrient loadings from the catchment and (2) the initial formation and long-term modifications to the basin morphometry by sedimentation. As sedimentation increases, a greater proportion of the basin sediments *per se* are brought into the photic zone and are available for colonisation by the extremely productive littoral and wetland communities discussed above. Greater productivity of small, shallow lakes is directly correlated with higher water–sediment interface area per water volume (i.e. lower mean depth).

Sedimentation rates are directly coupled with littoral and wetland productivity. Because phytoplanktonic productivity is so low, as discussed earlier, autochthonous productivity by littoral and wetland aquatic plants provides a major source of organic matter to lakes. On a global basis, the littoral zone dominates over the pelagic (giving low P:L ratios among most standing water ecosystems (Figure 8.4). Additions from the wetland

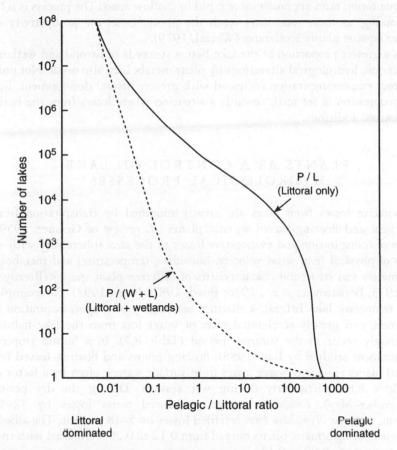

Figure 8.4 Estimated number of lakes in the world in relation to the ratios of pelagic (P) to littoral (L) and wetland (W) portions of aquatic ecosystems (after Wetzel, 1990). Printed with permission of the International Association of Theoretical and Applied Limnology.

components to the littoral zone result in low ratios (P/[W + L]). A similar very low ratio is apparent among rivers, because the floodplains are functioning as major wetland and littoral components of most river ecosystems.

Most of the particulate organic matter produced in the wetland–littoral complex of lake ecosystems is deposited within the land–water interface regions (Wetzel, 1983; 1990; 1992). Decomposition is incomplete because of anaerobic chemical and metabolic conditions, which reduce the availability of electron-acceptor compounds and enhance the production of fermentative metabolic products. The structural plant tissues of most productive higher aquatic plants are incompletely decomposed and result in an accumulation of relatively recalcitrant organic matter. Accumulation rates of sediments from higher plants is much greater than from planktonic sources. As a result, sedimentation rates are much more rapid in shallow areas. The process is self-reinforcing: as more sediments reach the photic zone, the productivity of higher aquatic plants accelerates (Wetzel, 1979).

As a greater proportion of the lake basins succeeds to littoral and wetland vegetation, hydrological alterations by plant metabolism also occur. Not only is direct evapotranspiration increased with greater littoral development, but the progression is set forth towards a situation where losses from the basin will exceed additions.

PLANTS AS A CONTROL ON LAKE HYDROLOGICAL PROCESSES

Evaporative losses from lakes are greatly modified by transpiration from emergent and floating-leaved aquatic plants (*cf.* review of Gessner, 1959). Rates of transpiration and evaporative losses to the atmosphere vary with an array of physical (e.g. wind velocity, humidity, temperature) and metabolic parameters and structural characteristics of different plant species (Brezny *et al.*, 1973; Bernatowicz *et al.*, 1976; Boyd, 1987; Jones, 1992). For example, in a temperate lake littoral, a distinct seasonality of evapotranspiration is common, and greatly accelerated rates of water loss from the lake habitat commonly occur in the summer period (Table 8.7). In a humid tropical environment studied by Rao (1988), floating plants and floating-leaved but rooted plants enhanced water losses from surface waters, often by a factor of 2 (Table 8.8), particularly during wet seasons. During the dry period (November–May), *Eichhornia crassipes* enhanced water losses by 32–51 percent, whereas *Nymphaea lotus* retarded losses by 5–18 percent. The albedo of these floating aquatic plants ranged from 0.12 to 0.20, compared with that of open water at 0.09 to 0.13.

Evapotranspiration rates of aquatic plants generally increase with increasing wind velocity up to modest speeds ($< c.$ 2 m s^{-1}) and with decreasing relative humidity (Gessner, 1959; Rao, 1988). Evapotranspiration

Table 8.7 Comparison of water loss from a stand of the emergent aquatic macrophyte *Phragmites australis* (= *communis*) (evapotranspiration) with that from open water (evaporation), Berlin, Germany, 1950.[a]

Date	Evapotranspiration ($kg\ m^{-2}\ day^{-1}$)[b]	Evaporation ($kg\ m^{-2}\ day^{-1}$)[b]	Ratio of evapotranspiration to evaporation
11th May	3.20	3.24	1.0
25th May	2.50	1.44	1.6
27th July	9.82	2.24	4.4
22nd August	16.01	2.29	7.0
17th October	2.79	0.79	3.9

[a] Modified from Gessner (1959).
[b] = mm day^{-1} or × 0.1 = g cm^{-2} day^{-1}.

rates are generally positively correlated with temperatures and solar irradiance on a seasonal basis. Evapotranspiration rates increase with increasing rates of photosynthesis up to a maximum rate that differs between species of aquatic plants. For example, the average photosynthetic rate of the emergent freshwater rush *Juncus effusus* L. increases with greater irradiance until *c.* 1000 μmol m^{-2} s^{-1} and then declines with further photo-inhibitory irradiance as photosynthetic efficiencies decline below 0.5 percent (Figure 8.5). Photosynthesis of this species is maximum at temperatures between 20 and 30 °C and declines precipitously at higher temperatures (Figure 8.5) as respiration rates increase markedly. As a result, stomatal conductance declines and evapotranspiration rates are reduced during midday periods of intense irradiance and maximum diurnal temperatures (e.g. Giurgevich and Dunn, 1982; Jones and Muthuri, 1984; Mann and Wetzel, submitted). These midday depressions in evapotranspiration suppress overall water losses to a greater extent than is counterbalanced by increased rates of evaporation from open water surfaces with the higher temperatures and lower relative humidity values that occur at midday.

Factors that suppress growth of aquatic plants will also reduce evapotranspiration. For example, growth and rates of photosynthesis of the common cattail (*Typha domingensis* Pers.) are markedly reduced by modest increases in the salinity of the water (Glenn *et al.*, 1995). Evapotranspiration (E_t) of *Typha* was 1.3 times pan evaporation (E_o) when the salinity of inflow water was 1.0‰, and E_t/E_o declined to 0.7 when the salinity was 3.2‰.

Plant growth and evapotranspiration are predominantly seasonal in lakes and river wetlands of exorheic regions. In subtemperate and tropical lakes, floodplains of rivers, and wetlands, however, many of the large perennial hydrophytes grow more or less continuously. In most situations, transport of water from the lake or river to the air is greatly increased by a dense stand of actively growing littoral vegetation, as compared with evaporation rates from

Table 8.8 Representative rates of evapotranspiration (E_t) by aquatic plants and comparison with rates of evaporation from open water (E_o).

Species	mm day^{-1}	E_t/E_o	Reference
Emergent:			
Typha domingensis Pers.	2.7–4.7	1.3	Glenn *et al.* (1995); Abtew (1996)
Typha latifolia L.	4–12	1.41 – 1.84	Snyder and Boyd (1987)
	4.8	3.7–12.5	Price (1994)
Carex lurida Wallenb.	4.0–6.3	1.33	Boyd (1987)
Panicum regidulum Nees	5.5–7.5	1.58	Boyd (1987)
Rice (*Oryza sativa* L.)	6–13		Humphreys *et al.* (1994)
Myriophyllum aquaticum (Vellozo) Verdcourt	0.2–1.0		Sytsma and Anderson (1993)
Juncus effusus L.	3.8–8.0	1.52	Boyd (1987)
Justicia americana (L.) Vahl.	2.2–6.4	1.17	Boyd (1987)
Alternanthera philoxeroides (Mart.) Grieseb.	4.0–6.3	1.26	Boyd (1987)
Willow carr (*Salix* spp.), Czech Republic	2.3–3.7		Pribán and Ondok (1986)
Sedge–grass marsh (*Carex*, *Calamagrostis*, *Glyceria*), Czech Republic	2.2–4.5		Pribán and Ondok (1985; 1986)
Lakeshore marsh (*Sagittaria*, *Pontederia*, *Panicum*, *Hibiscus* dominating), Florida	0.5–1.0	0.35–1.2	Dolan *et al.* (1984)
Carex-dominated marsh, Ontario subarctic	2.6–3.1	0.74–1.02	Lafleur (1990)
Floodplain forest, Florida	5.57		Brown (1981)
Reed (*Phragmites*) swamp, Czech Republic	1.4–6.9	1.03	Šmid (1975)
Arctic wetland, Canada	4.5 (2.2–7.3)		Roulet and Woo (1988)
Quaking Fen, Netherlands			
Typha latifolia L.	0.9–4.7	1.87	Koerselman and
Carex diandra Schrank	1.1–3.9	1.68	Beltman (1988)
Carex acutiformis/Sphagnum	1.0–3.7	1.65	
Floating-leaved rooted:			
Nymphaea lotus (L.) Willd.	2.5–6.0	0.82–1.35	Rao (1988)
Floating not rooted:			
Eichhornia crassipes (Mart.) Solms.	3.8–10.5	1.30–1.96	Rao (1988)
	6–11	1.45–2.02	Snyder and Boyd (1987)
		2.67	Lallana *et al.* (1987)
Salvinia molesta D.S. Mitchell	2.1–6.8	0.96–1.39	Rao (1988)
Pistia stratiotes L.	19.9	1.07	Brezny *et al.* (1973)
Azolla caroliniana Willd.	7.1	0.95	Lallana *et al.* (1987)
Lemna spp.		1.03	Brown (1981)

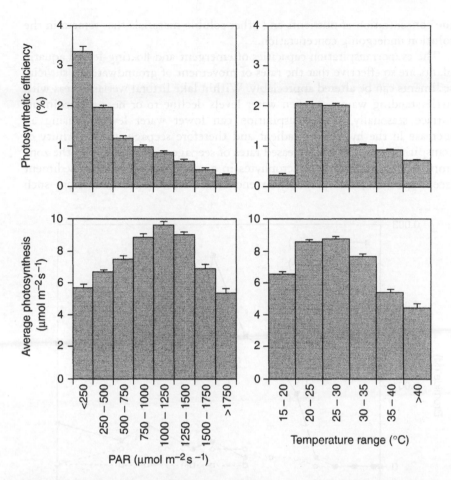

Figure 8.5 Annual average rates of photosynthesis and photosynthetic efficiencies of light utilisation by the emergent aquatic plant *Juncus effusus* under conditions of natural variations in irradiance (photosynthetically active radiation, PAR, 400–700 nm) and temperature in the Talladega Wetland Ecosystem, Hale County, Alabama, USA. Measurements were made *in situ* at biweekly intervals throughout the year; $n \geq 2500$. Bars = ± S.E. (from C. J. Mann and R. G. Wetzel, unpublished data).

open water (see Tables 8.7 and 8.8). Vegetation cover of sedge-dominated communities reduces the evaporating efficiency of the wet wetland sites and slightly increases evaporation efficiency of wetland sites in which the hydrosoils are not covered with standing water (Lafleur, 1990). Because most lakes are small and often possess a well-developed littoral and wetland flora, these communities contribute significantly to the water balance of many surface waters. Such evapotranspirative water loss can result in an increase in the concentration of nutrients and other soluble materials that remain in the solution undergoing concentration.

The evapotranspiration capacities of emergent and floating-leaved aquatic plants are so effective that the rates of movement of groundwater in surficial sediments can be altered appreciably. Within lake littoral wetland areas with little standing water, or when water levels decline to or near the sediment surface seasonally, evapotranspiration can lower water levels, causing an increase in the hydraulic gradient and therefore seepage in the vicinity of transpiring plants. Such increased rates of seepage were observed in the zone immediately adjacent to macrophytes in a study on a sandy clay sediment (see Figure 8.6) but probably extend for several metres away from such

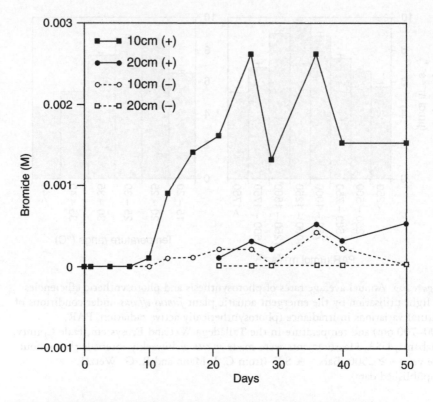

Figure 8.6 Variations in the transport of water laterally, as indicated by movements of a conservative bromide tracer, within sandy clay sediments of the Talladega Wetland Ecosystem, Hale County, Alabama. (–) = without macrophytes; (+) = with clumps of the emergent rush *Juncus effusus* L. at one end of the sediment transect. Suction sampler sampling depths of 10 and 20 cm, respectively. The root/rhizome bolus of *Juncus* extended from 0 to 10 cm into the sediments. Rates of tracer movement relate to a distance of 20 cm from the tracer injection site towards the macrophyte roots.

plants in many wetland environments. Where water overlying sediments is relatively shallow (e.g. several centimetres), as is common in littoral areas, the evapotranspiration losses result in a small but distinct diurnal periodicity of water levels (Figure 8.7). Reductions in level occur during the daylight period and recovery by seepage and shallow surface flow occurs during darkness.

Where appreciable depths of water overlie littoral sediments, sources of evapotranspirational water are largely from the overlying water of the immediate area. As noted by Gessner (1959) long ago, evapotranspiration losses from dense stands of emergent vegetation can be so large that these littoral regions function effectively as a major lake outflow. The evapo-transpiration of aquatic vegetation can cause water levels in littoral zones to fall below lake levels, resulting in accelerated seepage from the lake (*cf.* Meyboom 1966; Mann and Wetzel, submitted). During the growing season, evapotranspiration by plants causes a cone of depression such that water will seep to the cone from the lake. When plants are dormant, rainfall or larger-scale groundwater flows can raise lakeside water tables above the lake level, causing water to flow to the lake.

Under reduced summer water levels, evapotranspiration can dominate water losses and increase proportionally in relation to other losses in lakes of small size and mean depth. Measurements with piezometers within the lake bed (e.g. Lee, 1977), and closely spaced wells near lake boundaries indicate that flow reversals occur and are related to transpiration in a wide variety of climatic and geological settings (Winter, 1995). In addition to being influenced by changes in photosynthetic activity of the vegetation, these flows

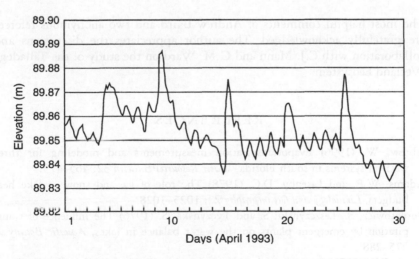

Figure 8.7 Diurnal fluctuations in surface water levels within the Talladega Wetland Ecosystem, Hale County, Alabama.

from the lake caused by transpiration from nearshore vegetation can be reversed quickly during or following periods of heavy rainfall.

CONCLUSIONS

The physical and biotic structure and resulting metabolism of a lake ecosystem are strongly coupled to the hydrology and resulting chemical loadings from its drainage basin. Lakes interact with adjacent surficial and shallow groundwater flow systems, which change continuously in response to a combination of local and regional flow systems. Plant evapotranspiration by nearshore littoral and wetland vegetation can cause highly dynamic, bidirectional seepage conditions over various periods of time. These loss processes alter nearshore recharge from shallow groundwater sources as well as hydrosoil water storage and fluctuations in water levels.

The high productivity of emergent littoral and wetland vegetation can rapidly increase rates of sedimentation of partially decomposed organic matter. The resulting decreases in basin volume, expanding habitat and emergent vegetation can result in gradual transitions to conditions where evapotranspiration losses exceed collective water inputs. As a result of biologically mediated water losses, eventual greater losses than inputs set forth a progressively accelerating transition of some shallow lakes to terrestrial ecosystems.

ACKNOWLEDGEMENTS

The most helpful comments of Andrew Baird and two anonymous referees are gratefully acknowledged. The author appreciates the discussions and collaboration with C.J. Mann and G.M. Ward on the study of the Talladega Wetland Ecosystem.

REFERENCES

Abtew, W. (1996) Evapotranspiration measurements and modeling for three wetland systems in south Florida, *Water Resources Bulletin* 32: 465–473.

Adams, W.P. and Lasenby, D.C. (1978) The role of ice and snow in lake heat budgets, *Limnology and Oceanography* 23: 1025–1028.

Bernatowicz, S., Leszczynski, S. and Tyczynska, S. (1976) The influence of transpiration by emergent plants on the water balance in lakes, *Aquatic Botany* 2: 275–288.

Boyd, C.E. (1987) Evapotranspiration/evaporation (E/E_o) ratios for aquatic plants, *Journal of Aquatic Plant Management* 25: 1–3.

294

Brezny, O., Metha, I. and Sharma, R.K. (1973) Studies on the evapotranspiration of some aquatic weeds, *Weed Science* 21: 197–204.

Brown, S. (1981) A comparison of the structure, primary productivity, and transpiration of cypress ecosystems in Florida, *Ecological Monographs* 51: 403–427.

Brutsaert, W.H. (1982) *Evaporation into the Atmosphere*, Dordrecht: Reidel, 299 pp.

Carter, S.M., Ward, G.M., Wetzel, R.G. and Benke, A.C. (in press) Growth, production, and senescence of *Nymphaea odorata* Aiton in a southeastern (U.S.A.) wetland, *Aquatic Botany*.

Coveney, M.F. and Wetzel, R.G. (1995) Biomass, production, and specific growth rate of bacterioplankton and coupling to phytoplankton in an oligotrophic lake, *Limnology and Oceanography* 40: 1187–1200.

Dodds, W.K. and Castenholz, R.W. (1988) The nitrogen budget of an oligotrophic cold water pond, *Archiv für Hydrobiologie/Supplement* 79: 343–362.

Dolan, T.J., Hermann, A.J., Bayley, S.E. and Zoltek, J., Jr (1984) Evapotranspiration of a Florida, U.S.A., freshwater wetland, *Journal of Hydrology* 74: 355–371.

Gessner, F. (1959) *Hydrobotanik. II. Stoffhaushalt*, Berlin: VEB Deutscher Verlag der Wissenschaften, Berlin, 701 pp.

Giurgevich, J.R. and Dunn, E.L. (1982) Seasonal patterns of daily net photosynthesis, transpiration and net primary productivity of *Juncus roemerianus* and *Spartina alterniflora* in a Georgia salt marsh, *Oecologia* 52: 404–410.

Glenn, E., Thompson, T.L., Frye, R., Riley, J. and Baumgartner, D. (1995) Effects of salinity on growth and evapotranspiration of *Typha domingensis* Pers., *Aquatic Botany* 52: 75–91.

Godshalk, G.L. and Wetzel, R.G. (1978) Decomposition of aquatic angiosperms. II. Particulate components, *Aquatic Botany* 5: 301–328.

Godshalk, G.L. and Wetzel, R.G. (1984) Accumulation of sediment organic matter in a hardwater lake with reference to lake ontogeny, *Bulletin of Marine Science* 35: 576–586.

Grimm, N. B. (1987) Nitrogen dynamics during succession in a desert stream, *Ecology* 68: 1157–1170.

Håkanson, L. (1981) *A Manual of Lake Morphometry*, New York: Springer-Verlag, 78 pp.

Håkanson, L. and Jansson, M. (1983) *Principles of Lake Sedimentology*, New York: Springer-Verlag, 316 pp.

Hobbie, J.E. (ed.) (1980) *Limnology of Tundra Ponds*, Stroudsburg: Hutchinson and Ross Inc., 514 pp.

Hostetler, S. W. (1995) Hydrological and thermal response of lakes to climate: description and modeling, in A. Lerman, D.M. Imboden, and J.R. Gat (eds) *Physics and Chemistry of Lakes*, Berlin: Springer-Verlag, pp. 63–82.

Humphreys, E., Meyer, W.S., Prathapar, S.A. and Smith, D.J. (1994) Estimation of evapotranspiration from rice in southern New South Wales: a review, *Australian Journal of Experimental Agriculture* 34: 1069–1078.

Hutchinson, G.E. (1957) *A Treatise on Limnology. I. Geography, Physics, and Chemistry*, New York: Wiley, 1015 pp.

Hutchinson, G.E. (1975) *A Treatise on Limnology. III. Limnological Botany*, New York: Wiley, 660 pp.

Imberger, J. and Patterson, J.C. (1990) Physical limnology, *Advances of Applied Mechanics* 27: 303–475.

Imboden, D.M. and Wüest, A. (1995) Mixing mechanisms in lakes, in A. Lerman, D.M. Imboden, and J.R. Gat (eds) *Physics and Chemistry of Lakes*, Berlin: Springer-Verlag, pp. 83–138.

Jones, F.E. (1992) *Evaporation of Water With Emphasis on Applications and Measurements*, Chelsea, Michigan: Lewis, 188 pp.

Jones, M.B. and Muthuri, F.M. (1984) The diurnal course of plant water potential, stomatal conductance and transpiration in a papyrus (*Cyperus papyrus* L.) canopy, *Oecologia* 63: 252–255.

Jordan, M.J., Likens, G.E. and Peterson, B.J. (1985) Mirror Lake – Biologic considerations. C. Organic carbon budget, in G. E. Likens (ed.) *An Ecosystem Approach to Aquatic Ecology: Mirror Lake and its Environment*, New York: Springer-Verlag, pp. 292–301.

Kajosaari, E. (1966) Estimation of the detention period of a lake, *Verhandlungen Internationale Vereinigung der Limnologie* 16: 139–143.

Keeney, D.R. (1972) *The Fate of Nitrogen in Aquatic Ecosystems. Literature Review 3*, Water Resources Center, University of Wisconsin, 59 pp.

Kilham, S.S. and Kilham, P. (1990) Tropical limnology: do African lakes violate the 'first law' of limnology? *Verhandlungen Internationale Vereinigung der Limnologie* 24: 68–72.

Koerselman, W. and Beltman, B. (1988) Evapotranspiration from fens in relation to Penman's potential free water evaporation (E_o) and pan evaporation, *Aquatic Botany* 31: 307–320.

Lafleur, P.M. (1990) Evapotranspiration from sedge-dominated wetland surfaces, *Aquatic Botany* 37: 341–353.

Lallana, V.H., Sabattini, R.A. and Lallana, M.C. (1987) Evapotranspiration from *Eichhornia crassipes*, *Pistia stratiotes*, *Salvinia herzogii* and *Azolla caroliniana* during summer in Argentina, *Journal of Aquatic Plant Management* 25: 48–50.

Lee, D.R. (1977) A device for measuring seepage flux in lakes and estuaries, *Limnology and Oceanography* 22: 155–163.

Lee, D.R. (1985) Method for locating sediment anomalies in lakebeds that can be caused by groundwater flow, *Journal of Hydrology* 79: 187–193.

Lee, D.R. and Cherry, J.A. (1978) A field exercise on groundwater flow using seepage meters and mini-piezometers, *Journal of Geological Education* 27: 6–10.

Lee, D.R. and Harvey, F.E. (1996) Installing piezometers in deepwater sediments, *Water Resources Research* 32: 1113–1117.

Likens, G.E. (1984) Beyond the shoreline: A watershed–ecosystem approach, *Verhandlungen Internationale Vereinigung der Limnologie* 22: 1–22.

Likens, G.E. (ed.) (1985) *An Ecosystem Approach to Aquatic Ecology: Mirror Lake and its Environment*, New York: Springer-Verlag, 516 pp.

Likens, G.E., Bormann, F.H., Pierce, R.S., Eaton, J.S. and Johnson, N.M. (1977) *Biogeochemistry of a Forested Ecosystem*, New York: Springer-Verlag, 146 pp.

Likens, G.E., Eaton, J.S., Johnson, N.M., and Pierce, R.S. (1985) Flux and balance of water and chemicals, in G. E. Likens (ed.), *An Ecosystem Approach to Aquatic Ecology: Mirror Lake and Its Environment*, New York: Springer-Verlag, pp. 135–155.

Lundqvist, G. (1927) Bodenablagerungen und Entwicklungstypen der Seen, *Die Binnengewässer* 2: 1–122.

Mann, C.J. and Wetzel, R.G. (submitted) Hydrology and subsurface water flows in a wetland ecosystem, *Wetland Ecology and Management*.

Mason, I.M., Guzkowska, M.A.J. and Rapley, C.G. (1994) The response of lake levels and areas to climatic change, *Climatic Change* 27: 161–197.

Meybeck, M. (1995) Global distribution of lakes, in A. Lerman, D.M. Imboden, and J.R. Gat (eds) *Physics and Chemistry of Lakes*, Berlin: Springer-Verlag, pp. 1–35.

Meyboom, P. (1966) Unsteady groundwater flow near a willow ring in hummocky moraine, *Journal of Hydrology* 4: 38–62.

Meyer, J.L., Likens, G.E. and Sloane, J. (1981) Phosphorus, nitrogen, and organic carbon flux in a headwater stream, *Archiv für Hydrobiologie* 91: 28–44.

Moeller, R.E. and Wetzel, R.G. (1988) Littoral vs profundal components of sediment accumulation: contrasting roles as phosphorus sinks, *Verhandlungen Internationale Vereinigung der Limnologie* 23: 386–393.

Moeller, R.G., Burkholder, J.M. and Wetzel, R.G. (1988) Significance of sedimentary phosphorus to a submersed freshwater macrophyte (*Najas flexilis*) and its algal epiphytes, *Aquatic Botany* 32: 261–281.

Moeller, R.E., Wetzel, R.G. and Osenberg, C.W. (1998) Concordance of phosphorus limitation in lakes: bacterioplankton, phytoplankton, epiphyte-snail consumers, and rooted macrophytes, in E. Jeppesen, Ma. Søndergaard, Mo. Søndergaard and K. Christoffersen (eds) *The Structuring Role of Submersed Macrophytes in Lakes*, Berlin: Springer-Verlag, pp. 318–325.

Naiman, R.J. and Melillo, J.M. (1984) Nitrogen budget of a subarctic stream altered by beaver (*Castor canadensis*), *Oecologia* 62: 150–155.

Neumann, J. (1959) Maximum depth and average depth of lakes, *Journal of Fisheries Research Board of Canada* 16: 923–927.

Pribán, K. and J.P. Ondok, (1985) Heat balance components and evapotranspiration from a sedge–grass marsh, *Folia Geobotanica et Phytotaxonomica Praha* 20: 41–56.

Pribán, K. and Ondok, J.P. (1986) Evapotranspiration of a willow carr in summer, *Aquatic Botany* 25: 203–216.

Price, J.S. (1994) Evapotranspiration from a lakeshore *Typha* marsh on Lake Ontario, *Aquatic Botany* 48: 261–272.

Rao, A.S. (1988) Evapotranspiration rates of *Eichhornia crassipes* (Mart.) Solms, *Salvinia molesta* D.S. Mitchell and *Nymphaea lotus* (L.) Willd. Linn. in a humid tropical climate, *Aquatic Botany* 30: 215–222.

Rich, P.H., Wetzel, R.G. and Thuy, N.V. (1971) Distribution, production and role of aquatic macrophytes in a southern Michigan marl lake, *Freshwater Biology* 1: 3–21.

Roulet, N.T. and Woo, M. (1988) Wetland and lake evaporation in the low Arctic, *Arctic and Alpine Research* 18: 195–200.

Sculthorpe, C.D. (1967) *The Biology of Aquatic Vascular Plants*, New York: St Martin's Press, 610 pp.

Serruya, C. and Pollinger, U. (1984) *Lakes of the Warm Belt*, Cambridge: Cambridge University Press, 569 pp.

Shaw, R.D. and Prepas, E.E. (1990) Groundwater–lake interactions. 2. Nearshore seepage patterns and the contribution of groundwater to lakes in central Alberta, *Journal of Hydrology* 119: 121–136.

Šmid, P. (1975) Evaporation from a reedswamp, *Journal of Ecology* 66: 299–309.

Snyder, R.L. and Boyd, C.E. (1987) Evapotranspiration by *Eichhorina crassipes* (Mart.) Solms and *Typha latifolia* L., *Aquatic Botany* 27: 217–227.

Street, F.A. (1980) The relative importance of climate and local hydrogeological factors in influencing lake level fluctuations, *Palaeoecology of Africa* 12: 137–158.

Sturrock, A.M., Winter, T.C. and Rosenberry, D.O. (1992) Energy budget evaporation from Williams Lake: a closed lake in north central Minnesota, *Water Resources Research* 28: 1605–1617.

Sytsma, M.D. and Anderson, L.W.J. (1993) Transpiration by an emergent macrophyte: Source of water and implications for nutrient supply, *Hydrobiologia* 271: 97–108.

Szesztay, K. (1974) Water balance and water level fluctuations of lakes, *Hydrological Sciences Bulletin* 19: 74–84.

Triska, F.J., Sedell, J.R., Cromack, K., Jr, Gregory, S.V. and McCorison, F.M. (1984) Nitrogen budget for a small coniferous forest stream, *Ecological Monographs* 54: 119–140.

Westlake, D.F. (1963) Comparisons of plant productivity, *Biological Reviews* 38: 385–425.

Westlake, D.F. (1965) Some basic data for investigations of the productivity of aquatic macrophytes, *Memorie dell'Istituto Italiano di Idrobiologia* 18 (supplement): 229–248.

Wetzel, R.G. (1965) Nutritional aspects of algal productivity in marl lakes with particular reference to enrichment bioassays and their interpretation, *Memorie dell'Istituto Italiano di Idrobiologia* 18 (supplement): 137–157.

Wetzel, R.G. (1966) Productivity and nutrient relationships in marl lakes of northern Indiana, *Verhandlungen Internationale Vereinigung der Limnologie* 16: 321–332.

Wetzel, R.G. (1970) Recent and post-glacial production rates of a marl lake, *Limnology and Oceanography* 15: 491–503.

Wetzel, R.G. (1979) The role of the littoral zone and detritus in lake metabolism, *Archiv für Hydrobiologie Beiheft Ergebnisse Limnologie* 13: 145–161.

Wetzel, R.G. (1983) *Limnology*, 2nd edition, Philadelphia: Saunders College Publishing, 860 pp.

Wetzel, R.G. (1990) Land–water interfaces: metabolic and limnological regulators, *Verhandlungen Internationale Vereinigung der Limnologie* 24: 6–24.

Wetzel, R.G. (1991) Extracellular enzymatic interactions in aquatic ecosystems: storage, redistribution, and interspecific communication, in R.J. Chróst (ed.) *Microbial Enzymes in Aquatic Environments*, New York: Springer-Verlag, pp. 6–28.

Wetzel, R.G. (1992) Gradient-dominated ecosystems: sources and regulatory functions of dissolved organic matter in freshwater ecosystems, *Hydrobiologia* 229: 181–198.

Wetzel, R.G. (1993) Microcommunities and microgradients: linking nutrient regeneration and high sustained aquatic primary production, *Netherlands Journal of Aquatic Ecology* 27: 3–9.

Wetzel, R.G. (1995) Death, detritus, and energy flow in aquatic ecosystems, *Freshwater Biology* 33: 83–89.

Wetzel. R.G. (in press) *Limnology: Lake, Reservoir, and River Ecosystems*, San Diego: Academic Press.

Wetzel, R.G. and Otsuki, A. (1974) Allochthonous organic carbon of a marl lake, *Archiv für Hydrobiologie* 73: 31–56.

Wetzel, R.G. and Likens, G.E. (1991) *Limnological Analyses*, 2nd edition, New York: Springer-Verlag, 391 pp.

Wetzel, R.G. and Howe, M.J. (in press) Continuous growth in herbaceous perennials: population dynamics, growth, and production of an aquatic rush, *Ecology*.

Whalen, S.C. and Cornwell, J.C. (1985) Nitrogen, phosphorus, and organic carbon cycling in an arctic lake, *Canadian Journal of Fisheries and Aquatic Sciences* 42: 797–808.

Winter, T.C. (1981) Uncertainties in estimating the water balance of lakes, *Water Resources Bulletin* 17: 82–115.

Winter, T.C. (1985) Mirror Lake and its watershed. A. Physiographic setting and geologic origin of Mirror Lake, in G.E. Likens (ed) *An Ecosystem Approach to Aquatic Ecology: Mirror Lake and Its Environment*, New York: Springer-Verlag, pp. 40–53.

Winter, T.C. (1995) Hydrological processes and the water budget of lakes, in A. Lerman, D.M. Imboden and J.R. Gat (eds) *Physics and Chemistry of Lakes*, Berlin: Springer-Verlag, pp. 37–62.

9

MODELLING

Andrew J. Baird

What I'm above all primarily concerned with is the substance
of life, the pith of reality. If I had to sum up my work, I
suppose that's it really: I'm taking the pith out of reality.
Alan Bennett (1966) 'The Lonely Pursuit',
On the Margin, BBC Television, UK

INTRODUCTION

Mathematical models are now widely used in hydrology and ecology and
could be regarded as essential tools for many aspects of both disciplines.
Nevertheless, to the non-modeller modelling is often seen as an arcane
discipline that has little relevance to real scientific problems. It is the business
of mathematicians and should be viewed with suspicion. To some extent this
suspicion is justified. Many models are presented in the academic literature
that have not been tested with field data. Many models seem alarmingly
complex, while others seem worryingly simplistic. In short, many models do
not seem to be *useful*. Hopefully, this chapter will show that modelling has an
important role to play within a science such as eco-hydrology, in particular in
theory testing. Hopefully, this chapter will also reveal, in an unjaundiced way,
some of the worst excesses of modellers while mapping out a clear way in
which we can avoid constructing useless and sometimes over-complex models.
Rather than reviewing specific models of plant–water relations it is better to
try to highlight what the main issues in modelling are; the context of this
chapter is, therefore, general rather than specific. As a result, the chapter is
divided into four main sections. After briefly defining what scientists mean
by models, the following issues are considered in detail: model complexity,
model construction, the relationship between modelling and mathematics,
and scale issues in modelling. Some readers may note that a detailed
discussion on methods of model testing, such as ways of assessing goodness of
fit between model predictions and measurements of the real system, has been
omitted. Model testing can be regarded as a fundamental issue in ecology
and hydrology, especially as a criticism of many modellers is that they do not

300

properly test their models. The reasons for such an omission are two-fold. First, there are some fundamental issues that need to be considered *before* testing a model, such as how simple or complex we should make our models and whether they are properly *testable*, and greater emphasis is given to these. Second, a discussion of goodness-of-fit tests could easily constitute a chapter in its own right. Other good reviews of model testing have been written recently, and these are noted later in this chapter.

WHAT IS A MODEL?

Before we consider the four main issues outlined above, we need to define what we mean by the term 'model'. Unfortunately, 'model' is used loosely to designate a variety of concepts in the arts, social sciences and sciences. The most banal definition of a model is that it is an abstraction or simplification of reality. Such a definition is too broad to be useful. To scientists, 'model' has more specific meanings, although even among scientists there is disagreement on what constitutes a model. Rather than attempt a single (and probably convoluted and unsatisfactory) definition it is perhaps better to define different types of model. Most texts on modelling differentiate between three broad model types.

Conceptual models

Conceptual models are statements or pictures of how we think a particular system behaves. A good example of a simple pictorial model is the familiar hydrological cycle, which shows the principal stores of water and routes of water movement at the global scale (Figure 9.1). Refinements of this conceptual model include the hydrological cascade, which shows routes of water movement through catchments or smaller sub-units of the landscape (Figure 9.2). Strictly, conceptual models, at best, can be only semi-quantitative. For example, line thickness on a flow diagram may be used crudely to differentiate principal from minor flows within a system. However, conceptual models can readily be made mathematical if the quantity of matter or energy moving through a system can be measured and described using equations (see below and Blackie and Eeles, 1985).

Physical and analogue models

Before the advent of high-speed computers, many modellers sought to model the real world using physical and analogue, rather than mathematical, models. Physical models, in particular, still have utility, as the example of a soil tank in Figure 9.3 shows. The tank shown in the photograph was used to simulate overland flow and erosion over semi-arid surfaces. Water could be

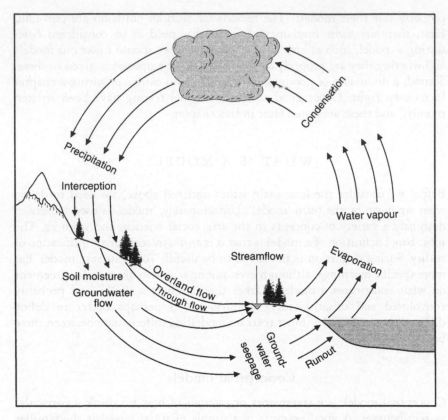

Figure 9.1 The hydrological cycle: a very simple conceptual model. Redrawn from Ward and Robinson (1990). Reproduced with permission of McGraw-Hill.

introduced directly from a trough at the top of the slope and by artificial rainfall. The advantage of such a scheme over field studies based on observing natural erosion events is that the soil type, the surface microtopography, rainfall characteristics and the amount of water arriving from upslope can be carefully controlled. Although not shown, it is also possible to grow plants in such a tank and analyse their effect on infiltration and overland flow. More information on these effects is given in Chapter 4.

Mathematical models

A simple mathematical model is the mass balance equation of a black box system. It is intuitive to expect the amount of mass flowing into a system to be equalled by the sum of (1) that flowing out, and (2) changes in storage in the system. Mathematically, this can be represented by

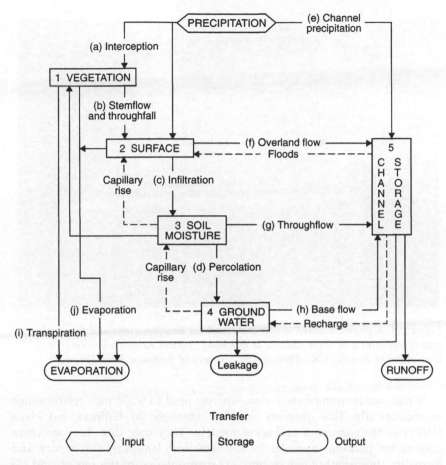

Figure 9.2 A conceptual model of hydrological stores and transfers at the catchment scale. Redrawn from Ward and Robinson (1990). Reproduced with permission of McGraw-Hill.

$$I - Q = \Delta S \tag{9.1}$$

where I is the amount of mass entering the system (M) during a given time period, Q is the mass lost from the system (M) and S is mass storage (M). In differential form this becomes

$$i - q = dS/dt \tag{9.2}$$

where i is the instantaneous rate of inflow (MT^{-1}), q is the instantaneous rate of discharge (MT^{-1}), and dS/dt is the rate of change of storage (MT^{-1}).

Figure 9.3 A physical model: a soil tank used for analysing overland flow and erosion on semi-arid slope surfaces at the Long Ashton Research Station, University of Bristol, UK. (Photograph courtesy of Professor John Thornes.)

It may not be immediately clear why we need to write such relationships mathematically. This question has been answered by Gillman and Hails (1997) in the context of ecological models. They note that there are three reasons for phrasing a model in mathematical language: (1) brevity and formality (precision) of description; (2) manipulation of the model; and (3) the discovery of emergent properties not apparent from non-mathematical reasoning. In terms of (1), equation (9.1) offers little advantage over the conceptual or 'word model' that precedes it. However, more complex models are difficult to phrase verbally but can be described relatively simply mathematically. Such equations can then be manipulated or rewritten, giving them a flexibility that goes well beyond a conceptual model phrased in ordinary language. To illustrate this, consider a saturated control volume of soil through which water is moving at a steady rate (i.e. the rate does not change over time) (Figure 9.4). The fluid moves in only one direction through the control volume. However, the actual fluid motion can be subdivided on the basis of the components of flow parallel to the three principal axes of a Cartesian coordinate system. Since the system is experiencing steady flow, the sum of the change in flow along the principal axes must equal zero, i.e. there is no change in storage. Formally, this is given by

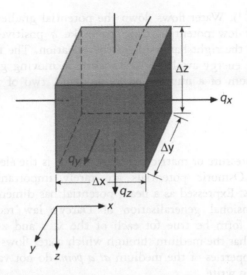

Figure 9.4 A saturated control volume for analysing flow through a soil.

$$\Delta q_x \Delta y \Delta z + \Delta q_y \Delta x \Delta z + \Delta q_z \Delta x \Delta y = 0 \qquad (9.3)$$

where $\Delta q_x = q_{xin} - q_{xout}$, $\Delta q_y = q_{yin} - q_{yout}$ and $\Delta q_z = q_{zin} - q_{zout}$, and where q is now given in units of volume discharge ($L^3 T^{-1}$), and *in* and *out* refer to flow into and out of the control volume. Dividing each term in equation (9.3) by $\Delta x \Delta y \Delta z$ and expressing in partial differential form gives

$$\frac{\partial q_x}{\partial x} + \frac{\partial q_y}{\partial y} + \frac{\partial q_z}{\partial z} = 0 \qquad (9.4)$$

In simple porous media, flow is dependent on the gradient of a flow potential (the total potential $-\phi$)* and is given by Darcy's law:

$$q = -K \frac{\Delta \phi}{\Delta l} \text{ or } q = -K \frac{d\phi}{dl} \qquad (9.5)$$

where l is distance in the direction of flow (L) and K, the proportionality constant linking flow rate with the gradient of potential, is the hydraulic

---

* The convention in groundwater hydrology is to use ϕ to denote flow or total potential. In keeping with this convention, this notation is used here. Note, however, that it differs from that used by Tyree (see Chapter 2), where ψ is used to denote total potential (equation (2.5)).

conductivity (LT^{-1}). Water flows down the potential gradient from areas of high to areas of low potential. Thus to make q positive, a minus sign is introduced on the right-hand side of the equation. The flow potential is a measure of the energy available to do work in moving groundwater and consists of the sum of a number of components, two of which are most important, i.e.

$$\phi = \phi_p + \phi_z + \ldots \tag{9.6}$$

where ϕ_p is the pressure or matric potential and ϕ_z is the elevation potential (Hillel, 1980). Osmotic potentials are rarely important in soil- and groundwater flow. Expressed as a head, potential has dimensions of length. The three-dimensional generalisation of Darcy's law requires that the one-dimensional form be true for each of the x, y and z components of flow. Assuming that the medium through which water flows is isotropic (i.e. the hydraulic properties of the medium *at a point* do not vary according to direction*) we can write

$$q_x = -K\frac{\partial\phi}{\partial x} \qquad q_y = -K\frac{\partial\phi}{\partial y} \qquad q_z = -K\frac{\partial\phi}{\partial z}$$

Substituting these into equation (9.4) gives

$$\frac{\partial}{\partial x}\left(-K\frac{\partial\phi}{\partial x}\right) + \frac{\partial}{\partial y}\left(-K\frac{\partial\phi}{\partial y}\right) + \frac{\partial}{\partial z}\left(-K\frac{\partial\phi}{\partial z}\right) = 0 \tag{9.7}$$

where $K = K(x, y, z)$. If K is assumed to be independent of x, y and z, i.e. the region is assumed to be homogeneous as well as isotropic, then equation (9.6) becomes

$$\frac{\partial^2\phi}{\partial x^2} + \frac{\partial^2\phi}{\partial y^2} + \frac{\partial^2\phi}{\partial z^2} = 0 \tag{9.8}$$

Equation (9.8) is known as the Laplace equation, after the eighteenth/nineteenth-century mathematician Pierre Simon Laplace, who discovered it. It is the governing equation for water flow through isotropic, homogeneous media under steady flows. It has wider applicability too and has been used in the analysis of steady-state heat conduction in solids, where the potential

---

* In some porous media, hydraulic conductivity is known to vary according to direction. The principal planes of the property are unusually the vertical and horizontal. It is often the case that the horizontal hydraulic conductivity is greater than the vertical hydraulic conductivity.

becomes temperature. The important point here is that such an important relationship simply could not be derived verbally. In addition, it contains properties that are not necessarily obvious from looking at the equation. For example, using a solution to equation (9.8) it is possible to model the distribution of total potential within a medium given a knowledge of the values of the flow potential along the boundaries of the volume of medium of interest (see Figure 9.5). For more information on this subject, the interested reader is referred to an introduction on groundwater modelling by Wang and Anderson (1982: pp. 11–39).

It is possible to define different types of mathematical models. Kirkby *et al.* (1993) distinguish between black box models (regression models and schematic models), process models, mass and energy balance models, and

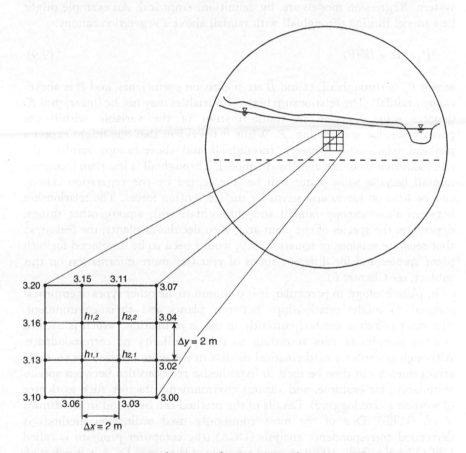

Figure 9.5 A cross-section of soil in a floodplain aquifer. The numbers at the nodes represent hydraulic head in metres above an arbitrary datum. The head at nodes (1,1), (1,2), (2,1) and (2,2) can be calculated using a solution to the Laplace equation provided that the head at the boundaries of the cross-section are known.

stochastic models. A more conventional breakdown of model types is given by Hillel (1977), who, along with other workers such as Addiscott and Wagenett (1985) and Watts (1997), considers the differences between empirical models, physically based (or mechanistic) models, deterministic models and stochastic models. The meaning of each of these terms is given below, after Hillel (1977).

Empirical models

Empirical models are concerned with prediction rather than explanation. They are based on observed quantitative relationships between variables and do not attempt to give an insight into the functional or causal operation of a system. Regression models are, by definition, empirical. An example might be a model linking throughfall with rainfall above a vegetation canopy:

$$P_{tf} = \alpha + \beta f(P) \tag{9.9}$$

where P_{tf} is throughfall, α and β are regression coefficients, and P is above-canopy rainfall. The relationship between variables may not be linear; thus $f()$ denotes some function or transformation of the variable within the parentheses, for example $\log_{10} P$. While it is evident that one might expect a physical relationship between throughfall and above-canopy rainfall, the mechanisms are not *explained* by the model. Throughfall is less than incoming rainfall because some water will be intercepted by the vegetation canopy and be held on leaves and stems by surface tension forces. The relationship between above-canopy rainfall and throughfall will, among other things, depend on the species of the plant and, with deciduous plants, the season, so that separate versions of equation (9.9) would need to be developed for each plant species and for different times of year (for more information on this subject, see Chapter 6).

In plant ecology, in particular, it is common to use other types of empirical method to model relationships between plants and their environment. The most common method currently in use is ordination, which is used to arrange samples of data according to their similarity or correspondence. Although not strictly mathematical models in the sense implied above, these arrangements can then be used to hypothesise relationships between species abundance, for example, and various environmental factors such as degree of wetness (waterlogging). Details of the method can be found in Goldsmith *et al.* (1986). One of the most commonly used ordination methods is detrended correspondence analysis (DCA) (the computer program is called DECORANA; Hill, 1979). A good example of the use of DCA is Busch *et al.* (1998), who looked at the relationship between plant communities and hydrological variables in the Florida Everglades, USA.

Physically based or mechanistic models

Physically based models are based on known mechanisms or processes that operate within the system being modelled. These mechanisms are described by the fundamental laws of physics and chemistry. It is important to note that 'physically based' does not necessarily mean that the model is entirely composed of fundamental physical laws, merely that it is based to a large extent on such laws. Models that combine significant elements of both the physical and empirical approaches are often called semi-empirical. A good example of a physically based model is the Penman–Monteith equation (Penman, 1948; Monteith, 1973 – see Campbell, 1981) used to predict evapotranspiration from a vegetated surface (see also Chapter 2). Physically based models have also been called *theoretical* models because they are essentially statements of how we think the real world behaves as opposed to an empirical model, which describes the behaviour of one variable in terms of another without any attempt to explain system behaviour. Theoretical models are widely used in research and represent formal statements of theories, which should then be tested with experimental data.

Deterministic models

With deterministic models exact relationships are postulated, and the output of the model is predicted with complete certainty by the input. Thus the water content of a length of river channel can be calculated with certainty for all times when the initial storage, as well as the inflows and outflows from the channel, is known. Deterministic models can be both empirical and physically based. In equation (9.9), the amount of throughfall is given exactly from a knowledge of above-canopy rainfall. This does not mean to say that the prediction of the model is correct, merely that for a given input, the model will produce a unique output.

Stochastic models

In stochastic models, a random or stochastic element is introduced into the model so that for a given input there is no longer a single output but a range or distribution of output values. For example, equation (9.9) could be rewritten

$$P_{tf} = \alpha + \beta f(P) + r \qquad (9.10)$$

where r is a random value or variable that has different values for each run of the model, even if the values of the model inputs remain the same. Thus, P_{tf} is no longer a unique function of P but also depends on r, which is randomly generated for each model run. r will usually be generated from a known probability distribution (Kirkby *et al.*, 1993). Sometimes, the term stochastic model is also used to describe Monte Carlo simulations using deterministic models. Monte Carlo simulations are used where there is some uncertainty in the input data. Thus input data sets are drawn from measured or assumed distributions of input parameters. The probability of different model outputs can then be determined by analysing the distribution of model outputs. For an example of a Monte Carlo simulation using an eco-hydrological model, see Franks and Beven (1997).

Other model types

Continuous models portray continuous processes, in contrast to *discrete models* which include discontinuous or abrupt phenomena. *Dynamic* models simulate time-dependent processes, as opposed to static systems or systems in which processes operate at a steady rate (*steady-state models* are then used). Two classes of model not explicitly discussed by Hillel (1977) are *lumped* and *distributed* models. In lumped models, no attempt is made to account for spatial variability of a given process. Many conceptual mathematical models are of the lumped type in which the river catchment is represented as a series of conceptual stores, with stores representing the soil system, the channel system and so on. Such a model, HYRROM (HYdrological Rainfall Runoff Model) (Blackie and Eeles, 1985), is illustrated in Figure 9.6, where it can be seen that the catchment is conceptualised as four lumped stores: the vegetation canopy, the soil, the groundwater store, and the channel or routing store. Flow from the soil, groundwater and routing stores is described mathematically as a function of the water content of each store. The flow equations used in the model are not explained here, and the interested reader should refer to the original paper. Distributed models explicitly account for spatial variations in parameter values and rates of process operation. Most account for such variation by 'chopping up' the system of interest, such as a river catchment or hill slope, into a series of stores. Thus for a catchment there may be hundreds or even thousands of soil stores, each representing different soil types or variations within a soil type, thousands of stores representing the river channel network and so on. This is illustrated below (see Figure 9.7).

WHAT IS AN ECO-HYDROLOGICAL MODEL?

In its broadest sense, an eco-hydrological model is any model used in an eco-hydrological study. Thus, for example, Wassen *et al.* (1996) use a groundwater flow model (FLOWNET – van Elburg *et al.*, 1991) to indicate possible pathways of water movement into and out of fens in order to understand better the distribution of wetland plant species. In a narrower sense, an eco-hydrological model is one that, *within* the model structure, accounts for

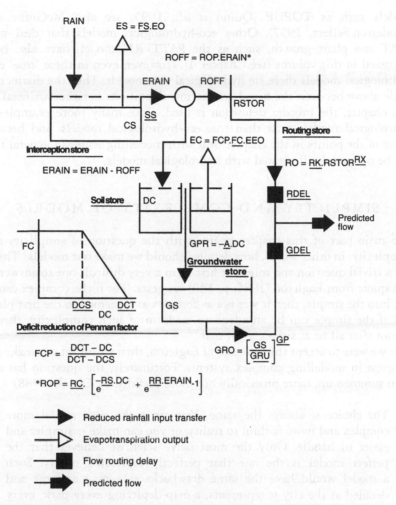

RAIN

ES = FS.EO

ROFF = ROP.ERAIN*

ERAIN   ROFF

SS

CS

Interception store

ERAIN = ERAIN - ROFF

RSTOR

EC = FCP.FC.EEO

Routing store

RO = RK.RSTOR^{RX}

Soil store   DC

RDEL

Predicted
flow

FC

GPR = –A.DC

GDEL

**Groundwater
store**

DCS   DCT   GS
    DC

Deficit reduction of Penman factor

GRO = $\left[\dfrac{GS}{GRU}\right]^{GP}$

$$FCP = \dfrac{DCT - DC}{DCT - DCS}$$

$$*ROP = RC \cdot \left[e^{\dfrac{-RS.DC}{}} + e^{\dfrac{RR.ERAIN}{}} {-1}\right]$$

▶ Reduced rainfall input transfer

△ Evapotranspiration output

■ Flow routing delay

▶ Predicted flow

Figure 9.6 A conceptual mathematical model, HYRROM, of catchment runoff
(redrawn from Blackie and Eeles, 1985). HYRROM is a lumped model in which
the catchment is divided into four stores: the interception store, the soil store, the
groundwater store and the routing store. Reproduced with permission from John
Wiley & Sons Ltd.

eco-hydrological processes. SVAT (soil–vegetation–atmosphere transfer)
models, which are the subject of much current research, fit into this class
because they look directly at how plants affect rates of water loss from the
Earth's surface. SVAT models can range from the relatively complex such as
SWIM v.2.1 (Verburg *et al.*, 1996), which can cater for up to four vegetation
types, different root distributions in the soil, multiple-layer soils, Darcy–
Richards-type soil water flow and soil solute transport, to relatively simple

models such as TOPUP (Quinn *et al.*, 1995; see also McGuffie and Henderson-Sellers, 1997). Other eco-hydrological models that deal with SVAT and plant growth, such as the PATTERN model, have also been discussed in this volume (see Chapter 4). However, even in these 'true' eco-hydrological models there are hydrological sub-models. Thus the distinction made above between the broad and narrow definitions is rather artificial. In this chapter, the broader definition is used, since many more examples of hydrological models exist than true eco-hydrological models, and because many of the points in the chapter deal with modelling issues in general that can be adequately illustrated with hydrological models.

SIMPLICITY AND COMPLEXITY OF MODELS

The main part of this chapter begins with the question of simplicity and complexity: in other words, how detailed should we make our models? This is not a trivial question and might at first seem a very difficult one to answer, as this quote from Eagleton (1986: p. 149) suggests: 'For if the complex can be put into the simple, then it was not as complex as it seemed in the first place; and if the simple can be an adequate medium of such complexity, then it cannot after all be as simple as all that.'

If we were to accept the cynicism of Eagleton, then we might not make any progress in modelling complex systems. Fortunately, the question has also been summed up, more prosaically but more usefully, by Gleick (1988):

> The choice is always the same. You can make your model more complex and more faithful to reality, or you can make it simpler and easier to handle. Only the most naive scientist believes that the perfect model is the one that perfectly represents reality. Such a model would have the same drawbacks as a map as large and detailed as the city it represents, a map depicting every park, every street, every building, every tree, every pothole, every inhabitant, and every map. Were such a map possible, its specificity would defeat its purpose: to generalise and abstract.

Gleick's message is clear: models should always attempt to simplify and we should be wary of complexity in model design, because with a complex model we can lose the power of abstraction. In any case, as workers on deterministic chaos have demonstrated, even simple models can display complex behaviour and we should, perhaps, try to understand these simple models before we attempt to build more complex ones.

If the purpose of a model is, as Gleick suggests, to generalise and abstract, it is perhaps surprising to see so many complex models in the academic and professional literature. Put harshly, it is almost as if we are scared to leave something out of our models in case we miss some important aspect of system behaviour. There is also the possibility of 'model snobbery': 'My model is bigger than yours and has more "governing" equations; therefore, it must be a better representation of reality.' However, before we become too jaundiced it is important to note that it seems intuitive to expect that the more complex or detailed the model, the better it should be able to represent reality. To use Gleick's analogy with a map, it (the model) will give a better depiction of reality if it includes the potholes. However, this is not necessarily true, as the well-known SHE (Système Hydrologique Européen) model can be used to show.

The SHE model (Abbott *et al.*, 1986a and b) is regarded, even by its detractors, as the most complete theoretical representation of catchment behaviour currently available. In other words, it is a compilation of our state-of-the-art knowledge of how water 'cascades' through catchments. Any diagram would be a poor representation of such complexity. However, that shown in Figure 9.7 (after Beven, 1985) gives a flavour of what processes are represented in the model. It should be noted that a number of versions of SHE exist, including versions that predict soil erosion (see, for example, Wicks and Bathurst, 1996). Therefore, Figure 9.7 shows just some of the processes that

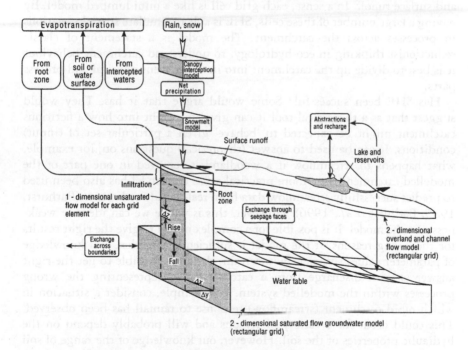

Figure 9.7 A physically based, distributed catchment model: Système Hydrologique Européen (SHE) (redrawn from Beven, 1985). Reproduced with permission from John Wiley & Sons Ltd.

can be simulated using the SHE package. From the figure, it can be seen that the model deals with almost every aspect of the catchment cascade. Incoming precipitation in the form of either rain or snow can be intercepted by a vegetation canopy or arrive directly at the ground surface. Snow can accumulate at the ground surface or, depending on meteorological variables, can melt. Water is lost directly from the catchment through the processes of evapotranspiration. Water accumulating at the soil surface can infiltrate into the soil and move through the unsaturated zone either as throughflow or vertically into the deeper groundwater zone. If infiltration is insufficient to prevent ponding of water at the soil surface, or if the water table rises to the surface to cause ponding, overland flow or surface runoff will occur. This water will eventually enter the channel network and flow to the catchment outlet. Deeper groundwater flow is also simulated. A concise review of these processes is given by Baird (1997a), while a good detailed treatment is given by Dingman (1994). The SHE model is also a distributed model, as can be seen clearly from Figure 9.7. The catchment is divided into a grid, and for every grid 'cell', each of the processes listed above is, where relevant, simulated. Thus for a given cell the following processes could be simulated: canopy interception, evapotranspiration, soil water flow, groundwater flow and surface runoff. In a sense, each grid cell is like a mini lumped model. By having a large number of these cells, SHE is able to simulate spatial variations in processes across the catchment. The model is a statement of classic reductionist thinking in eco-hydrology: to understand catchment behaviour it is best to divide up the catchment into its component physical and process parts.

Has SHE been successful? Some would argue that it has. They would suggest that as a theoretical tool it can give an insight into how a fictitious catchment might be expected to 'behave' given a particular set of (input) conditions. It can be used to answer (theoretical) questions on, for example, what happens to streamflow if a woodland is removed in one part of the modelled catchment. In a more practical sense the model has also been used to predict successfully streamflow data from real-world catchments (Bathurst, 1986; Bathurst et al., 1996). However, this is where we can identify weaknesses in the model. It is possible for a complex model to give the right results for the wrong reasons. If the model is sufficiently complex, and knowledge of parameter values sufficiently uncertain, then it is possible to get the right answer (stream discharge from a catchment) by representing the wrong processes within the modelled system. For example, consider a situation in which rapid catchment (streamflow) response to rainfall has been observed. This could occur for a variety of reasons and will probably depend on the hydraulic properties of the soil. However, our knowledge of the range of soil hydraulic conductivities in the real catchment is likely to be scanty at best. Such data are time-consuming and expensive to collect. As an alternative to collecting data on hydraulic conductivity from *every* part of a catchment and for *every* layer of soil, researchers will *estimate* values of the parameter. Estimation is, by definition, an uncertain procedure, so ranges of *possible* values for the parameter will be used. The values of many parameters in a model such as SHE will be estimated. The range of uncertainty for each parameter is often sufficiently large that it is possible for a model such as SHE to be run with quite different sets of *apparently* realistic input conditions and give the same or very similar results in terms of water discharging from the catchment. For example, a rapid streamflow response could be predicted when high-permeability subsoils allow rapid throughflow to the channel in the model catchment. A very similar discharge time series could be predicted when the model catchment contains low-permeability subsoils in the vicinity of stream channels. During rainfall, these will prevent downward percolation of water through the soil profile, causing a rapid rise in the (perched) water table. If the water table intercepts the soil surface, overland flow may occur, causing rainwater to flow rapidly to the channel system. We therefore have a situation where quite different processes (overland flow and rapid through-flow) occurring within the model catchment give very similar outcomes in terms of streamflow response to rainfall. This is *equifinality*, or more strictly *model equifinality*, as defined by Beven (1996) and discussed in the context of other hydrological and eco-hydrological models models by Grayson *et al.* (1992) and Franks *et al.* (1997). The important point here is that, as modellers, we have been unable to identify what is going on in the real world: we do not know whether overland flow or rapid throughflow is the reason for the observed rapid streamflow response to rainfall. Both are plausible given our uncertain knowledge about input conditions. This is the problem of model and parameter *identifiability* or *indeterminacy* (Kennedy, 1985). We have developed a model that purports to represent the real world to a high level of physical detail. In practice, we have developed a very expensive computer code that, although it can be used to obtain accurate predictions of streamflow, cannot necessarily be used to give insights into what is going on physically in the real catchment. The model has, therefore, failed to do what it was set up for. It is likely that we could achieve similar levels of predictive accuracy using a much simpler conceptual mathematical store and flow model with just a few parameters. To find out why the real catchment shows the response it does would require detailed field observations to determine whether or not overland flow and rapid throughflow occur. It is important to realise that just because a physically based model accurately predicts some aspect of real system behaviour does not guarantee that it has done so by representing the real system accurately.

The famous mathematician and physicist Roger Penrose has classified theories as 'superb', 'useful', 'tentative' and 'misguided' (Gardner, 1989). Such a four-fold classification could also be applied to models. Some would argue that SHE and like models are superb; others, myself included, would suggest that they are tentative.

A BLUEPRINT FOR BETTER MODELLING?

If models such as SHE are tentative, how can we set about producing better models? Is it possible to set up a blueprint for better modelling? The answer to this is probably a qualified 'yes', although, as will be shown below, model construction and testing are more difficult than they first appear. Among others, Kirkby *et al.* (1993) have suggested that choice of an appropriate model should be centred around a number of related questions. With some modifications, these are:

1 What is the purpose of the model?

Models can be used for many purposes. Simple or even simplistic models are often used as teaching tools in order to illustrate the basic features of a system. Many models are used for prediction, and a growing number are used as research tools. In the latter case, the predictive value of the model may not be of particular interest. Often a research model is used for 'numerical experimentation'. Here, the behaviour of an often fictitious system is analysed. The modeller can change the value of input variables and parameters used in the model and see how these changes affect system behaviour; in other words, they are conducting a type of sensitivity analysis. Such heuristic 'what if?' modelling could involve, for example, assessing the impact of removing vegetation on soil water regime by assuming different amounts of non-vegetated ground from full cover to a cover of, say, 25 percent. Similarly, the effect of changing the hydraulic conductivity of subsoils on the amount of overland flow could be simulated. A brief discussion of heuristic numerical experiments of this type, looking at deforestation and desertification, is given in Chapter 4. As noted in Chapter 4, such numerical experiments are often used in conjunction with, or to direct, field investigations into system behaviour. In addition, they can be used to identify the most sensitive model parameters, and the properties of the system that a sensitive parameter represents can be singled out for more thorough investigation.

Numerical experimentation represents a very important use of models and one that should not be underestimated. In such applications, it is clear that the model is being used as a tool for developing more refined theories of system behaviour to be tested with field data, if possible. Such models can also provide provisional estimates of system behaviour where field data are expensive or impossible to collect, or to provide tentative predictions of future changes in an ecosystem. A good example of the latter is Mulligan's (1996a, b) PATTERN model described in Chapter 4, which is used to predict possible future changes in eco-hydrological processes in the Mediterranean. Thus numerical experimentation, if used carefully, can provide useful insights into how a system *might* behave. However, numerical experimentation is not without its pitfalls and may give misleading information on system behaviour, as

316

was explained above with the example of SHE (see also below), in which case it is important to test the internal behaviour/predictions of the model and not just its main output.

Some ecologists distinguish between 'tactical' and 'strategic' models (May, 1973, and Holling, 1966; cited in Gillman and Hails, 1997). The former are detailed models that attempt to descibe all the processes operating within a system. As the following extract from Gillman and Hails (1997: p. 4) shows, tactical models are similar to the SHE model described above and are thought to suffer from similar problems of complexity:

> At the tactical end of the spectrum we attempt to measure all the relevant factors and determine how they interact with the target population or community. For example, in producing a model of change in plant numbers with time we might find that the plants are affected by 12 factors, including summer and winter rainfall, spring temperature and herbivores. All the information is combined into a computer program, initial conditions are set (e.g. the number of plants at time 1), values for the variables entered (e.g. level of summer rainfall) and the model run. . . . After a certain number of generations or time periods the number of plants is recorded from the program output.
>
> Now comes the tricky part. We have produced a 'realistic' model in the sense that it mimics closely what we believe is happening in the field. However, we do not really know why it produces the answer that it does. The model is intractable (and perhaps unpredictable) owing to its complexity. Tweaking a variable such as rainfall may radically change the output but we do not know why. In other words we have created a black box which receives variables and spews out population dynamics.

Strategic models are deliberately simple and attempt to look at only a few aspects of system behaviour. Frequently, they consist of one or two governing equations. Numerical experimentation with such simple models can give insights into how a system might behave without its complicating factors.

2 Should the model be process-based?

Many empirical models that do not take explicit account of processes often give accurate predictions of system behaviour. Thus if accurate predictions are required, and sufficient data on system behaviour exist to construct an empirical model, then there may be no need to construct a process-based model. If, on the other hand, the model is to be used as a theoretical tool for investigating and explaining system behaviour, it will need to be process-based. There are many shades of model between these two extremes, and terms such as 'grey box', 'semi-empirical' and 'hybrid' are used to describe models that combine empirical and physically based approaches.

3 What resolution in space and time is required?

As noted above, models can be lumped, distributed, transient or steady-state. The spatial and temporal resolution of the model should be tailored to its use. Clearly, if a model for predicting floods is required there is no sense in using a model that gives streamflow averaged over monthly time steps. Often, models used as predictive tools are lumped, while models used as research tools are distributed, although this is by no means always the case. More models are transient than steady-state, although steady-state models are still widely used in some branches of hydrology such as groundwater hydrology, where the Laplace equation (equation (9.7)) is used to predict spatial patterns of steady groundwater flow (for recent examples of the application of a steady-state flow model in an eco-hydrology study see Gilvear *et al.*, 1993 and 1997).

4 Given the first three questions, is the model capable of being tested?

It could be suggested that if a model cannot be tested, there is no means of assessing its worth, and if there is no means of assessing its worth it is, to all intents and purposes, worthless. This is an extreme view. Many modellers use strategic models knowing that they are naive representations of the real world. The purpose behind such use is to identify aspects of real system behaviour that *may* exist and be worthy of further investigation. Thus, although the model is not being tested *per se*, it is being used to direct further (field)work that will elucidate system behaviour. Despite this, many complex models are presented in the literature that cannot readily be tested; SHE is a good example. The issue of testability is often conveniently forgotten by many modellers. A complex model may be plausible, it may contain what is thought to be a reasonable representation of the real world but, if it cannot be tested, then its results must be treated with great caution regardless of the claims made for it by its authors. As Kirkby *et al.* (1996: p. 352) note in a discussion of an eco-hydrological model of plant growth, hydrology and erosion on Mediterranean hill slopes:

> Models may be able to exist in isolation but their credibility can only
> be built on existing field studies. . . . Good models have the power to
> generalize from a small number of field studies, but good fieldwork
> can destroy a model overnight, and send us all back to the drawing
> board!

Nevertheless, model testing can be very difficult, and the problem of model falsification *sensu* Popper (1968; 1972) has been raised by a number of authors. Most models are in a form that does not allow clear falsification, and the issue then is how we should aim to produce models that are essentially testable (see Weiner, 1995). This problem is dealt with in more detail below.

Rather than deal with each of the above questions generically, it is perhaps better to illustrate how and, indeed, whether, they can be used to guide better modelling practice with an eco-hydrological example. The example chosen is that of water table fluctuations within a floodplain. The real system (Figure 9.8) could be represented pictorially, as shown in Figure 9.9. Note that this is not a conceptual model in the scientific sense, because no attempt is made to convey how we think the system works (*cf.* the picture of the hydrological cycle shown in Figure 9.1).

Case 1

Let us suppose that we want a model of such a system that gives good predictions of water tables and that we are not concerned about whether or not the model can represent water flow processes within the system. The answer to the first question is, therefore, that the model should be a predictive

Figure 9.8 River and adjacent floodplain. (Photograph courtesy of Professor Rob Ferguson.)

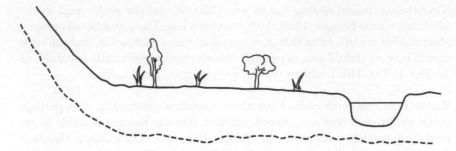

Figure 9.9 Pictorial representation of a cross-section through the floodplain shown in Figure 9.8.

tool. The answer to the second question is that we are not interested in process representation, i.e. the model is atheoretical, it is simply a 'calculation tool' (Weiner, 1995). In answer to the third question on spatial and temporal resolution we want a model that can give a broad indication of daily water table position across the floodplain. The simplest model that could meet each of these requirements would be a simple empirical model. An example is illustrated in Figure 9.10. Water tables are expressed as a function of previous water tables and other variables, for example rainfall, i.e.

$$h_t = \beta_1 P_t + \ldots + \beta_n h_{t-1} + \ldots \tag{9.11}$$

where h is water table elevation above a given datum, P is rainfall, β_1 to β_n are regression coefficients, t is time, and $t-1$ denotes a time a given interval before t. The choice of the time interval depends on the system being modelled. Clearly, in the example given, h_t is more likely to be a function of

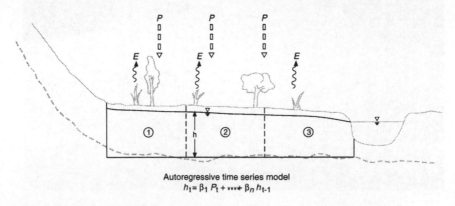

Autoregressive time series model
$$h_t = \beta_1 P_t + \cdots + \beta_n h_{t-1}$$

Figure 9.10 Empirical model of water table behaviour in the floodplain shown in Figure 9.8.

recent water tables within the last few hours to days rather than water tables within the last few weeks or months. Models similar to equation (9.11) have been used by Rennolls *et al.* (1980) and Armstrong (1988), the latter using a Box–Jenkins (1976) type model to simulate water table behaviour in mole-drained agricultural land. To give a representation of water table variations across the floodplain, empirical models could be developed for each of the zones denoted by 1, 2, and 3 in Figure 9.10, and, possibly, for different seasons. Empirical models are, in a sense, already tested, because they have been developed from existing data. Their continued reliability and accuracy can be checked (tested) periodically by collecting more data of system behaviour and assessing the goodness of fit between predicted and observed.

Case 2

Now let us suppose that we want a model that is more transportable and that has a theoretical or physical basis, i.e. one that 'explains' the real system. Nevertheless, we want a model still capable of being used as a predictive tool but not one that is 'parameter-hungry'; i.e. the model should remain more strategic than tactical. The model should be capable of simulating water table positions every few metres across the floodplain for time intervals of hours rather than days. Such a model is illustrated in Figure 9.11. Here we model water table behaviour using a simplification of Darcy's law for groundwater flow that, when combined with the continuity equation, is called the Boussinesq equation (Boussinesq, 1904):

$$\frac{\partial h}{\partial t} = \frac{K}{s} \frac{\partial}{\partial x} \left(h \frac{\partial h}{\partial x} \right) - S(t) \tag{9.12}$$

where K is hydraulic conductivity (LT^{-1}), s is the specific yield defined as the ratio of the volume of water that a soil will yield by gravity drainage to the volume of soil from which it drains (dimensionless), and S is a sink term representing evapotranspiration (rainfall is treated as a negative sink). The Boussinesq equation is based on the Dupuit–Forchheimer approximation, which states that in a shallow groundwater system where the flow is approximately horizontal, the potential or energy gradient driving flow can be approximated by the gradient of the water table (see pp. 96–98 of McWhorter and Sunada, 1977, and pp. 188–189 of Freeze and Cherry, 1979, for a general treatment; and Childs and Youngs, 1968, Kirkham, 1967, and Murray and Monkmeyer, 1973, for more detailed discussions of the approximation). Equation 9.12, because it is non-linear, can often only be solved for real-world cases using numerical techniques such as the finite-difference method (see 'Modelling and mathematics', below). Hence the floodplain is represented in the model as a series of finite-difference cells or nodes. To obtain a solution to the equation we need a knowledge of what is happening at the system

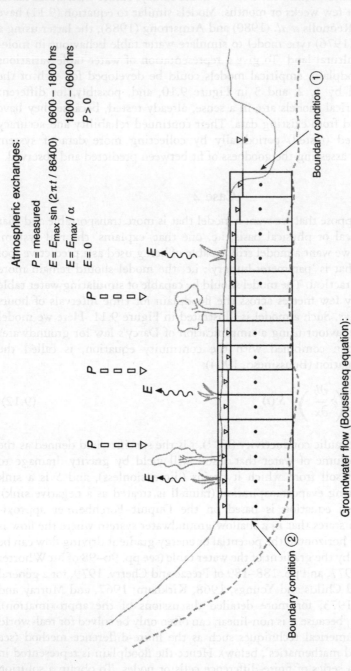

Atmospheric exchanges:

P = measured

$E = E_{max} \sin(2\pi t / 86400)$ 0600 – 1800 hrs

$E = E_{max} / \alpha$ 1800 – 0600 hrs

$E = 0$ $P > 0$

Groundwater flow (Boussinesq equation):

$$\frac{\partial h}{\partial t} = \frac{K}{s} \frac{\partial}{\partial x}\left[h \frac{\partial h}{\partial x} \right] - S(t)$$

Boundary condition ①

Boundary condition ②

P

E

Figure 9.11 A simple physically based model of water table behaviour in the floodplain shown in Figure 9.8.

boundary (the boundary conditions), together with a knowledge of the position of the water table in each computational cell at the beginning of the simulation (the initial conditions). In Figure 9.11, it is assumed that the floodplain soils are underlain by relatively impermeable soils at a depth shown by the broken line. The river channel represents one boundary. Here the level of water in the river needs to be specified. This level can be held constant or can be allowed to vary over time. The second model boundary is where the hill slope meets the floodplain. This could be represented as either a given water table elevation, as a given flux of water across the boundary, or a flux dependent on the water table within the model. In modelling jargon, these are known as Dirichlet, von Neumann and Cauchy conditions, respectively (see Wang and Anderson, 1982). It can already be seen that the model is substantially more complex than the previous empirical model. However, in many ways it still represents quite a severe simplification of the real system. In the form of the Boussinesq equation given in equation (9.12) it is assumed that the hydraulic conductivity and specific yield of the soil can be given by a single representative value when in fact it is known that both parameters, the former in particular, can show considerable spatial variability in real systems (see *inter alia* Nielsen *et al.*, 1973; Ragab and Cooper, 1993; Rogowski, 1972; Warrick and Nielsen, 1980; and Viera *et al.*, 1981). Rainfall and evapotranspiration are also assumed to be uniform across the floodplain, when the latter can be expected to show some variability depending on the distribution of different vegetation types and water table depths. In addition, no attempt is made to account for the effects of the unsaturated zone on water table behaviour. Thus rainfall is added directly to the water table and is not taken up by storage in the unsaturated zone. This is probably a reasonable assumption during winter, when the water table is close to the ground surface, but it could be a source of error in model predictions during low water table conditions in summer months. Evapotranspiration is predicted using a simple sine-wave model in which losses of water to the atmosphere vary as a function of the maximum (midday) measured evapotranspiration (see Hillel *et al.*, 1976):

$$E = E_{max} \sin (2\pi t/86400) \quad \text{between 0600 and 1800 hrs}$$

$$E = E_{max}/\alpha \quad \text{between 1800 and 0600 hrs} \quad (9.13)$$

$$E = 0 \quad \text{when } P > 0$$

where E is evapotranspiration rate (m s^{-1}), E_{max} is maximum midday evapotranspiration rate (m s^{-1}), t is time (s), and α is an empirically determined parameter (dimensionless). Obviously, such a model is very simple and does not take explicit account of changes in atmospheric controls on the process. In nature, evapotranspiration rates are controlled principally by (1) the provision at the evaporating surface of sufficient energy for the latent heat of vaporisation and the operation above the surface of diffusion and mixing processes, which remove water vapour produced by evaporation, and (2) by the properties of the floodplain vegetation (since these will affect wind flow in the boundary layer) and the depth of the water table (Gardner, 1957). Plants will also exert a control on evapotranspiration through their stomatal resistance to vapour flux. For more information on these processes, see Chapters 2 and 6.

How could such a model be tested? If one purpose of the model is to predict or represent water table behaviour across the floodplain, there is obviously a need to measure water tables at strategic locations away from the river. Lined auger holes could be used for such a purpose. Water will flow into or out of these in order to attain equilibrium with the free water surface in the soil. If the time to attain equilibrium is relatively short, the auger holes should provide a good measure of actual water table fluctuations. Having measured water tables, it is then necessary to measure or estimate K and s and E_{max}. Once this has been done, the predicted water tables could be compared with measured values to ascertain the goodness of fit between the two. However, it is possible, even in the simple model presented, that errors in the evapo-transporation sub-model could cancel out errors in the groundwater flow model, giving a good fit between observed and predicted water tables. Equally, a poor fit between observed and predicted water tables could be due to only one, and not both, of the sub-models. In order to resolve this problem, the sub-models should be tested separately (this is known as internal validation). For example, the groundwater model could be tested separately for periods when evapotranspiration is known to be very low, while modelled rates of evapotranspiration could be compared with measured rates using methods such as the Bowen ratio and eddy correlation (see Chapter 6). The model uses a single value of K and a single value of s. The validity of using a single representative value of each variable can be assessed from the comparison of modelled and measured water tables. If the two show close correspondence, then it could be concluded that the use of a single value of K and s is justified, although it should be noted that such 'effective' parameters have less physical meaning than measured point values.

Since the model is also theoretical, it is important to test the assumptions of the governing equation and to demonstrate that they are met in the real system. One such assumption, that of using a single K and s, has already been discussed above. Some workers do not regard this as testing *per se* (Hornberger, personal communication), yet it is important to ensure that the model is not being applied wrongly. Noting the problems with SHE discussed above, it is possible to get the right results for the wrong reasons. Hence, a simple goodness-of-fit test between modelled and measured water tables may not be sufficient to demonstrate the validity of the model, although in simple models like equations (9.12) and (9.13) equifinality is much less of a problem than it is in very complex models such as SHE. Part of this problem can be overcome by separate testing of component models as noted above.

To illustrate such assumption testing, consider the assumption that groundwater flow is largely horizontal and can be described by the Dupuit–Forchheimer approximation. Two lines of evidence could be used to test the validity of this approximation. The first, which has already been discussed, is to see if there is a good fit between observed and predicted water tables. If there is, this would *suggest* that the Boussinesq equation provides an adequate description of real system behaviour, i.e. the assumption of horizontal flow appears to be reasonable because the model gives good predictions. Second, piezometers could be used to measure potentials and hydraulic conductivity at different depths in the groundwater flow field to see if there is a substantial vertical flow in the real system. If significant amounts of vertical flow were found then it would have to be concluded that the model was wrong and would have to be regarded as empirical rather than theoretical. The second line of evidence would be definitive; the first is only suggestive. Nevertheless, deciding on what is substantial vertical flow and, therefore, when the model is falsified is very difficult. The issue of falsification is discussed further below.

Case 3

If we change our emphasis to a model that is more theoretical, while retaining the other requirements of Case 2, a more complex representation of the real system results. Such a model is shown in Figure 9.12. Water movement through the soil is now represented as a combination of saturated and unsaturated flow, and vertical as well as horizontal flow is simulated, i.e.

$$c\left(\phi_p\right)\frac{\partial\phi_p}{\partial t} = \frac{\partial}{\partial x}\left(K\left(\phi_p\right)\frac{\partial\phi_p}{\partial x}\right) + \frac{\partial}{\partial z}\left(K(\phi_p)\left(\frac{\partial\phi_p}{\partial z} + 1\right)\right) - S(t, x, z)$$

(9.14)

where ϕ_p is the pressure/matric potential, $c(\phi_p)$ is the specific water capacity or the change in water content in a unit volume of soil per unit change in pressure/matric potential given by $\partial\theta/\partial\phi_p$ (where θ is the volumetric water content $[L^3 L^{-3}]$), z is the vertical ordinate with positive upwards, and S is a sink term representing losses from the system due to evapotranspiration. This equation is called the Darcy–Richards equation, since it is based on Darcy's law and Richards' (1931) extension of the law to unsaturated flow (cited and explained in detail in Hillel, 1980). For a derivation of equation (9.14), the reader should consult Freeze and Cherry (1979: pp. 64–67). Solving this equation in the form given above (also in Figure 9.12) represents a much more major undertaking than solving the Boussinesq equation. The method of solution is similar, in that the real system is represented as a system of finite-difference nodes or cells. However, many more nodes/cells are now needed to

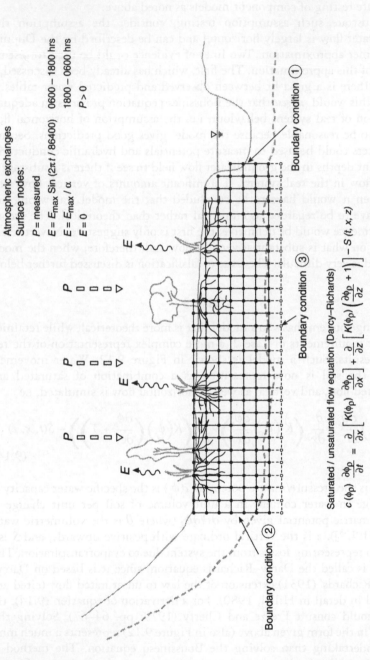

Atmospheric exchanges

Surface nodes:

P = measured

$E = E_{max} \, Sin \, (2\pi t \, / \, 86400)$ 0600 – 1800 hrs

$E = E_{max} \, / \, \alpha$ 1800 – 0600 hrs

$E = 0$ $P > 0$

Boundary condition ①

Boundary condition ③

Boundary condition ②

Saturated / unsaturated flow equation (Darcy–Richards)

$$c(\phi_\rho) \, \frac{\partial \phi_\rho}{\partial t} = \frac{\partial}{\partial x} \left[K(\phi_\rho) \, \frac{\partial \phi_\rho}{\partial x} \right] + \frac{\partial}{\partial z} \left[K(\phi_\rho) \left(\frac{\partial \phi_\rho}{\partial z} + 1 \right) \right] - S \, (t, x, z)$$

Figure 9.12 A Darcy–Richards model for predicting water tables in the floodplain shown in Figure 9.8.

represent vertical flow and the variability of flow parameters across the floodplain. The Darcy–Richards equation contains many more parameters than the Boussinesq equation. To solve equation (9.14), the following need to be measured or estimated: (1) the relationship between unsaturated hydraulic conductivity and matric potential and how this relation varies in space across the real system, i.e. $K(\phi_p)$, although properly this should be $K(\phi_p, x, z)$ (in other words we need to know $K(\phi_p)$ for each node of the model); and (2) the relationship between matric potential and soil water content for each node of the model. Clearly, collecting data on these parameters would be a major undertaking and involve the collection of hundreds of samples of floodplain soil. As noted in the discussion of the SHE model, such a measurement programme is rarely, if ever, attempted, and values of parameters at each model node are often estimated.

The model shown in Figure 9.12 also contains a more detailed sub-model describing losses of water to the atmosphere. The relatively simple sine-wave model is retained, but instead of evapotranspiration losses being removed from the water table they are removed from all model nodes 'in' the rooting zone of the floodplain soil. The precise form of such a root extraction model is not given and is merely represented as the last term in the Darcy–Richards equation. Various models exist, and most follow the principles of cohesion–tension theory outlined in Chapter 2. Empirical models are also available that describe losses to roots. For an example of the use of a root extraction model in a soil water model, readers should consult Hillel *et al.* (1976) or Verburg *et al.* (1996). Semi-empirical root extraction models are reviewed in Feddes *et al.* (1988). Both physical and semi-empirical models of root extraction require parameterisation and again make the application of the model shown in Figure 9.12 far more difficult than the model based on the Boussinesq equation.

Case 4

It is possible to design and construct a (tactical) model that is as full a statement as possible of water transfers into, through and out of a floodplain (no attempt has been made here at illustrating such a model). The main purpose of such a model would probably be for heuristic testing/numerical experimentation, as identified above. However, such models must be tested in order to have any credibility. In addition to Darcy–Richards-type soil water fluxes, such a model could consider macropore flow and the role of larger soil pipes in transferring water through the floodplain. To account for situations when the river is in flood or the water table has risen to the soil surface, an overland flow routine could be included. Evapotranspiration could be simulated using an energy balance model driven by data collected using an automatic weather station (for an example of such an application see Spieksma *et al.*, 1997).

Such a model would involve a considerable extension of the already very complex model shown in Figure 9.12 and, in view of earlier comments on the SHE model, would seem to be somewhat misguided, for it is unlikely that all of the processes could be measured accurately in the field in order to test the model. Indeed, it is unlikely that all of the processes included in the model are important all the time in the real system. The very complexity of the model would also render it very difficult to use, and interpretation of 'what if?' testing would be confounded by the problem of parameter variability; i.e. different credible parameter sets could be used in a numerical experiment involving change of the floodplain vegetation, but each set may give quite different results for the same vegetation change treatment. Thus it is difficult to see how the model could be used to make even basic statements about how the real system might behave. Franks and Beven (1997) have explored this issue in an analysis of a SVAT model (TOPUP – Quinn *et al.*, 1995) and conclude that the inclusion of a greater number of parameters/processes in a model leads to more input parameter sets than could be considered realistic and consequently to greater uncertainty in model outputs or predictions. On the other hand, simpler models allow more robust parameter estimation but have greater uncertainty associated with their more basic representation of certain processes.

Two options are open to the modeller when confronted with the problem of which processes to include in a model and which to ignore. To illustrate this, consider the construction of a floodplain model but one in which the emphasis is now on predicting water flow from the floodplain into the river rather than water table rise and fall within the floodplain. The first approach could be called a *top-down* modelling approach. In this, the complex model, including all of the water transfer processes likely to occur in a floodplain (Darcian soil water and groundwater flow, pipe flow, overland flow, etc.), is constructed. A sensitivity analysis of the model is then used to indicate which processes are most important in controlling system behaviour. An example of such an analysis would involve altering the values of hydraulic conductivity of the floodplain soil. The value of K will affect groundwater flow rates through the soil according to Darcy's law (equation (9.5)). If changes in the value of K had little impact on model predictions, i.e. the amount of water entering the river channel from the floodplain, then K would be described as an insensitive parameter. It is probably the case that for some modelling scenarios the model output would be sensitive to the value of K, while in others it would be insensitive. If, for example, water tables in the floodplain are near the ground surface, ponding of water could occur on the floodplain shortly after rainfall, giving rise to overland flow and rapid water movement to the river channel. Under such conditions, the much slower rates of flow associated with ground-water movement would be insignificant. Thus the model would be insensitive to groundwater flow. One outcome of such an analysis might be the construction of simpler models for different flow scenarios. Thus a model that represents only overland flow could be used for predicting flow into the river during storms in late winter, when water tables might be expected to be near the ground surface. This simpler model would then be tested with field observations. Kirkby *et al*. (1993: p. 172) have described this as a desirable outcome resulting from the use of complex models, although it is not clear whether they advocate constructing complex models in the first place in order to produce simpler models. The problem with this approach is that several possible model scenarios could still produce essentially the same output. For example, groundwater flow to a dense network of natural soil pipes could also produce rapid flows into the river from the floodplain, and in this situation the model could be expected to be sensitive to K because of its control on (shallow) groundwater flow into the soil pipes. An additional problem with this approach is that the sensitivity analysis of a complex model would be difficult (perhaps impossible) to design and extremely time-consuming.

An alternative, and possibly more efficient, way of proceeding (see below) could be called a *bottom-up* approach. In this approach, use is made of a limited amount of field data *before* a model is constructed. For example, field measurements of river flow may indicate a very rapid response to late winter rainfall. Such a rapid response could arise only if water was conveyed rapidly to the river channel. A reconnaissance field survey would give basic information on the soil type and whether or not features such as soil pipes are present. If the soil was clayey and soil pipes were absent it might be concluded that the floodplain would be prone to winter waterlogging. From this fairly basic information it might then be postulated that overland flow occurs in winter, when the water table is very near the soil surface, and that this mechanism is the main explanation of the rapid increase in river discharge during and after winter rain. To test such a postulate, a simple model could be constructed. A sensible starting model would then be one in which the floodplain soil is treated as impermeable and rainfall is routed to the channel as overland flow; in other words, Darcian flow through the floodplain soil is ignored, even though it undoubtedly occurs. In this example, basic field information has been used to produce a simple model that can then tested with more detailed field data.

Both the top-down and bottom-up approaches can be used to construct simple models that retain the power of abstraction referred to by Gleick. Generally, however, the bottom-up approach is likely to prove the more efficient. In the above example this is clearly the case. The top-down approach involves the expense of first constructing the complex model and then the problem of deciding which alternative sub-model to test first, assuming, of course, that the sensitivity analysis is not prohibitively expensive in terms of time or too difficult. In addition, if basic field data have not been collected, it is possible that both of the alternatives (pipes and overland flow) may be inappropriate for simulating flow to the river during winter, because it may be the case that rapid increases in river discharge during and after rainfall never occur!

Does the blueprint work?

The blueprint based on the four questions above would appear to work in that, as is shown in the four cases, it minimises modelling effort and 'forces' modellers into producing models that can be readily used and tested. In following the blueprint, pitfalls, such as developing complex physically based models in order to predict water tables on a daily basis, are avoided. Essentially, the blueprint forces the modeller to consider developing the simplest model that meets her/his requirement, and therefore encourages efficiency in model development. The question of testability is perhaps the most contentious in the blueprint, since, from the argument presented above, it is suggested that complex models are more difficult to test than simple models and are consequently less useful. Hatton *et al.* (1997) reach a similar conclusion on testability and simplicity. While recognising that complex models have a role to play in eco-hydrology, they suggest that relatively simple, scale-independent models such as those developed by Eagleson (1978a–g) are a better way forward to understanding eco-hydrological behaviour in real systems, even though such models are based on physically less meaningful 'effective parameters'. Nevertheless, as has been shown, testing can be difficult and can be subjective. Even with simple models, model predictions may not match the scale at which measurements can be taken (see Chapter 3), or the model output may not equate with a simple physical parameter but rather a compound representation of several real-world phenomena that is not capable of being easily measured/calculated. Similarly, in Case 2 it was suggested that deciding when a model is falsified is not a straightforward matter.

The issue of testability has also been discussed by Weiner (1995) who notes that while most ecologists accept the need for testable theories, some central models in ecology, such as the Lotka–Volterra population models, show little sign of producing testable predictions. Some ecologists have even argued that the importance of testability has been exaggerated (Fagerström, 1987, cited in Weiner, 1995) and that attributes such as the elegance and beauty of a model are equally important. Weiner (*ibid.*) dimisses such suggestions, noting that they would lead to 'a sort of nihilistic postmodern view of ecology where there is no truth, only stories, and the choice among stories is a question of individual taste (and power).' Weiner concludes that the production of testable mathematical models has to occupy the core of ecology, for without the production of explanatory and testable theories ecology will be little more than a discipline, like history, based on causal narratives. Testing is not easy, and currently much effort is expended in eco-hydrology on testing method-ology (for example, GLUE – generalised likelihood uncertainty estimation:

see Beven and Binley, 1992, and Freer et al., 1996). This must continue as a central theme within the discipline. However, it is important to stress that an important part of such a methdology should be the production, in the first instance, of testable models.

Although the discussion here has been largely on (eco-)hydrological models, it is worth noting that workers in other areas of the environmental sciences have reached similar conclusions about simplicity/complexity and testability. Climatologists use models more than most environmental scientists. McGuffie and Henderson-Sellers (1997) place climate models into a scale of increasing complexity from one-dimensional energy balance models, which predict the variation of Earth surface temperature with latitude, to general circulation models (GCMs), in which the three-dimensional nature of the atmosphere/oceans is simulated. Interestingly, the latter often contain SVAT sub-models which, as noted earlier, can be regarded as eco-hydrological. McGuffie and Henderson-Sellers (ibid.: p. 55) note that by 1980 most research funding went into producing complex three-dimensional models. They continue:

> by the mid- to late 1980s a series of occurrences of apparently correct results being generated for the wrong reason by these highly non-linear and highly complex models prompted many modelling groups to move backward, in an hierarchical sense, in order to try and isolate the essential processes responsible for the results which are observed from the more complex models.

Later in their discussion they stress the need for developing models that are both relatively simple and capable of being tested. Similar conclusions have been reached by Oreskes et al. (1994) in a review of modelling in the Earth sciences.

In conclusion, model construction is, or should be, the first stage of model testing. A model must be designed in a way that renders it capable of being tested. To reiterate the point at the beginning of this section, a complex model, no matter how plausible, cannot be considered credible unless it has been tested or is capable of being tested. It is also important to stress again that testing is often not a simple procedure that is used to declare that a model is 'valid' or 'invalid'. Deciding when a model gives a good representation of the real system is both a qualitative and quantitative exercise, and often the conclusion reached on model validity is partly a matter of opinion.

MODELLING AND MATHEMATICS

As the focus of this chapter is mathematical models, it is appropriate to consider briefly the role of mathematics in modelling. Many mathematicians are modellers and have the same concerns as outlined in 'A blueprint for better modelling?' above. Mathematicians are also interested in developing solutions to the equations used in models. For simple empirical models such as those shown in Case 1 (see Figure 9.10), the parameter values in the equation are derived from field data using statistical fitting methods. Once values of these parameters have been calculated, the equations can be used directly to calculate water table elevation. With physically based models, the relationships between variables are usually non-linear and are posed as partial differential equations. Such equations can be solved using flexible but approximate numerical solutions or, in a limited number of cases, traditional exact analytical methods. The difference between analytical and numerical methods and their advantages and disadvantages are important considerations in modelling. Each approach is explained below.

Analytical (exact) solutions

An analytical solution to a problem results in a closed-form equation that can be used directly to obtain the value of a variable. If we consider the study of infiltration, the form of an analytical solution can be made clear. Other examples could be used to illustrate the analytical approach, but these usually involve formidable mathematical detail. For this reason too, the derivation of the analytical solution given below is not given.

A standard test in soil hydrology used for estimating the physical properties of surface layers of soil involves ponding water in a cylinder inserted into the soil surface. The rate of water uptake by the soil, the infiltration rate, is measured as a function of time. This rate of uptake depends on physical soil properties and the potential gradient driving flow down through the soil profile, and it can be expressed using empirical and physically based equations. An example of an analytical infiltration equation derived from physical theory is that of Philip (1957), given by

$$i = A + Bt^{-0.5} \tag{9.15}$$

where i is infiltration rate (LT^{-1}), A and B are constants that can, in principle, be derived from soil properties (A has dimensions of LT^{-1} and B has dimensions of $LT^{-0.5}$) and t is time (T). It can be seen that the equation can be used directly to obtain a value of infiltration rate providing that A, B and t are known. The equation is thus described as closed-form.

Numerical (approximate) solutions

Some infiltration equations have been developed that describe the rate of infiltration as a function of the total amount of water stored in the soil and only indirectly as a function of time. An example is the Green and Ampt (1911) equation (as cited by Kirkby, 1985):

$$i = A + \frac{C}{S(t)} \tag{9.16}$$

where C is an empirically derived constant (L^2T^{-1}) and $S(t)$ is the time-dependent soil water storage expressed as a depth of water (L). Although an analytical solution, it can be seen that equation (9.16), unlike equation (9.15), does not explicitly consider changes in infiltration rate over time. The equation can, however, be solved iteratively so that i is expressed as a function of time.* Computationally, the procedure (after Baird, 1997b) is as follows:

1 Specify S at the beginning of the period of infiltration to be modelled ($t = 0$). This is the *initial condition* of the model.
2 Calculate i using equation (9.16). This rate is assumed to be constant or steady for a given (small) time step (Δt).
3 Update S i.e., $S = S + i\Delta t$.
4 Update the model's internal clock, i.e. $t = t + \Delta t$
5 Simulation over?
   If YES then stop.
   If NO then go to step 2.

Steps 1 to 5 represent a numerical solution to equation (9.16) known as an explicit finite-difference solution. 'Finite-difference' refers to the fact that a process that is continuous in time has been divided up into a series of finite steps. Finite-differencing is also used to represent processes or phenomena that are continuous in space as well as time, such as groundwater flow in the models in Figures 9.11 and 9.12. Numerical solutions are approximate. The solution here is approximate because the values of i calculated for different elapsed times (t) will depend on the value of Δt. If relatively large time steps are used, the continuous nature of the infiltration process is crudely represented. The finite-difference solution assumes that a process that is continuous in time (infiltration) can be approximated as a series of steady values of infiltration. Calculations of infiltration at one time interval will affect calculations in succeeding iterations, as can be seen from stages 1 to 5

---

* It should be noted that analytical solutions to forms of this equation are also possible. The numerical development is given to illustrate the finite-difference method.

above. This can result in an unsatisfactory solution to equation (9.16), as shown in Figure 9.13, if large time steps are used. In general, the numerical solution will converge with the exact solution as the value of Δt tends to zero.

Many workers still regard analytical solutions as superior to numerical solutions because of their exactness. This is despite the fact that most differential equations of interest to environmental and eco-hydrological modellers cannot be solved using analytical methods without sometimes considerable simplification. This point has been stressed by Remson *et al.* (1971: p. 56) in a treatise on the use of numerical methods in groundwater hydrology:

> The determination of exact analytic [*sic*] solutions is highly desirable. A variety of analytical solutions are available for saturated flow and unsaturated flow. Unfortunately, many of the governing equations are non-linear and the auxillary conditions of most natural systems are extremely complicated. For such situations, exact analytic [*sic*] solutions are not available and recourse must be made to the use of numerical methods of approximation for the solution of differential equations. Thus numerical methods are powerful tools for the solution of *realistic* [*my italics*] mathematical models of complicated natural subsurface hydrologic systems.

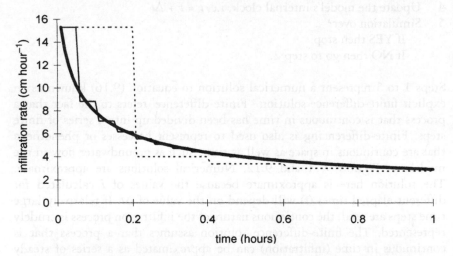

Figure 9.13 Rates of ponded infiltration as a function of time calculated using equation (9.16). $A = 0.3$ cm hour^{-1}; $C = 15$ cm^2 hour^{-1}; $S = 1$ cm. Rates calculated using time steps (Δt) of 0.01 hour (thick solid line); 0.05 hour (thin solid line) and 0.2 hour (broken line). For each time step in the finite difference solution, i is considered to be constant or steady. Large values of Δt result in gross approximations to the exact solution, which in the figure is close to the thick solid line.

To illustrate the point, consider the Boussinesq equation (equation (9.12)) given earlier. Because it is non-linear, a full solution to the equation is not possible unless numerical techniques are used. Analytical solutions can be obtained if certain simplifying assumptions are made. For the example of groundwater flow in a floodplain given above (Case 2), an analytical solution can sometimes be obtained if (1) a function describing boundary condition 1 (river water level) can be specified; (2) water table fluctuations become negligible with increasing distance away from the river (towards the hill slope) i.e. $\partial h/\partial t \to 0$, $x \to \infty$; and (3) a linearised form of the equation is used in which the saturated thickness of the aquifer is assumed constant (d) rather than being the thickness of the local water table, i.e. the Boussinesq equation is replaced by the diffusion equation

$$\frac{\partial h}{\partial t} = \frac{Kd}{s} \frac{\partial^2 h}{\partial x^2} \tag{9.17}$$

The simplifications described in points (1) to (3) could sometimes lead to a much poorer representation of the real system than a full solution to the equation. For example, the saturated thickness of the aquifer can show great variability temporally between summer and winter, violating condition (3). It is also possible that water table fluctuations at the boundary between the floodplain and the hill slope will show considerable fluctuation, depending on the amount of water flowing through the hill-slope soils (violating conditions (2) and (3)). Finally, the river level boundary condition may be complex and may not be adequately described in the analytical solution. Thus, although an analytical solution will be exact, it will also be an *exact approximation*, because it is a solution to an approximation of the original differential equation to be solved. This is different from numerical approximations in that the approximation comes *before* the solution. In such circumstances, which method is worse? Sometimes, exact analytical solutions are poorer representations of the real world. In the above example, a number of numerical solutions to the full Boussinesq equation exist. Used properly, with carefully chosen time steps and sensible spatial discretisation of the floodplain, these solutions will involve very little approximation to the orginal equation and are thus to be preferred to exact analytical approximations.

Analytical solutions are often regarded as more 'elegant' than numerical solutions, presumably because of their exactness, and sometimes, if they are relatively simple, they can reveal the structure of the solution to the differential equation. However, as shown here, elegance in a solution to an equation is not always important, whereas 'elegance' in the design and construction of a model is, as noted earlier in this chapter, essential. The two should not be confused. The same holds true for choosing between numerical solutions. Many numerical methods exist, and some are more elegant than others. For the modeller, it is not generally important which method is chosen as long as it is stable and can be shown to converge with the exact solution.

Mathematical onanism

A number of modellers indulge in 'mathematical onanism'. If 'onanism' is defined as 'to stimulate the genital organs of (oneself or another) to achieve sexual pleasure' (Hanks, 1986), then mathematical onanism in eco-hydrological modelling is 'the search for new solutions to "governing" equations to achieve more publications' (my definition). This definition is not meant as a slur on mathematicians. They are quite rightly concerned with how equations are solved and the elegance and brevity of solution methods. However, some modellers seem more concerned with presenting new solutions (analytical solutions in particular) to equations for which perfectly good solutions (often numerical) already exist. This is best left to mathematicians. Arguably, modellers should concern themselves with whether a differential equation is an adequate description of the real system and with model design and testing. In a similar vein, Wilby (1997: p. 330) notes, 'Rather than continuing the proliferation of new, "improved" models, greater understanding of processes could arise from critical assessments of existing theories.' It is, then, disappointing to see so many papers in the ecological and hydrological literature that do not address these fundamental problems but rather present new solutions to equations. Examples of mathematical onanism are abundant, as a look through recent volumes of any international hydrological journal such as *Journal of Hydrology* and *Water Resources Research* will reveal. It is perhaps invidious to highlight examples of the practice. Nevertheless, two equations within hydrology seem to have been the subject of most mathematical onanism: the Darcy–Richards equation and the Boussinesq equation. Recent examples in just one journal (*Journal of Hydrology*) of new solutions to the Boussinesq equation when one did not appear to be needed are Guo (1997), Rai and Singh (1995), Ram *et al.* (1994), Govindaraju and Koelliker (1994) (see also Hogarth *et al.*, 1997), and Su (1994).

SCALE, PROCESS, AND GOVERNING EQUATIONS IN ECO-HYDROLOGICAL MODELS

The issue of scale in eco-hydrology has already been dealt with in Chapter 3. However, some points relevant to the construction of eco-hydrological models are worth revisiting here. An increasing number of authors have given prominence to the idea that nature is subject to different physical laws, or at least different forms of the same physical laws, at different scales (for a brief review see Baird, 1997a; see also Turner, 1990, and Bloschl and Sivapalan, 1995). Yet the idea that the mathematical description of a given phenomenon may depend on the scale at which it is being studied is not new. A classic example of how physical laws are scale-dependent is groundwater flow through a porous medium. At many scales, Darcy's law can be used to describe water flow through porous media. However, the law is a macroscopic law and applies only to flow through a representative volume of soil, called the representative elementary volume (REV), and not individual pores. The concept of the REV was introduced in a seminal paper by Hubbert (1940), although the term was apparently coined by Bear (1972), who defined it as a volume of sufficient size that there are no longer any statistical variations in the value of a particular property (in this case porosity) with the size of the element. The REV depends on the textural and structural properties of the soil and is shown in Figure 9.14.

At the scale of an individual pore, water flow is virtually impossible to describe mathematically unless the pore is conceptualised as a perfect cylinder. In cylindrical pores laminar flow can be described by the Hagan–Poiseuille equation

$$Q = \frac{\gamma \pi r^4}{8\mu} \left(\frac{P_1 - P_2}{L} + \sin\phi \right) \qquad (9.18)$$

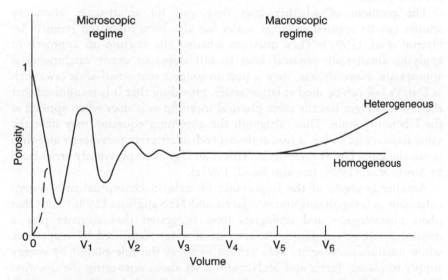

Figure 9.14 Definition of the representative elementary volume (REV). The REV in this case is V3. Source: Hubbert (1956) in Domenico and Schwartz (1990) (redrawn). Reproduced with permission of the American Institute of Mining, Metallurgical and Petroleum Engineers.

337

where Q is pore discharge (L^3T^{-1}), γ is the weight density of water ($ML^{-2}T^{-2}$), r is the radius of the pore (L), μ is the dynamic viscosity of the fluid flowing through the pore ($ML^{-1}T^{-1}$), P is the hydrostatic pressure head acting on the face of the cylinder (L) (subscript 1 refers to the inflow point and 2 to the outflow), L is the length of the pore between 1 and 2 (L), and ϕ is the angle of the pore from the horizontal ($\sin\phi$ represents the gradient of the elevation head or potential). In ideal situations, equation (9.18) could be applied to larger scales than the individual pore. Flow through a porous medium could be conceptualised as a bundle of capillary pores. The mathematics of this approach are explained in Dingman (1984: pp. 294–296) and give rise to an equation analogous to Darcy's law. Unfortunately, such an approach is not generally possible in real soils. Soil pores are not uniform, smooth tubes; they are highly irregular in shape, and flow through them is limited by numerous constrictions (Hillel, 1980). This microscopic complexity defies effective description using laws similar to equation (9.18), and macroscopic laws that ignore the microscopic detail are needed. There are upper limits to the applicability of Darcy's law. For example, Beven (1989) *inter alia*, has questioned whether it is appropriate to use Darcy's law in finite-difference models with computational cells with dimensions of tens or hundreds of metres across. Is there any meaning in a hydraulic gradient calculated between two large cells when in the real system the gradient is changing significantly over much smaller scales, partly due to the (relatively small-scale) heterogeneity of the real system?

The problem of whether laws developed for small-scale laboratory studies can be applied at larger scales has also been discussed recently by Hatton *et al.* (1997). They question whether the bottom-up approach of applying small-scale physical laws to hill slopes or event catchments is appropriate. Nevertheless, they appear to suggest that small-scale laws such as Darcy's law can be used at larger scales, providing that it is recognised that the law no longer has the same physical meaning as it does when applied at the laboratory scale. Thus, although the governing equation may have the same form across scales, it uses scale-dependent effective parameters and thus is not the same law at every scale. This is an argument previously articulated by Smith *et al.* (1994) (see also Baird, 1997a).

Another example of the importance of scale in conceptualising system behaviour is (evapo)transpiration. Jarvis and McNaughton (1986) note that plant physiologists and ecologists have suggested that stomata play a dominant role in regulating the amount of water transpired by vegetation, while micrometeorologists have tended to stress the role played by energy supply to plants. Jarvis and McNaughton set about answering the question 'To what extent do stomata control transpiration?' and show that the answer depends on the scale of investigation. At the scale of the leaf or individual plant, transpiration is determined by vapour pressure deficit and stomatal processes, whereas at regional scales it is controlled by net radiation (for a detailed explanation of this see the original paper or Chapter 2). In the example of groundwater flow the basic processes describing the phenomenon are the same regardless of whether Darcy's law or Poiseuille's equation are being used. In the case of evapotranspiration, it is the governing processes as well as the mathematical description that change depending on the scale of investigation.

The message from this brief section is clear: the suitability of a model for describing a particular phenomenon will depend on the scale at which it is being applied. This point should always be uppermost in the modeller's mind when constructing a model and choosing which processes to include in a description of system behaviour.

THE VALUE OF MATHEMATICAL MODELS

This chapter has, in places, tended to dwell on the deficiencies in existing models and approaches to modelling while giving less attention to the value of mathematical models. Therefore, it is worth re-emphasising that mathematical models are an essential tool in many branches of eco-hydrology. In many cases (i.e. with physically based models), models are our formal encodings of theories and if coded properly allow the theories to be tested and the discipline to advance. Without testing, a theory lacks credibility. It is not enough simply to suggest that the theory appears to represent the essential patterns of a system; this must be shown to be the case. Most science is quantitative and predictive (although see the impassioned and well-argued plea for descriptive science given by Gould, 1989). Indeed, prediction is often used to test the validity of our understanding of nature. Assessing the value of quantitative prediction is not easy and involves subjectivity; science is, after all, partly a social process. It is, however, far easier to compare two sets of data, one modelled, the other measured, than it is to assess the validity of a nebulous, qualitative 'prediction' or a feeling that a model or theory captures in some way the essential dynamics of a system. In addition, although empirical models have a place within eco-hydrology, without theory testing there is the danger that eco-hydrology will become a discipline that deals with site-specific causal narratives (Weiner, 1995) rather than true explanations of process. Thus physically based mathematical models will remain fundamental tools of researchers in eco-hydrology, but they must be used sensibly.

ACKNOWLEDGEMENTS

This chapter has benefited enormously from the critical comments on an earlier version of the manuscript of George Hornberger, Rob Wilby and

Simon Gaffney. As usual, any errors or inconsistencies remain the fault of the author. The diagrams were prepared by Paul Coles and Graham Allsopp of Cartographic Services at the Department of Geography, University of Sheffield.

REFERENCES

Abbott, M.B., Bathurst, J.C., Cunge, J.A., O'Connell, P.E. and Rasmussen. J. (1986a) An introduction to the European Hydrological System – Système Hydrologique Européen, 'SHE'. 1: History and philosophy of a physically-based, distributed modelling system, *Journal of Hydrology* 87: 45–59.

Abbott, M.B., Bathurst, J.C., Cunge, J.A., O'Connell, P.E. and Rasmussen. J. (1986b) An introduction to the European Hydrological System – Système Hydrologique Européen, 'SHE'. 2: Structure of a physically-based, distributed modelling system, *Journal of Hydrology* 87: 61–77.

Addiscott, T.M., and Wagenet, R.J. (1985) Concepts of solute leaching in soils: a review of modelling approaches, *Journal of Soil Science* 36: 411–424.

Armstrong, A.C. (1988) Modelling rainfall/watertable relations using linear response functions, *Hydrological Processes* 2: 383–389.

Baird, A.J. (1997a) Continuity in hydrological systems, in R.L. Wilby (ed.) *Contemporary Hydrology: Towards Holistic Environmental Science*, Chichester: Wiley, pp. 25–58.

Baird, A.J. (1997b) Overland flow generation and sediment mobilisation by water, in D.S.G. Thomas (ed.) *Arid Zone Geomorphology: Process, Form and Change in Drylands*, Chichester: Wiley, 713 pp.

Bathurst, J.C. (1986) Physically-based distributed modelling of an upland catchment using the Système Hydrologique Européen, *Journal of Hydrology* 87: 79–102.

Bathurst, J.C., Kilsby, C. and White, S. (1996) Modelling the impacts of climate and land-use change on basin hydrology and soil erosion in Mediterranean Europe, in Brandt, C.J. and Thornes, J.B. (eds) *Mediterranean Desertification and Land Use*, Chichester: Wiley, pp. 355–387.

Bear, J. (1972) *Dynamics of Fluids in Porous Media*, Elsevier, New York, 764 pp.

Beven, K. (1985) Distributed models, in M.G. Anderson and T.P. Burt (eds) *Hydrological Forecasting*, Chichester: Wiley, pp. 405–435.

Beven, K. (1989) Changing ideas in hydrology – the case of physically based models, *Journal of Hydrology* 137: 149–163.

Beven, K. (1996) Equifinality and uncertainty in geomorphological modelling, in B.L. Rhoads and C.E. Thorn (eds) *The Scientific Nature of Geomorphology: Proceedings of the 27th Binghamton Symposium in Geomorphology*, held 27–29 September 1996, Chichester: Wiley.

Beven, K. and Binley, A.M. (1992) The future of distributed models: model calibration and uncertainty prediction, *Hydrological Processes* 6: 279–298.

Blackie, J.R. and Eeles, C.W.O. (1985) Lumped catchment models, in M.G. Anderson and T.P. Burt (eds) *Hydrological Forecasting*, Chichester: Wiley, pp. 311–345.

Bloschl, G. and Sivapalan, M. (1995) Scale issues in modelling: a review, in J.D. Kalma and M. Sivapalan (eds) *Scale Issues in Hydrological Modelling*, Chichester: Wiley, pp. 71–88.

Boussinesq, J. (1904) Recherches théoriques sur l'écoulement des nappes d'eau infiltrées dan le sol et sur le débit des sources, *Journal de Mathématiques Pures et Appliquées* 10: 5–78.

Box, G.E.P. and Jenkins, G.M. (1976) *Time Series Analysis: Forecasting and Control*, San Francisco: Holden-Day.

Busch, D.E., Loftus, W.F., and Bass, O.L. Jr (1998) Long-term hydrologic effects on marsh plant community structure in the southern Everglades, *Wetlands* 18: 230–241.

Campbell, G.S. (1981) Fundamentals of radiation and temperature relations, in *Encyclopedia of Plant Physiology* New Series Vol. 12A, New York: Springer-Verlag, pp. 11–40.

Childs, E.C. and Youngs, E.G. (1968) Note on the paper 'Explanation of paradoxes in Dupuit–Forchheimer seepage theory', by Don Kirkham, *Water Resources Research* 4: 219–220.

Dingman, S.L. (1984) *Fluvial Hydrology*, New York: Freeman, 383 pp.

Dingman, S.L. (1994) *Physical Hydrology*, New York: Macmillan.

Domenico, P.A. and Schwartz, F.W. (1990) *Physical and Chemical Hydrogeology*, New York: Wiley, 824 pp.

Eagleson, P.S. (1978a) Climate, soil, and vegetation. 1. Introduction to water balance dynamics, *Water Resources Research* 14: 705–712.

Eagleson, P.S. (1978b) Climate, soil, and vegetation. 2. The distribution of annual precipitation derived from observed storm sequences, *Water Resources Research* 14: 713–721.

Eagleson, P.S. (1978c) Climate, soil, and vegetation. 3. A simplified model of soil water movement in the liquid phase, *Water Resources Research* 14: 722–730.

Eagleson, P.S. (1978d) Climate, soil, and vegetation. 4. The expected value of annual evapotranspiration, *Water Resources Research* 14: 731–739.

Eagleson, P.S. (1978e) Climate, soil, and vegetation. 5. A derived distribution of storm surface runoff, *Water Resources Research* 14: 741–748.

Eagleson, P.S. (1978f) Climate, soil, and vegetation. 6. Dynamics of the annual water balance, *Water Resources Research* 14: 749–764.

Eagleson, P.S. (1978g) Climate, soil, and vegetation. 7. A derived distribution of annual water yield, *Water Resources Research* 14: 765–776.

Eagleton, T. (1986) The critic as clown, in Eagleton, T., *Against the Grain: Essays 1975–1985*, London: Verso, p. 149.

Fagerström, T. (1987) On theory, data and mathematics in ecology, *Oikos* 50: 258–261.

Feddes, R.A., Kabat, P., van Bakel, P.J.T., Bronswijk, J.J.B. and Halbertsma, J.

(1988) Modelling soil water dynamics in the unsaturated zone – state of the art, *Journal of Hydrology* 100: 69–111.

Franks, S.W. and Beven, K.J. (1997) Bayesian estimation of uncertainty in land surface–atmosphere flux predictions, *Journal of Geophysical Research* 102 D20: 23991–23999.

Franks, S.W., Beven, K.J., Quinn, P.F. and Wright, I.R. (1997) On the sensitivity of soil–vegetation–atmosphere transfer (SVAT) schemes: Equifinality and the problem of robust calibration, *Agricultural and Forest Meteorology* 86: 63–75.

Freer, J., Beven, K.J. and Ambroise, B. (1996) Bayesian estimation of uncertainty in runoff prediction and the value of data: an application of the GLUE approach, *Water Resources Research* 32: 2161–2173.

Freeze, R.A. and Cherry, J.A. (1979) *Groundwater*, Englewood Cliffs, New Jersey: Prentice-Hall, 604 pp.

Gardner, M. (1989) Foreword, in R. Penrose, *The Emperor's New Mind: Concerning Computers, Minds, and the Laws of Physics*, Oxford: Oxford University Press.

Gardner, W.R. (1957) Some steady-state solutions of the unsaturated moisture flow equation with application to evaporation from a water table, *Soil Science* 85: 228–232.

Gillman, M. and Hails, R. (1997) *An Introduction to Ecological Modelling: Putting Practice into Theory*, Oxford: Blackwell, 202 pp.

Gilvear, D.J., Andrews, R., Tellam, J.H., Lloyd, J.W. and Lerner, D.N. (1993) Quantification of the water balance and hydrogeological processes in the vicinity of a small groundwater-fed wetland, East Anglia, UK, *Journal of Hydrology* 144: 311–334.

Gilvear, D.J., Sadler, P.J.K., Tellam, J.H. and Lloyd, J.W. (1997) Surface water processes and groundwater flow within a hydrologically complex floodplain wetland, Norfolk Broads, UK, *Hydrology and Earth System Sciences* 1: 115–135.

Gleick, J. (1988) *Chaos: Making a New Science*, London: Heinemann.

Goldsmith, F.B., Harrison, C.M. and Morton, A.J. (1986) Description and analysis of vegetation, in Moore, P.D. and Chapman, S.B. (eds) *Methods in Plant Ecology*, Oxford: Blackwell, pp. 437–524.

Gould, S.J. (1989) *Wonderful Life: The Burgess Shale and the Nature of History*, London: Hutchinson Radius, 347 pp.

Govindaraju, R.S. and Koelliker, J.K. (1994) Applicability of linearized Boussinesq equation for modeling bank storage under uncertain aquifer parameters, *Journal of Hydrology* 157: 349–366.

Grayson, R.B., Moore, I.D. and McMahon, T.A. (1992) Physically based hydrologic modeling. 2. Is the concept realistic? *Water Resources Research* 28: 2659–2666.

Green, W.H. and Ampt, G.A. (1911) Studies on soil physics. 1. The flow of water through soils. *Journal of Agricultural Science* 4.1: 1–24.

Guo, W. (1997) Transient groundwater flow between reservoirs and water-table aquifers, *Journal of Hydrology* 195: 370–384.

Hanks, P. (ed.) (1986) *Collins English Dictionary*, Glasgow: Collins, 1771 pp.

Hatton, T.J., Salvucci, G.D. and Wu, H.I. (1997) Eagleson's optimality theory of an ecohydrological equilibrium: quo vadis? *Functional Ecology* 11: 665–674.

Hill, M.O. (1979) *DECORANA, a FORTRAN Program for Detrended Correspondence Analysis and Reciprocal Averaging*, Ithaca, New York: Microcomputer Power.

Hillel, D. (1977) *Computer Simulation of Soil Water Dynamics: A Compendium of Recent Work*, Ottawa: International Development Research Center.

Hillel, D. (1980) *Fundamentals of Soil Physics*, New York: Academic Press.

Hillel, D., Talpaz, H. and Van Keulen, H. (1976) Macroscopic-scale model of water uptake by a non-uniform root system and of water and salt movement in the soil profile, *Soil Science* 121: 242–255.

Hogarth, W.L., Govindaraju, R.S., Parlange, J.-Y., and Koelliker, J.K. (1997) Linearised Boussinesq equation for modelling bank storage – a correction, *Journal of Hydrology* 198: 377–385.

Holling, C.S. (1966) The strategy of building models in complex ecological systems, in K.E.F. Watt (ed.) *Systems Analysis in Ecology*, New York: Academic Press.

Hubbert, M.K. (1940) The theory of groundwater motion, *Journal of Geology* 48: 785–944.

Hubbert, M.K. (1956) Darcy's law and the field equations of the flow of underground fluids, *Transactions of the American Institute of Mining, Metallurgical and Petroleum Engineers* 207: 222–239.

Jarvis, P.G. and McNaughton, K.G. (1986) Stomatal control of transpiration: scaling up from the leaf to the region, *Advances in Ecological Research* 15: 1–49.

Kennedy, B.A. (1985) Indeterminancy, in A. Goudie (ed.) *The Encyclopaedic Dictionary of Physical Geography*, Oxford: Blackwell, 528 pp.

Kirkby, M.J. (1985) Infiltration, in A. Goudie (ed.) *The Encyclopaedic Dictionary of Physical Geography*, Oxford: Blackwell, 528 pp.

Kirkby, M.J., Baird, A.J., Diamond, S.M., Lockwood, J.G., McMahon, M.L., Mitchell, P.L., Shao, J., Sheehy, J.E., Thornes, J.B. and Woodward, F.I. (1996) The Medalus slope catena model: a physically based process model for hydrology, ecology and land degradation interactions, in Brandt, C.J. and Thornes, J.B. (eds) *Mediterranean Desertification and Land Use*, Chichester: Wiley, pp. 303–354.

Kirkby, M.J., Naden, P.S., Burt, T.P. and Butcher, D.P. (1993) *Computer Simulation in Physical Geography*, Chichester: Wiley.

Kirkham, D. (1967) Explanation of paradoxes in Dupuit–Forchheimer seepage theory, *Water Resources Research* 3: 609–622.

May, R.M. (1973) *Stability and Complexity in Model Ecosystems*, New Jersey: Princeton University Press.

McGuffie, K. and Henderson-Sellers, A. (1997) *A Climate Modelling Primer*, Chichester: Wiley, 253 pp.

McWhorter, D.B. and Sunada, D.K. (1977) *Ground-Water Hydrology and Hydraulics*, Fort Collins: Water Resources Publications.

Monteith, J.L. (1973) *Principles of Environmental Physics*, London: Edward Arnold.

Mulligan, M. (1996a) *Modelling hydrology and vegetation change in a degraded semi-arid environment*, unpublished PhD thesis, King's College, University of London.

Mulligan, M. (1996b) Modelling the complexity of land surface response to climatic variability in Mediterranean environments, in Anderson, M.G. and Brooks, S.M. (eds) *Advances in Hillslope Processes*, Chichester: Wiley, pp. 1099–1149.

Murray, W.A. and Monkmeyer, P.L. (1973) Validity of Dupuit–Forchheimer equation. *Journal of the Hydraulics Division (American Society of Civil Engineers)* 97(HY9): 1573–1583.

Nielsen, D.R., Biggar, J.W. and Erh, K.T. (1973) Spatial variability of field measured soil water properties, *Hilgardia* 42: 215–259.

Oreskes, N., Shrader-Frechette, K. and Belitz, K. (1994) Verification, validation and confirmation of numerical models in the earth sciences, *Science* 263: 641–646.

Penman, H.L. (1948) Natural evaporation from open water, bare soil and grass, *Proceedings of the Royal Society, Series A* 193: 120–145.

Philip, J.R. (1957) The theory of infiltration. 4: Sorptivity and algebraic infiltration equations, *Soil Science* 84: 257–264.

Popper, K. (1968) *Conjectures and Refutations: The Growth of Scientific Knowledge*, New York: Harper & Row.

Popper, K. (1972) *Objective Knowledge: An Evolutionary Approach*, Oxford: Oxford University Press.

Quinn, P.F., Beven, K.J. and Culf, A. (1995) The introduction of macroscale hydrological complexity into land surface–atmosphere transfer models and the effect of planetary boundary layer development, *Journal of Hydrology* 166: 421–444.

Ragab, R. and Cooper, J.D. (1993) Variability of unsaturated zone water transport parameters: implications for hydrological modelling. 1: In situ measurements. *Journal of Hydrology* 148: 109–132.

Rai, S.N. and Singh, R.N. (1995) Two dimensional modelling of water table fluctuation in response to localised transient recharge, *Journal of Hydrology* 167: 167–174.

Ram, S., Jaiswal, C.S. and Chauhan, H.S. (1994) Transient water table rise with canal seepage and recharge, *Journal of Hydrology* 163: 197–202.

Remson, I., Hornberger, G.M. and Molz, F.J. (1971) *Numerical Methods in Subsurface Hydrology with an Introduction to the Finite Element Method*, New York: Wiley, 389 pp.

Rennolls, K., Carnell, R. and Tee, V. (1980) A descriptive model of the relations between rainfall and soil water table, *Journal of Hydrology* 47: 103–114.

Richards, L.A. (1931) Capillary conduction of liquids through porous mediums, *Physics* 1: 318–333.

Rogowski, A.S. (1972) Watershed physics: soil variability criteria, *Water Resources Research* 8: 1015–1023.

Smith, R.E., Goodrich, D.R., Woolhiser, D.A. and Simanton, J.R. (1994) Comment on 'Physically based hydrologic modeling. 2. Is the concept realistic?' by R.B. Grayson, I.D. Moore, and T.A. McMahon, *Water Resources Research* 30: 851–854.

Spieksma, J.F.M., Moors, E.J., Dolman, A.J. and Schouwenaars, J.M. (1997) Modelling evaporation from a drained and rewetted peatland, *Journal of Hydrology* 199: 252–271.

Su, N. (1994) A formula for computation of time-varying recharge of groundwater, *Journal of Hydrology* 160: 123–135.

Turner, M.G. (1990) Spatial and temporal analysis of landscape patterns, *Landscape Ecology* 4: 21–30.

van Elburg, H., Engelen, G.B. and Hemker, C.J. (1991) *FLOWNET Version 5.1: User's Manual. Microcomputer Modeling of Two-dimensional Steady State Groundwater Flow in a Rectangular Heterogeneous Anisotropic Section of the Subsoil*, Amsterdam: Institute of Earth Sciences, Free University of Amsterdam.

Verburg, K., Ross, P.J. and Bristow, K.L. (1996) *SWIMv2.1 User Manual*, Divisional Report of the CSIRO Division of Soils No. 130, 107 pp.

Viera, S.R., Nielsen, D.R. and Biggar, J.W. (1981) Spatial variability of field-measured infiltration rates, *Proceedings of the Soil Science Society of America* 45: 1040–1048.

Wang, H.F. and Anderson, M. P. (1982) *Introduction to Groundwater Modelling: Finite Difference and Finite Element Methods*, San Diego: Academic Press, 237 pp.

Ward, R.C. and Robinson, M. (1990) *Principles of Hydrology*, third edition, London: McGraw-Hill.

Warrick, A.W. and Nielsen, D.R. (1980) Spatial variability of soil physical properties in the field, in D. Hillel (ed.) *Applications of Soil Physics*, London: Academic Press.

Wassen, M.J., van Diggelen, R., Wolejko, L. and Verhoeven, J.T.A. (1996) A comparison of fens in natural and artificial landscapes, *Vegetatio* 126: 5–26.

Watts, G. (1997) Hydrological Modelling in Practice, in R.L. Wilby (ed.) *Contemporary Hydrology: Towards Holistic Environmental Science*, Chichester: Wiley, pp. 151–193.

Weiner, J. (1995) On the practice of ecology, *Journal of Ecology* 83: 153–158.

Wicks, J.M. and Bathurst, J.C. (1996) SHESED: A physically based, distributed erosion and sediment yield component for the SHE hydrological modelling system, *Journal of Hydrology* 174: 213–238.

Wilby, R.L. (1997) Beyond the river catchment, in R.L. Wilby (ed.) *Contemporary Hydrology: Towards Holistic Environmental Science*, Chichester: Wiley, pp. 317–346.

10

THE FUTURE OF ECO-HYDROLOGY

Robert L. Wilby

INTRODUCTION

> One of the defining social themes of the decade has been ecological awareness. Consequently, ecology now stands at the interface between science and public policy. Pressure groups, citizens, and policy-makers can draw on ecological research to form opinions on the 'big questions' facing the planet: what to do about climate change, biodiversity, population control, and other pressing matters. So the issue of how ecologists gather information, and how that information can be applied, becomes tremendously important.
>
> Gallagher *et al.* (1995)

The same could also be said of hydrology. Water, in the form of vapour, liquid, snow and ice, is critical to the present functioning and future changes in the global climate. Water-borne nutrients, soil moisture deficits and surpluses are often limiting factors affecting the temporal and spatial dynamics of vegetation systems. Water resources for domestic supplies, irrigation, navigation and industry are subject to increasing competition as populations and economies continue to grow. Water is involved in many human-induced problems, such as accelerated soil erosion, environmental pollution and ecosystem degradation. So the issue of how hydrologists gather information, and how that information can be applied, becomes tremendously important.

The preceding chapters on plant–water relations have demonstrated how hydrology can be applied within ecology, and how knowledge of ecology is important to our understanding of hydrological processes. Three underlying themes emerged, namely the importance of considering scale effects, the value of well-designed field and laboratory experiments, and the need for appropriate mathematical models in eco-hydrological research. As Tyree (Chapter 2) and Wilby and Schimel (Chapter 3) indicated, plant–water relations can be considered at a multitude of scales, ranging from individual cells in plants to the planetary scale. Indeed, the issue of scale is receiving much attention as hydrologists seek to unify processes operating at different temporal and spatial scales (*cf.* Blöschl, 1997), and as ecologists seek to formulate a new understanding of multi-scale biotic and abiotic inter-connections (*cf.* Root and Schneider, 1995).

Chapters 4 to 8 examined the dominant plant–water relations operating within specific environments. A common theme emerging from all the chapters is the difficulty of isolating causes and effects in plant–water rela-tionships. For example, Wainwright, Mulligan and Thornes (Chapter 4) demonstrated that in arid and semi-arid environments the distribution of plants has a profound effect on overland flow processes and erosion. The focusing of overland flow in inter-shrub areas leads to the development of desert pavements, a process, perhaps triggered by overgrazing, that is hydrologically self-perpetuating and ultimately results in the replacement of grassland by scrub. Wheeler (Chapter 5) discussed the elusive definition of 'typical' wetland species, emphasising the hydrological controls exerted on plant assemblages. Although the 'wetness' processes affecting individual plant growth are well understood from laboratory experiments, it was shown that plant communities are responsive to fluctuating water levels, that indi-vidual species' ranges are often broad and, because of this, it is often difficult to specify exact environmental preferences. Roberts (Chapter 6) demonstrated on the one hand the role of plants in the storage and routing of water movement via the canopy and soils, and on the other, the regulation in some cases, of forest evapotranspiration by soil moisture variations.

The importance of considering dynamic, three-dimensional hydrological controls on plants was discussed with reference to streams and rivers in Large and Prach (Chapter 7). It was demonstrated that ephemeral and subsurface flows complicate the concept of the river corridor, and that the two-way exchange of water and nutrients between the river and floodplain is critical to the regulation of plant communities in these ecotones. Similarly, Wetzel (Chapter 8) showed that the productivity of emergent macrophytes at the land–water interface of lake margins shapes, and is shaped by, hydrological processes regulating nutrient and sediment stores. Plants in lakes are effectively a 'metabolic wick', altering the patterns of flow, rates of evapo-transpiration and water balances of the ecosystem.

Finally, Baird (Chapter 9) critically examined the use and abuse of mathematical models as a tool for representing the complex eco-hydrological processes discussed in the previous chapters. It was shown that it is preferable to apply simpler models, in which the key processes are well prescribed and appropriate, than over-parameterised 'replicas' of the real system. This is due to the problems associated with parameter identifiability and the equifinality of model results that can arise when a reductionist approach to model building is employed.

As the following sections indicate, mathematical models will continue to figure prominently as a means of exploring plant–water relations in a multitude of terrestrial environments and across a wide spectrum of temporal and spatial scales. However, this presupposes the existence of carefully designed field and laboratory experiments, as well as the assimilation of new data types, in particular from remote-sensing technology. Furthermore, there is considerable scope for coordinated fieldwork campaigns, and for the inter-comparison of models using common data sets. Increasingly, human impacts on eco-hydrological processes are attaining global dimensions: these concerns are reflected by the ongoing development of integrated atmosphere–biosphere–hydrosphere models, and by a growing appreciation of the significance of plant–water–climate feedbacks. At the landscape scale, field experimentation and model developments could herald a new era in hydrology: an age in which eco-hydrological processes may be deliberately manipulated in order to bring about environmental restoration, remediation or recovery. But before this can happen there are a number of environmental uncertainties that require our urgent attention.

EMBRACING ENVIRONMENTAL UNCERTAINTIES

There are few if any terrestrial and, to a lesser extent, marine environments that have not been marked by the human fingerprint: all organisms modify their environment, and humans are no exception. As Vitousek *et al.* (1997) point out, human activities have transformed the Earth's land surface, reduced biodiversity, altered marine ecosystems (particularly at coastal margins), regulated most major rivers, modified biogeochemical cycles and changed the composition of soils, water and the atmosphere (Figure 10.1). In other words, there are few, if any, pristine environments or landscapes for eco-hydrologists to study. Therefore, any serious analysis of the functioning of plant–water systems should incorporate human factors. Fortuitously, hydrology has a long history of pure and applied research (James, 1991) and is well placed to address the 'big questions' of the twenty-first century, because hydrological processes transcend traditional disciplinary divides and provide scope for the development of a holistic environmental science (Wilby, 1997). The challenge that confronts us now is to infuse eco-hydrological research with the same principles.

Inevitably, the significance of human impacts and the Earth's capacity to support human populations is determined both by natural constraints and by societal choices concerning economics, politics, culture and demography (Cohen, 1995). A good example of the inherent uncertainty surrounding these choices is the recent debate concerning economic and environmental options for the stabilisation of atmospheric greenhouse gas concentrations, a necessary

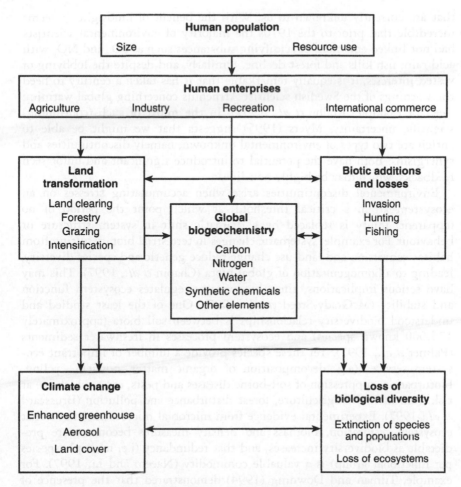

Figure 10.1 A conceptualisation of humanity's direct and indirect effect on the Earth's ecosystems (after Vitousek *et al.*, 1997). Permission applied for.

action to prevent or reduce dangerous anthropic interference with the climate system. By viewing the stabilisation issue as a carbon budget allocation problem, Wigley *et al.* (1996) argue that emission reductions should be deferred for economic and technical reasons. In the medium term, both the stabilisation pathway and climatic response are highly uncertain, with concomitant implications for global ecosystems. Since future economic and technological developments remain uncertain, this means that we would be unwise to make any assumptions about the stability of the climatic boundary conditions affecting long-term plant–water relations.

In a recent review article, Myers (1995) explored the possibility that among the environmental problems ahead, the most important ones could be those that are currently unknown to us. With the benefit of hindsight, it seems incredible that prior to the 1970s the majority of environmental scientists had not linked emissions of acidifying substances such as SO_2 and NO_x with acid rain, fish kills and forest decline. Similarly, and despite the lobbying of vested interests, it is equally remarkable that it has taken a century to heed the warnings of the Swedish scientist Arrhenius concerning global warming (see, for example, Rodhe et al., 1997). In the midst of such (continuing) scientific uncertainty, Myers (1995) suggests that we might be able to anticipate two types of environmental unknown, namely discontinuities and synergisms. Both have the potential to introduce significant and unforeseen feedbacks, which may be highly non-linear.

Environmental discontinuities arise when accumulating stresses on an ecosystem attain a critical threshold, at which point the period of no (apparent) injury is replaced by a dramatic shift in system structure or behaviour. For example, systematic changes in terrestrial biota resulting from habitat conversion and land-use change reduce genetic and species diversity, leading to a homogenisation of global biota (Chapin et al., 1997). This may have serious implications, since biodiversity regulates ecosystem function and stability (McGrady-Steed et al., 1997). One of the least studied and understood biodiversity relationships is between soil biota (approximately 175,000 known species) and ecosystem processes in freshwater sediments (Palmer et al., 1997). Yet these species provide a number of important eco-system services (e.g. decomposition of organic matter, nutrient cycling, bioturbation, suppression of soil-borne diseases and pests, etc.), which are at risk from intensive agriculture, forest disturbance and pollution (Brussaard et al., 1997). Experimental evidence from microbial microcosms shows that ecosystem respiration, biomass and density measures become more pre-dictable as biodiversity increases, and that redundancy (i.e. multiple species per functional group) is a valuable commodity (Naeem and Li, 1997). For example, Tilman and Downing (1994) demonstrated that the presence of drought-tolerant species in diverse grasslands maintains higher productivity under drought conditions than grasslands with lower diversity. This is because high diversity within functional groups also provides greater resilience in response to environmental change, since decreases in the abundance of one species will be compensated for by increases in other functionally similar species.

Therefore, the fewer species there are in any assemblage, the more likely it is that further extinctions will dramatically affect ecosystem processes. In particular, the introduction or removal of predators or diseases that have large 'keystone' effects can result in ecosystem responses that are substantially greater than might have been anticipated from the biomass of the species alone. For example, the removal by humans of elephants or other keystone herbivores can lead to the encroachment of woody plants into savannahs; conversely, the concentration of high densities of elephants in national parks can result in significant environmental degradation (Chapin *et al.*, 1997). Thus subtle changes in ecosystem resource dynamics, trophic structure or disturbance regimes can ultimately have major implications for plant–water processes.

Nonetheless, threshold or discontinuity effects are thought to characterise both natural and human-impacted systems. For example, the climate system is known to have more than one equilibrium state, and intrinsic or extrinsic perturbations can trigger rapid transitions between states. Evidence from Greenland ice cores and deep ocean sediment records suggest that although the Holocene has been relatively stable for the last 9000 years, the previous interglacial began and ended with abrupt changes in deep-water flow, in as little as 400 years (Adkins *et al.*, 1997). It has also been shown that modifications to the global carbon cycle, as a consequence of human activities, have the potential to weaken the thermohaline circulation of the North Atlantic Ocean, which in turn could significantly affect future oceanic CO_2 uptake. Stocker and Schmittner (1997) found that a modelled increase to 750 parts per million by volume CO_2 within 100 years (in line with current emission growth rates) is sufficient to permanently shut down the thermohaline circulation. This would have severe repercussions for precipitation and temperature regimes across northwest Europe and Scandinavia (Hurrell, 1995), for the generation of freshwater runoff from these regions, and for the salinity and thermal profiles of the North Atlantic (Manabe and Stouffer, 1995). Such dramatic changes in regional climate could also have the potential to trigger further discontinuities in vegetation cover, water resources, agriculture and soil erosion (see, for example, Watson *et al.*, 1998).

The second type of uncertainty affecting terrestrial ecosystems might be environmental synergisms – a multiplier effect that arises when two or more environmental processes interact. For example, the inclusion of CO_2 fertilisation effects with those due to direct temperature and/or precipitation changes result in higher modelled biomass and maize yields compared with values obtained using the effects of climate change alone (Dhakwa *et al.*, 1997). A further example of a likely synergistic effect is the phenomenon of forest decline. Forest damage, forest decline and forest dieback that cannot be attributed to biotic factors alone are currently occurring on a global scale. Explanations of the causes of the phenomena include air pollution (acid deposition and ozone), nitrogen saturation or CO_2 fertilisation (inducing nutrient imbalances), magnesium deficiency, episodic drought, epidemic diseases, pathogen attack, stand senescence, vulcanism, and cyclones (Huettl and Mueller-Dombois, 1993). In practice, forest decline seldom results from a single cause, but rather one factor (such as drought or disease) predisposes the forest to the devastating effect of a second impact such as acid deposition. What is clear is that there are many indirect damage pathways and that these are often site-specific, soil-mediated processes, with repeated droughts in particular acting as the final trigger mechanism. Furthermore, Worrall (1994) has reported that leaf and stem rusts, mildews and other foliage diseases can actually show a decrease in polluted air, whereas abiotic diseases such as frost damage and winter injury increase in polluted situations. Similarly, Wilby (1996) demonstrated that the critical loads of acidic deposition that can be tolerated by ecosystems without exhibiting stress are highly dependent upon the assumed climate and hydrological regimes (Figure 10.2).

From the preceding discussion, it is evident that both discontinuities and synergistic effects, and their associated feedbacks, may have profound effects on future plant–water relations. The question then arises as to what are the most appropriate research experiments and models for addressing the phenomenon of environmental unknowns given the uncertainty surrounding the future status of most human-dominated ecosystems (due to resource exploitation, land transformation, biodiversity reduction, climate changes, etc.). The following sections consider the complementary roles of well-designed field experiments and mathematical modelling.

ECO-HYDROLOGICAL EXPERIMENTS

The river catchment has long been used as the fundamental unit for hydrological research, because runoff at the basin outlet is an integration of all the upstream atmospheric and terrestrial processes. The paired-catchment concept also has a long tradition in hydrology as a vehicle for comparing

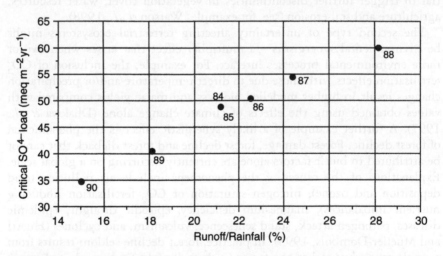

Figure 10.2 Critical sulphate loads for the Beacon Hill catchment, Charnwood Forest, Leicestershire, UK, obtained from the MAGIC model and derived using observed rainfall–runoff data for 1984–1990. Note that 1989 and 1990 were years of exceptionally severe drought and returned the lowest critical loads. From Wilby (1996). Permission applied for.

runoff and water quality from two contrasting ecosystems. The classic experiment of this type was conducted at Hubbard Brook Experimental Forest, where an undisturbed forest was compared with one in which the trees were cut and regrowth was inhibited by the application of herbicide (Likens et al., 1978; Bormann and Likens, 1979). Such experiments underline the value of unmanipulated (or 'benchmark') ecosystems and long-term observations as a means of establishing equilibrium conditions (Burt, 1994). However, despite their clear scientific value, long-term monitoring networks are coming under increasing financial pressure (for example, there have been recent moves to 'rationalise' the long-standing Plynlimon catchment experiments in Wales, UK). Yet, to date, experimental catchments have been used to investigate many important research issues such as the chemical and biological functions of the hydrological cycle, the scaling of dynamic ecosystem behaviour, land surface–atmosphere interactions, and the hydrological effects of human activity (Farrell, 1995).

Ecologists have also been conducting controlled laboratory and field manipulations since the 1960s in the style of 'hard science' (which some commentators have unfairly attributed to 'physics envy'). Laboratory systems have many advantages, such as the replicability of experiments and the control of a limited number of parameters. For example, closed chambers have been used to investigate the specific issue of the sequestration of carbon by the terrestrial biosphere in response to rising concentrations of atmospheric CO_2 (Hungate et al., 1997). Again, there has been a strong argument for careful experimental design and the augmentation of laboratory results with long periods of field observation – what some researchers refer to as 'muddy boots biology'. However, it has been argued that too many ecological experimenters reduce nature to oversimplified caricatures of the real world in their quest for cause and effect (Roush, 1995). The remedy for such reductionism is to embrace more complexity and multiple causality in experimental designs, and to set up multiple working hypotheses that can be subjected to unambiguous statistical testing. There has also been a call for a broadening of the spatial scales of ecological experiments, because many natural processes such as mobility, dispersal and species interaction can only be observed from a macro-perspective (ibid.).

Controlled environment facilities, whether in the laboratory or the field, are halfway houses between the simplicity and abstraction of mathematical models and the full complexity of the environment (Lawton, 1995). Here hydrologists and ecologists have found common ground through experimental manipulations of entire lake, catchment, terrestrial and marine ecosystems. One of the most prominent of such experiments was the RAIN (Reversing Acidification In Norway) study of the effects of acid deposition on soil and river water quality (Wright et al., 1988). At Risdalsheia, Norway, acid rain was excluded from the whole catchment by means of a roof, whereas at Sogndal the acid load was artificially increased. The respective experiments demonstrated that acid deposition effects are reversible, and that the onset of surface water acidification can occur after only a few years of acid deposition. The scientific principles underpinning whole-catchment experiments have subsequently been applied to studies of liming and other measures to mitigate acidification (Warfvinge and Sverdrup, 1988), to the manipulation of forest nutrient balances (Tietema *et al.*, 1997), and to studies of the effects of elevated CO_2 concentrations and temperature on catchments (Schindler and Bayley, 1993). Manipulation experiments have also been used to study nutrient enrichment, biogeochemical cycling and food webs in lake, ocean and terrestrial ecosystems (see Carpenter *et al.*, 1995).

A significant feature of the most recent experiments has been the increasing level of international coordination and cooperation. This enables the repetition of important experiments and the rationalisation of effort, as well as checks on the generality of results. For example, the Boreal Ecosystem–Atmosphere Study (BOREAS) was a large-scale international investigation that focused on improving understanding of the exchanges of radiative energy, sensible heat, water, CO_2 and other radiatively active trace gases between the boreal forest and the lower atmosphere (Sellers *et al.*, 1997). The field campaign was conducted between 1993 and 1997 and adopted a nested multi-scale measurement strategy (Figure 10.3) to integrate observations and

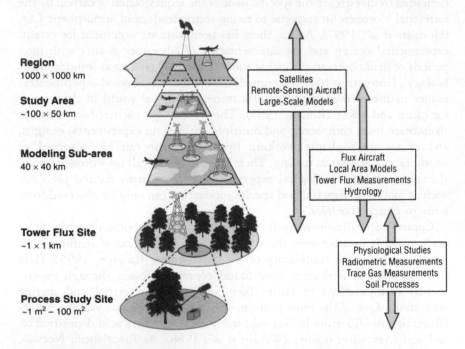

Figure 10.3 Multi-scale measurement strategy used in BOREAS (after Sellers *et al.*, 1997). Permission applied for.

process models over five scales ranging from the process study site (1 m² to 100 m²) up to the region (1,000,000 km²). The measurement and modelling activities were undertaken by eighty-five scientific teams organised into six disciplinary groups (airborne fluxes and meteorology, tower fluxes, terrestrial ecology, trace gas biogeochemistry, hydrology, and remote-sensing science). An important outcome of such experiments is the development of common data sets for model development and inter-comparison, as in the case of the Project for Intercomparison of Land–Surface Parameterization Schemes (PILPS) (Chen et al., 1997) or the Vegetation/Ecosystem Modeling and Analysis Project (VEMAP) (VEMAP members, 1995). However, the value of 'big science' field and/or model data sets is truly maximised when the results are made freely available to the wider scientific community. In this respect, the Climate Impacts LINK Project (Viner and Hulme, 1997) has been an outstanding success. By the free dissemination of results from the UK Meteorological Office Hadley Centre climate change experiments, the 'impacts community' has been able to conduct investigations using state-of-the-art climate scenarios, while returning important information concerning the climate model's realism. For more details about the project, interested readers should refer to the world wide web site: http://www.cru.uea.ac.uk/link.

Large-scale ecosystem experiments are also vital for the validation of existing mathematical models (Oreskes et al., 1994). A good example of an integrated field manipulation and modelling experiment involved the evaluation of MAGIC (Model of Acidification of Groundwater In Catchments) (Wright et al., 1990). This model has a well-documented pedigree and has been widely used to study soil and river water acidification in North America, Europe and Scandinavia. The most recent field results suggest that MAGIC does not adequately simulate observed inter-annual variations in water quality (Figure 10.4). This was attributed to the lumped structure of the hydrological sub-model and poor representations of aluminium dynamics (Cosby et al., 1995). Nonetheless, these shortfalls provide a valuable focus for *both* future field experimentation and model refinement. However, such integrated experimentation presupposes the existence of dedicated field sites, with sustained funding, in a diversity of habitats. The importance of such work should not be understated given that ecosystem experiments are the most direct means of relating cause(s) and effect(s) at the actual scale of many environmental impacts (Carpenter et al., 1995).

PLANETARY SCALE ECO-HYDROLOGY

As has been shown, integrated field and modelling experiments afford considerable opportunities for theory building and model testing. However,

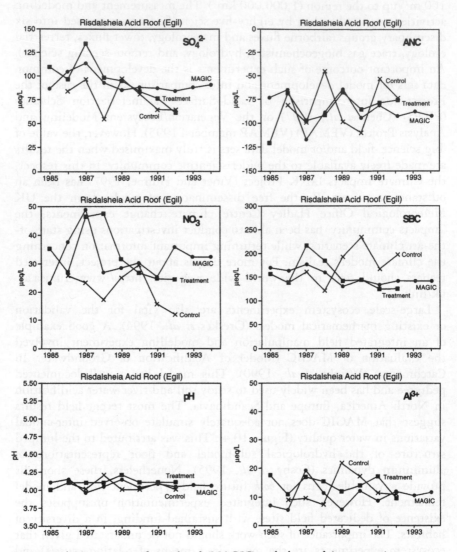

Figure 10.4 Time series of simulated (MAGIC) and observed (treatment) volume-weighted annual average concentrations of selected variables in runoff of the Egil catchment (roof, collected acid rain applied) of the RAIN project. The observed (control) values of the control catchment Rolf (no roof, background acid rain) are also shown. ANC corresponds to the acid neutralising capacity, and SBC to the sum of base cations. Adapted from Cosby *et al.* (1995). Permission applied for.

lake watershed experiments deal essentially with point to catchment-scale processes, have relatively limited sampling capabilities, and often involve the broad-scale extrapolation of laboratory or field measurements to studies of global change. Clearly, there are many eco-hydrological problems that are beyond the scope of conventional field monitoring systems. Thus, studies of landscape- to regional-, or regional to global-scale plant–water relations have largely depended upon, and benefited enormously from, remote-sensing research. As Table 10.1 suggests, there have been considerable advances in the capabilities of remote-sensing platforms to map hydrologically and ecologically relevant surface variables at regional scales and beyond. For example, Moulin et al. (1997) used NOAA/AVHRR satellite measurements to assess phenological stages of canopy development at the global scale. Rhythms in vegetation succession exert a strong control over seasonal energy and water exchanges between the Earth's surface and the atmosphere. Similarly, Hall et al. (1995) used microwave data to measure soil moisture quantitatively under a variety of topographic and vegetation cover conditions. Such measurements have formed the basis of much interesting research into biogeochemical cycles, and hydrological and land-surface changes.

For example, Asner et al. (1998) suggest that at the ecosystem to biome scales (1 m–1 km resolution), remote sensing can provide valuable information on the structural and biophysical attributes of individual

Table 10.1 Surface parameters derived from remotely sensed data (after Schmullius and Evans, 1997).

Discipline	Research topic/operational application
Ecology	Vegetation mapping
	Biomass estimation
	Canopy geometry
	Monitoring flooded forests
	Non-forested wetland flood cycles
	Agricultural land use and crop monitoring
	Tundra monitoring
	Frozen/thawed vegetation
Hydrology	Soil moisture and texture
	Salinity mapping
	Soil surface roughness
	Land–water boundaries and flooding
	Snowpack extent
	Snow wetness and water equivalent
	Mountain glacier extent
	Sea ice discrimination
	Ice sheet and shelves
	Iceberg monitoring

canopies (such as leaf and stem area indices (LAI) or the fraction of absorbed photosynthetically active radiation), which are needed for CO_2 and trace gas exchange estimation, and for carbon, water and nutrient cycle modelling. At continental to global scales (1–4 km data resolution), studies of global biogeochemistry and ecosystem dynamics, energy, water and CO_2 exchanges, employ structural parameters such as *effective* LAI or *plant functional types* derived from grid cell aggregation of land-surface properties. This is not a straightforward procedure, and there has been considerable debate within the literature about how best to aggregate or parameterise heterogeneous land-surface properties (e.g. Arain *et al.*, 1997; Avissar, 1995; Chen *et al.*, 1997; Franks and Beven, 1997). Nonetheless, the development of high-resolution global land data sets such as the 1 km International Geosphere–Biosphere Programme Data and Information System (IGBP-DIS) has proved to be a valuable resource for studies of whole ecosystems, biomes and the globe (Loveland and Belward, 1997). Although this data set is currently undergoing 'ground truthing', six continental data bases (Africa, Antarctica, Australia, Eurasia, North America and South America) have been constructed and are now accessible through the World Wide Web at: http://edcwww.cr.usgs.gov/landdaac/.

High-resolution remote-sensing products have also contributed significantly to the development of theory and research into landscape ecology (the study of the reciprocal effects of spatial pattern on ecological processes) (Pickett and Cadenasso, 1995). Spatial heterogeneity in ecological systems can influence important ecosystem functions, such as population structure and ecosystem processes. Edge effects, for example at forest margins, govern fluxes of moisture and energy, thereby modifying local and regional climates. Such local climate change may be one of the environmental consequences of the piecemeal development of the National Forest in the East Midlands, UK (Wilby, in press).

Continental-scale land-surface changes detected by Earth Observation Systems (EOS) may have profound effects on regional hydrological and biogeochemical processes (see, for example, the summaries in Pitman *et al.*, 1993, or Wilby, 1995). The most popular modelling experiments are those to replace tropical forests with pastures (e.g. Henderson-Sellers *et al.*, 1993), or to replace vegetation with desert (e.g. Xue and Shukla, 1993). This is typically accomplished within general circulation models (GCMs) by altering the surface albedo and roughness, or gross soil properties, and then comparing the results with a control simulation. Most GCMs treat the hydrological cycle in terms of a single, or overlying set of soil 'buckets'; relatively few incorporate lateral water exchanges, groundwater or climatically responsive surface vegetation (Miller *et al.*, 1994; Polcher and Laval, 1994). For example, Lean and Rowntree (1997) found that the complete removal of the Amazonian forest would produce area mean decreases in evaporation of 0.76 mm day^{-1} (18 percent) and rainfall of 0.27 mm day^{-1} (4 percent) and a rise in surface temperature of 2.3 °C. The rainfall changes were attributed to increased albedo, resulting in reduced atmospheric moisture convergence and ascent, while changes in both rainfall and evaporation were found to be very sensitive to one soil parameter – the maximum infiltration rate. On the other hand, Bonan (1997) asserts that the replacement of the natural forests and grasslands of the United States has contributed to summer cooling, especially in the east, which has actually offset any warming due to greenhouse forcing.

Changes in the function and structure of terrestrial ecosystems may also influence the climate by means of biogeochemical feedbacks that regulate land–atmosphere exchanges of radiatively active gases such as carbon dioxide, methane and nitrous oxide, as well as exchanges of energy and water vapour (Melillo et al., 1996). Estimating the magnitude of the terrestrial carbon sink is highly problematic, because the rate of carbon sequestration is a function of CO_2 and nitrogen fertilisation effects, changes in the area of agricultural land, the age structure of biomes, species composition, ecosystem metabolism, susceptibility to pathogens, and limiting factors such as drought or the frequency of fire. Terrestrial ecosystems are also significant sources of methane and nitrous oxide, the production of which is highly sensitive to soil moisture, temperature and oxygen levels. Similarly, as biomes shift in response to global warming, there will be latitudinal changes in surface characteristics such as albedo, rooting depth and surface roughness, which in turn govern surface energy and water fluxes. Furthermore, physiological adjustments such as increasing stomatal resistance in the presence of higher atmospheric CO_2 concentrations, without changes in leaf area, could lead to increased soil moisture storage and additional surface warming over the land (Martin et al., 1989). However, some have argued that at the *planetary* scale, land-surface changes are of less significance to the climate than the thermal regime of the oceans and the spatial distribution of snow and ice (Henderson-Sellers, personal communication). Thus the emphasis of research on terrestrial processes is more often a reflection of the needs and concerns of society than of scientific imperative.

While the focus of this text has been on terrestrial and freshwater ecosystems, it is appropriate that during the International Year of the Ocean (1998) we should consider the role of plants in the remaining two-thirds of the planet, that is, the marine ecosystem. The ocean ecosystem constitutes a significant component of the carbon cycle and produces chemical compounds that ultimately influence the global hydro-climate (Bigg, 1996), making it a legitimate research arena for eco-hydrologists. Most biological activity in the ocean occurs in the surface layers (<200 m), where phytoplankton take up CO_2 and emit sulphurous trace gases. Carbon is sequestered by phytoplankton during primary production, and calcium carbonate is fixed during photosynthesis by hard-shelled marine biota. A shift in the ratio of major seawater nutrients (i.e. C_{org}:N:P) – due to, for example, a change in ocean circulation – could modify the export of organic carbon to the deep ocean, with the potential to influence atmospheric CO_2 concentrations. Additionally, micro-organisms regulate the oceanic component of the nitrogen and phosphorus cycles and are sensitive to anthropic inputs of both these nutrients, particularly at continental margins (Cornell *et al.*, 1995; Galloway *et al.*, 1995). Primary production by phytoplankton may also be fertilised by the addition of iron stimulates, especially in the tropical Pacific region (Watson *et al.*, 1994). However, much more research is needed to ascertain the role of indirect climate–biogeochemical feedbacks in modulating the production and degradation of greenhouse gases (Denman *et al.*, 1996).

Phytoplankton are also significant sources of other climatically active trace gases. For example, total emissions of sulphur gases from natural sources are estimated to be 16 percent of the Northern Hemisphere total atmospheric burden and 58 percent in the Southern Hemisphere (Bates *et al.*, 1992). In terms of climatic influence, however, hydrogen sulphide and dimethyl sulphide (DMS) are the two most important gases (Figure 10.5). When oxidised, both contribute to the formation of cloud condensation nuclei, which can reduce direct radiative forcing at the Earth's surface by clear-sky back-scattering and by higher albedo due to cloud cover (Wigley, 1989). Recent research into DMS chemistry has focused on remote marine regions (e.g. over the Southern Ocean and Antarctic) because of concerns that increased UV-B radiation resulting from the Antarctic ozone hole could significantly affect marine phytoplankton and hence sulphur emissions

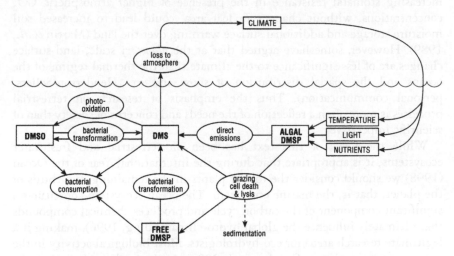

Figure 10.5 The marine biogeochemical cycle of DMS (dimethyl sulphide): production of DMSP (dimethyl sulphonioproprionate) by phytoplankton; transformation by bacteria to DMS and by bacteria or photo-oxidation to DMSO (dimethyl sulphoxide); and alternative utilisation pathways that may influence the quantity of DMS lost to the atmosphere (from Malin *et al.*, 1992). Permission applied for.

(Berresheim and Eisele, 1998). Furthermore, species-specific sensitivities to UV-B by phytoplankton could lead to shifts in community structure and associated changes in the ability of high-latitude ecosystems to act as carbon sinks (Denman *et al.*, 1996).

From the preceding discussion, it is evident that considerable resources are now at the disposal of researchers investigating global-scale eco-hydrology. In particular, these include high-resolution, remotely sensed data sets with global coverage, the results of integrated field and model validation campaigns, and the products of coupled land–atmosphere–ocean GCMs. Returning to the original quotation, the final question is now raised as to how such eco-hydrological information might be applied.

APPLIED ECO-HYDROLOGY

The significance of plants to human well-being has long been recognised, as a South American tribal legend testifies: 'the tropical rainforest supports the sky; cut down the trees and disaster follows' (cited in Park, 1992). And yet the degradation of the Earth's terrestrial vegetated surface by human activities is occurring at unprecedented levels. Daily (1995) estimated that the total area affected by soil, dryland and tropical moist forest degradation is 5.0 billion hectares, or 43 percent of the Earth's vegetated surface (Figure 10.6). Despite this shocking statistic, there has been a growing belief among environmental scientists that a deeper understanding of the functioning and significance of eco-hydrological relations can result in the restoration or rehabilitation of degraded terrestrial ecosystems. In other words, eco-hydrological science can be applied in a positive way to protect or enhance landscapes, rather than negatively to exploit them still further (Petts, 1990). This optimism is reflected in the publication of texts such as *The Ecological Basis for River Management* (Harper and Ferguson, 1995), *Restoration of Aquatic Ecosystems* (National Research Council, 1992) and *The Rivers Handbook: Hydrological and Ecological Principles* (Calow and Petts, 1992). The belief that environmental damage is reversible may also have intense psychological as well as practical significance in the face of compounding human impacts on the biosphere.

Current approaches to the restoration of degraded environments rely upon the identification of limiting factors, and artificial interventions to accelerate natural processes of primary succession and ecological restoration, which might otherwise occur over decades or centuries (Table 10.2). All such techniques stress the importance of retaining biodiversity via nature reserves or areas of undamaged ecosystems, which then provide the natural colonists for the regeneration/reseeding of degraded landscapes (Dobson *et al.*, 1997). Restoration efforts are particularly successful if treatments can mimic natural recovery mechanisms in the substrate, the ecosystem community structure, and in pollutant removal.

A

B

Figure 10.6 Examples of degraded landscapes in A: Moroccan dryland, and B: Malaysian tropical moist forest.

Table 10.2 The timescales for biological and physical processes involved in the development of ecosystems on a newly produced bare area (after Dobson *et al.*, 1997).

Timescale (years)	Biological process	Timescale (years)	Physical process
1–50	Immigration of appropriate plant species	1–1000	Accumulation of fine material by rock weathering or physical deposition
1–50	Establishment of appropriate plant species		
1–10	Accumulation of fine materials captured by plants	1–1000	Decomposition of soil minerals by weathering
1–100	Accumulation of nutrients by plants from soil minerals	1–100	Improvements of soil available water capacity
1–100	Accumulation of N by biological fixation and from atmospheric inputs	1–1000	Release of mineral nutrients from soil minerals
1–20	Immigration of soil flora and fauna supported by accumulating organic matter		
1–20	Changes in soil structure and organic matter turnover due to plant, soil micro-organism and animal activities		
1–20	Improvements in soil water-holding capacity due to changes in soil structure	10–10,000	Leaching of mobile materials from surface to lower layers
10–1000	Reduction in toxicities due to accumulation of organic matter	100–10,000	Formation of distinctive horizons in the soil profile

Soil-mediated limiting factors often hinder initial ecosystem development or recovery, paticularly at sites where the degraded material has a low organic matter content and is nitrogen-deficient. In the short term, the addition of artificial or organic fertilisers may be appropriate, particularly if a controlled-availability fertiliser is used to increase nutrient use efficiency and to minimise leaching (Yanai *et al.*, 1997). In the long term, nitrogen deficiency may be overcome by the introduction of nitrogen-fixing plant species such as the herbaceous legumes *Trifolium* and *Lespedeza* or tree species such as *Casuarina* and *Acacia*. The effectiveness of such measures presupposes a thorough knowledge of the fixing species' plant community, soil and climate preferences. A more radical solution to the problems of soil structure and biological properties is simply to remove the topsoil prior to the disturbance and to replace it afterwards, a practice that is well established for surface mining in most developed nations. As Table 10.3 suggests, there are a wide range of other options for dealing with soil-related problems in ecological restoration. However, the extent to which some of these solutions are sustainable management options is debatable. For example, the addition of lime to acidified soils or hydrological source areas may bring about the rapid recovery of soil and stream water quality (pH) in upland catchments impacted by acid rain (Jenkins *et al.*, 1991), but there can also be a wide range of undesirable side effects for the native, acidophilous biota (Hildrew and Ormerod, 1995).

Assuming that key soil characteristics have been restored, the next stage in ecological restoration often requires the introduction of a community of plants species that are important for restoring ecosystem function, the main

Table 10.3 Short-term and long-term approaches to soil problems in ecological restoration (after Dobson *et al.*, 1997).

Category	Problem	Immediate treatment	Long-term treatment
Texture	coarse	add organic matter or fines	vegetation
	fine	add organic matter	vegetation
Structure	compact	rip or scarify	vegetation
	loose	compact	vegetation
Stability	unstable	stabiliser or multiple crop	regrade or vegetation
Moisture	wet	drain	drain
	dry	irrigate or mulch	tolerant vegetation
Macronutrients	nitrogen	fertiliser	N-fixing species
	others	fertiliser and lime	fertiliser and lime
Micronutrients	deficient	fertiliser	
Toxicity	low	lime	lime or tolerant species
pH	high	pyritic waste or organic matter	weathering
Heavy metals	high	organic matter or tolerant plants	inert covering or bioremediation
Organics	high	inert covering	microbial breakdown
Salinity	high	weathering or irrigate	weathering or tolerant species

component of the final ecosystem, or an element in the final biodiversity of the ecosystem (Dobson *et al.*, 1997). Insights into the 'rules' for assembling such communities and natural rates of recolonisation have often come from studies of chemical waste heaps or from violent volcanic eruptions such as those of Mount St Helens and Krakatau (Plage and Plage, 1985). However, it is unclear whether efforts should be directed towards ecosystem *restoration* (of all that was previously present) or towards *replacement* (by a new but functionally equivalent system). What is clear is that the ecological engineering should be undertaken at the landscape scale in order that natural immigration and dispersal mechanisms may be incorporated. For example, Noble and Dirzo (1997) suggest that in forests used for logging, whole-landscape management is essential to ensure that areas of intensive use are interspersed with areas of conservation and river catchment protection.

There is considerable scope for the manipulation of eco-hydrological processes for the purpose of phyto-volatilisation, phyto-stabilisation or phyto-extraction of soils and groundwaters contaminated by heavy metals and certain organic compounds. Phyto-extraction involves the use of plants to take up toxic metals such as Zn, Ni, Cd, Pb, Cu and Co. For example, Robb and Robinson (1995) proposed the use of natural marshlands within former mining districts as a sink for heavy metals. The biomass is then harvested and, in some cases, metals may be commercially recovered from incineration products. Examples of plants endemic to metalliferous soils include *Alyssum* (Ni and Cr), *Thlaspi* (Zn, Cd and Pb), and *Brassica* (Pb). Phyto-extraction can also be used to remove excess nutrients such as in nitrogen uptake via riparian buffer zones of poplar (Haycock *et al.*, 1993) or through the judicious planting and management of new forests (Wilby, 1998). However, the efficiency of all such techniques is ultimately governed by rates of plant growth relative to biomass and the bio-availability (both chemically and hydrologically) of the contaminant. Furthermore, where the phyto-extraction involves the planting of large stands it is important that secondary conflicts do not occur, for example with flood control or water resource management (Large *et al.*, 1993).

The latter point underlines the necessity of undertaking eco-hydrological manipulations within a holistic environmental framework that embraces both terrestrial and aquatic ecosystems: in other words, integrated catchment management (Burt and Haycock, 1992). For example, EC Directive 91/676 compels member states to designate areas that are hydrologically active, steeply sloping, waterlogged, frozen or adjacent to water courses as nitrate 'vulnerable zones'. Within these areas, the amount and timing of nitrate fertiliser applications are restricted, while in surrounding areas codes of good agricultural practice must be implemented. This is because modern methods of floodplain drainage are highly efficient at coupling nitrate-rich runoff sources to river channels, relative to undrained floodplain soils (Haycock and Burt, 1993). Similarly, Reynolds *et al.* (1995) have demonstrated the potential for mitigating forest harvesting impacts on downstream rivers by enabling the runoff from treeless moorland areas to dilute the nitrate pulse from felled hillsides.

In other cases, ecological restoration may require the conjunctive manipulation of multiple eco-hydrological processes. For example, the management of surface waters affected by algae often requires the simultaneous regulation of point inputs from water treatment works, of *in situ* hydrochemical stores and fluxes, and the control of diffuse nutrient loading (Table 10.4). Reservoir sediments containing high concentrations of bio-available phosphate may continue to enrich the overlying water column long after the original source of enrichment has been controlled. River water concentrations of nitrate and phosphate are also highly sensitive to the prevailing hydro-climate. Thus the contemporaneous decline in *Ranunculus* and increased filamentous algae growth in the River Test, UK, is indirectly linked to a long-term reduction in summer low flows and flushing events arising from a combination of

Table 10.4 Rehabilitation techniques for surface waters affected by algae.

Treating the cause(s)	Technique
Limiting nutrient inputs to surface water bodies	Advanced sewage treatment by biological phosphate removal
	Sewage or effluent diversion away from sensitive watercourses
	Land cultivation and runoff control
Biomanipulation or biomelioration	Nutrient control by promoting growth of macrophytes
	Bankside storage, dilution and sedimentation
Chemical methods	Phosphate stripping using iron and aluminium salts
	Sediment conditioning and sealing
Physical methods	Removal of nutrient-rich sediments by dredging
	Artificial destratification of water body
	Forced aeration of the hypolimnion
	Controlling reservoir draw-down and water depth

Treating the effect(s)	Technique
Application of algicides	Approved proprietary algicides
	Release of phytotoxic organic compounds from decomposing barley straw
Biological cropping	Fishery manipulation by increasing predatory species
	Introduction of filter-feeding fish
Physical methods	Selective draw-off of water of most favourable quality

Source: Nick Everall, senior assessor (biology), Severn Trent Water Ltd, UK (personal communication).

adverse climate, land-use changes and/or aquifer de-watering (Wilby *et al.*, in press). The suggested options for remediation involve the re-evaluation of wider catchment management issues such as the licensing of ground and surface water abstractions, the regulation of undesirable algae growth via both direct and indirect flow manipulation as well as by stringent water quality standards, and the use of sympathetic river engineering as a means of manipulating in-stream hydraulics (such as channel gradient, bankside shading, etc.) at critical sites. All these suggestions presuppose a management perspective that looks beyond the immediate confines of the river margin or floodplain and, in some cases, even the river catchment (Wilby, 1997).

CONCLUSION

It is evident from this and preceding chapters that human intervention in plant–water relations has attained global dimensions. What then can we do to stop the 'sky falling on our heads'? At one level, the scale and diversity of the impacts demands political, social and economic reforms at an international level. At the same time, eco-hydrologists are well equipped to inform policy makers of the potential environmental thresholds, synergisms and feedbacks that may lie ahead. Data from remote-sensing systems allow us to observe changes in the complex interactions between the global biosphere and hydrosphere at high resolutions. (With the launch of satellites such as the Lunar Prospector, which will scan the surface of the Moon for the presence of frozen water, there is even scope for the development of extraterrestrial hydrological science!) In the meantime, integrated field, laboratory and mathematical modelling experiments allow us to focus resources and expertise on critical unknowns at the scale of many human impacts. At the other end of the spectrum, coupled ocean–atmosphere–hydrosphere models allow us to test hypotheses about the sensitivity of eco-hydrological systems to different scenarios of global environmental change. Finally, knowledge of the functioning of plant–water systems allows us to harness natural recovery processes and to restore degraded landscapes. However, these insights will increasingly require a holistic outlook and cooperation between scientists working in allied disciplines. Eco-hydrology provides a valuable example of the benefits that can accrue from such integration.

REFERENCES

Adkins, J.F., Boyle, E.A., Keigwin, L. and Cortijo, E. (1997) Variability of the North Atlantic thermohaline circulation during the last interglacial period, *Nature* 390: 154–156.

Arain, A.M., Shuttleworth, W.J., Yang, Z.-L., Michaud, J. and Dolman, J. (1997) Mapping surface-cover parameters using aggregation rules and remotely sensed cover classes, *Quarterly Journal of the Royal Meterorological Society* 123: 2325–2348.

Asner, G.P., Braswell, B.H., Schimel, D.S. and Wessman, C.A. (1998) Ecological research needs from multiangle remote sensing data. *Remote Sensing of Environment* 63: 155–165.

Avissar, R. (1995) Scaling of land–atmosphere interactions: an atmospheric modelling perspective, in Kalma, J.D. and Sivapalan, M. (eds) *Scale Issues in Hydrological Modelling*, Chichester: Wiley, pp. 435–452.

Bates, T.S., Lamb, B.K., Guenther, A., Dignon, J. and Stoiber, R.E. (1992) Sulfur emissions to the atmosphere from natural sources, *Journal of Atmospheric Chemistry* 14: 315–337.

Berresheim, H. and Eisele, F.L. (1998) Sulfur chemistry in the Antarctic Troposphere Experiment: an overview of project SCATE, *Journal of Geophysical Research* 103: 1619–1627.

Bigg, G.R. (1996) *The Oceans and Climate*, Cambridge: Cambridge University Press.

Blöschl, G. (1997) Preface to the special section on scale problems in hydrology, *Water Resources Research* 33: 2881.

Bonan, G.B. (1997) Effects of land use on the climate of the United States, *Climatic Change* 37: 449–486.

Bormann, F.H. and Likens, G.E. (1979) *Pattern and Process in a Forested Ecosystem: Disturbance, Development and the Steady State Based on the Hubbard Brook Ecosystem Study*, New York: Springer-Verlag, 253 pp.

Brussaard, L., Behan-Pelletier, V.M., Bignell, D.E., Brown, V.K., Didden, W., Folgarait, P., Fragoso, C., Freckman, D.W., Gupta, V.V.S.R., Hattori, T., Hawskworth, D.L., Klopatek, C., Lavelle, P., Malloch, D.W., Rusek, J., Söderström, B., Tiedje, J.M. and Virginia, R.A. (1997) Biodiversity and ecosystem functioning in soil, *Ambio* 26: 563–570.

Burt, T.P. (1994) Long-term study of the natural environment – perceptive science or mindless monitoring? *Progress in Physical Geography* 18: 475–496.

Burt, T.P. and Haycock, N.E. (1992) Catchment planning and the nitrate issue: a UK perspective, *Progress in Physical Geography* 16: 379–404.

Calow, P. and Petts, G.E. (eds) (1992) *The Rivers Handbook: Hydrological and Ecological Principles*, Oxford: Blackwell Scientific Publications.

Carpenter, S.R., Chisholm, S.W., Krebs, C.J., Schindler, D.W. and Wright, R.F. (1995) Ecosystem experiments, *Science* 269: 324–327.

Chapin III, F.S., Walker, B., Hobbs, R.J., Hooper, D.U., Lawton, J.H., Sala, O.E. and Tilman, D. (1997) Biotic control over the functioning of ecosystems, *Science* 277: 500–504.

Chen, T.H., Henderson-Sellers, A., Milly, P.C.D., Pitman, A.J., Beljaars, A.C.M., Polcher, J., Abramopoulos, F., Boone, A., Chang, S., Chen, F., Dai, Y., Desborough, C.E., Dickinson, R.E., Dümenil, L., Ek, M., Garratt, J.R., Gedney, N., Gusev, Y.M., Kim, J., Koster, R., Kowalczyk, E.A., Laval, K., Lean, J., Lettenmaier, D., Liang, X., Mahfouf, J.-F., Mengelkamp, H.-T., Mitchell, K., Nasanova, O.N., Noilhan, J., Robock, A., Rosenzweig, C., Schaake, J., Schlosser,

C.A., Shulz, J.-P., Shao, Y., Shmakin, A.B., Verseghy, D.L., Wetzel, P., Wood, E.F., Xue, Y., Yang, Z.-L. and Zeng, Q. (1997) Cabauw experimental results from the Project for Intercomparison of Land-Surface Parameterization Schemes, *Journal of Climate* 10: 1194–1215.

Cohen, J.E. (1995) Population growth and Earth's human carrying capacity, *Science* 269: 341–346.

Cornell, S., Rendell, A. and Jickells, T. (1995) Atmospheric inputs of dissolved organic nitrogen to the oceans, *Nature* 376: 243–246.

Cosby, B.J., Wright, R.F. and Gjessing, E. (1995) An acidification model (MAGIC) with organic acids evaluated using whole-catchment manipulations in Norway, *Journal of Hydrology* 170: 101–122.

Daily, G.C. (1995) Restoring value to the world's degraded lands, *Science* 269: 350–354.

Denman, K., Hofmann, E. and Marchant, H. (1996) Marine biotic responses to environmental change and feedbacks to climate, in Houghton, J.T., Meira Filho, L.G., Callander, B.A., Harris, N., Kattenberg, A. and Maskell, K. (eds) *Climate Change 1995: The Science of Climate Change*. Contribution of Working Group I to the Second Assessment Report of the Intergovernmental Panel on Climate Change, Chapter 10, Cambridge: Cambridge University Press.

Dhakhwa, G.B., Campbell, C.L., LeDuc, S.K. and Cooter, E.J. (1997) Maize growth: assessing the effects of global warming and CO_2 fertilization with crop models, *Agricultural and Forest Meteorology* 87: 253–272.

Dobson, A.P., Bradshaw, A.D. and Baker, A.J.M. (1997) Hopes for the future: restoration ecology and conservation biology, *Science* 277: 515–522.

Farrell, D.A. (1995) Experimental watersheds: a historical perspective, *Journal of Soil and Water Conservation* 50: 432–437.

Franks, S.W. and Beven, K.J. (1997) Estimation of evapotranspiration at the landscape scale: a fuzzy disaggregation approach, *Water Resources Research* 33: 2929–2938.

Gallagher, R.B., Fischman, J. and Hines, P.J. (1995) Big questions for a small planet, *Science* 269: 283.

Galloway, J.N., Schlesinger, W.H., Levy II, H., Michaels, A.F. and Schnoor, J.L. (1995) Nitrogen fixation: anthropogenic enhancement–environmental response. *Global Biogeochemical Cycles* 9: 235–252.

Galy-Lacaux, C., Delmas, R., Jambert, C., Dumestre, J-F., Labroue, L., Richard, S. and Gosse, P. (1997) Gaseous emissions and oxygen consumption in hydro-electric dams: a case study in French Guiana, *Global Biogeochemical Cycles* 11: 471–483.

Hall, F., Townsend, J. and Engman, T. (1995) Status of remote sensing algorithms for estimation of land surface state parameters, *Remote Sensing of Environment* 51: 138–156.

Harper, D.M. and Ferguson, A., (eds) (1995) *The Ecological Basis for River Management*, Chichester: Wiley.

Haycock, N.E. and Burt, T.P. (1993) Role of floodplain sediments in reducing the nitrate concentration of subsurface runoff: a case study in the Cotswolds, UK, *Hydrological Processes* 7: 287–295.

Haycock, N.E., Pinay, G. and Walker, C. (1993) Nitrogen retention in river corridors: European perspective, *Ambio* 22: 340–346.

Henderson-Sellers, A., Dickinson, R.E., Durbridge, T.B., Kennedy, P.J., McGuffie, K. and Pitman, A.J. (1993) Tropical deforestation: modelling local to regional-scale climate change, *Journal of Geophysical Research* 98: 7289–7315.

Hildrew, A. and Omerod, S.J. (1995) Acidification, causes, consequences and solutions, in Harper, D.M. and Ferguson, A. (eds) *The Ecological Basis for River Management*, Chichester: Wiley.

Huettl, R.F. and Mueller-Dombois, D. (1993) *Forest Decline in the Atlantic and Pacific Region*, Berlin: Springer-Verlag.

Hungate, B.A., Holland, E.A., Jackson, R.B., Stuart-Chapin, F., Mooney, H.A. and Field, C.B. (1997) The fate of carbon in grasslands under carbon dioxide enrichment, *Nature* 388: 576–579.

Hurrell, J.W. (1995) Decadal trends in the North Atlantic oscillation: regional temperature and precipitation, *Science* 269: 676–679.

James, L.D. (1991) Hydrology: infusing science into a demand-driven art, in Bowles, D.S. and O'Connell, P.E. (eds) *Recent Advances in the Modelling of Hydrologic Systems*, the Netherlands: Kluwer Academic Publishers, pp. 31–43.

Jenkins, A., Waters, D. and Donald, A. (1991) An assessment of terrestrial liming strategies in upland Wales, *Journal of Hydrology* 124: 243–261.

Large, A.R.G., Petts, G.E., Wilby, R.L. and Greenwood, M.T. (1993) Restoration of floodplains: a UK perspective, *European Water Pollution Control* 3: 44–53.

Lawton, J.H. (1995) Ecological experiments with model systems, *Science* 269: 328–331.

Lean, J. and Rowntree, P.R. (1997) Understanding the sensitivity of a GCM simulation of Amazonian deforestation to the specification of vegetation and soil characteristics, *Journal of Climate* 10: 1216–1235.

Likens, G.E., Bormann, F.H., Pierce, R.S. and Reiness, W.A. (1978) Recovery of a deforested ecosystem, *Science* 199: 492–496.

Loveland, T.R. and Belward, A.S. (1997) The IGBP-DIS global 1 km land cover data set, DISCover: first results, *International Journal of Remote Sensing*, 18: 3289–3295.

Malin, G., Turner, S.M. and Liss, P.S. (1992) Sulfur: the plankton/climate connection, *Journal of Phycology* 28: 590–597.

Manabe, S. and Stouffer, R.J. (1995) Simulation of abrupt climate change induced by freshwater input to the North Atlantic Ocean, *Nature* 378: 165–167.

Martin, P., Rosenberg, N.J. and McKenney, M.S. (1989) Sensitivity of evapo-transpiration in a wheat field, a forest and a grassland to changes in climate and the direct effects of carbon dioxide, *Climatic Change* 14: 117–151.

McGrady-Steed, J., Harris, P.M. and Morin, P.J. (1997) Biodiversity regulates ecosystem predictability, *Nature* 390: 162–165.

Melillo, J.M., Prentice, I.C., Farquhar, G.D., Schulze, E.-D. and Sala, O.E. (1996) Terrestrial biotic responses to environmental change and feedbacks to climate, in

Houghton, J.T., Meira Filho, L.G., Callander, B.A., Harris, N., Kattenberg, A. and Maskell, K. (eds) *Climate Change 1995: The Science of Climate Change.* Contribution of Working Group I to the Second Assessment Report of the Intergovernmental Panel on Climate Change, Chapter 9, Cambridge: Cambridge University Press.

Miller, J.R., Russell, G.L. and Caliri, G. (1994) Continental scale river flow in climate models, *Journal of Climate* 7: 914–928.

Moulin, S., Kergoat, L., Viovy, N. and Dedieu, G. (1997) Global-scale assessment of vegetation phenology using NOAA/AVHRR satellite measurements, *Journal of Climate* 10: 1154–1170.

Myers, N. (1995) Environmental unknowns, *Science* 269: 358–360.

Naeem, S. and Li, S. (1997) Biodiversity enhances ecosystem reliability, *Nature* 390: 507–509.

National Research Council (1992) *Restoration of Aquatic Ecosystems*, Washington: National Academy Press.

Noble, I.R. and Dirzo, R. (1997) Forests as human-dominated ecosystems, *Science* 277: 522–525.

Oreskes, N., Shrader-Frechette, K. and Belitz, K. (1994) Verification, validation and confirmation of numerical models in the earth sciences, *Science* 263: 641–646.

Palmer, M., Covich, A.P., Finlay, B.J., Gibert, J., Hyde, K.D., Johnson, R.K., Kairesalo, T., Lake, S., Lovell, C.R., Naiman, R.J., Ricci, C., Sabater, F. and Strayer, D. (1997) Biodiversity and ecosystem processes in freshwater sediments, *Ambio* 26: 571–577.

Park, C.C. (1992) *Tropical Rainforests*, London: Routledge.

Petts, G.E. (1990) Water, engineering and landscape: development, protection and restoration, in Cosgrove, D. and Petts, G.E. (eds) *Water, Engineering and Landscape*, London: Belhaven Press.

Pickett, S.T.A. and Cadenasso, M.L. (1995) Landscape ecology: spatial heterogeneity in ecological systems, *Science* 269: 331–334.

Pitman, A.J., Durbridge, T.B. and Henderson-Sellers, A. (1993) Assessing climate model sensitivity to prescribed deforested landscapes, *International Journal of Climatology* 13: 877–898.

Plage, D. and Plage, M. (1985) Java's wildlife returns, *National Geographic* 167: 750–771.

Polcher, J. and Laval, K. (1994) The impact of African and Amazonian deforestation on tropical climate, *Journal of Hydrology* 155: 389–405.

Reynolds, B., Stevens, P.A., Hughes, S., Parkinson, J.A. and Weatherby, N.S. (1995) Stream chemistry impacts of conifer harvesting in Welsh catchments, *Water, Air and Soil Pollution* 79: 147–170.

Robb, G.A. and Robinson, J.D.F. (1995) Acid drainage from mines, *Geographical Journal*, 161: 47–54.

Rodhe, H., Charlson, R. and Crawford, E. (1997) Svante Arrhenius and the greenhouse effect, *Ambio* 26: 2–5.

Root, T.L. and Schneider, S.H. (1995) Ecology and climate: research strategies and implications, *Science* 269: 334–341.

Roush, W. (1995) When rigour meets reality, *Science* 269: 313–315.

Schindler, D.W. and Bayley, S.E. (1993) The biosphere as an increasing sink for atmospheric carbon: estimates from increased nitrogen deposition, *Global Biogeochemical Cycles* 7: 717–733.

Schmullius, C.C. and Evans, D.L. (1997) Synthetic aperture radar (SAR) frequency and polarization requirements for applications in ecology, geology, hydrology, and oceanography: a tabular status quo after SIR-C/X-SAR, *International Journal of Remote Sensing* 18: 2713–2722.

Sellers. P.J., Hall, F.G., Kelly, R.D., Black, A., Baldocchi, D., Berry, J., Ryan, M., Ranson, K.J., Crill, P.M., Lettenmaier, D.P., Margolis, H., Cihlar, J., Newcomer, J., Fitzjarrald, D., Jarvis, P.G., Gower, S.T., Halliwell, D., Williams, D., Goodison, B., Wickland, D.E. and Guertin, F.E. (1997) BOREAS in 1997: Experiment overview, scientific results and future directions, *Journal of Geophysical Research* 102: 28731–28769.

Stocker, T.F. and Schmittner, A. (1997) Influence of CO_2 emission rates on the stability of the thermohaline circulation, *Nature* 388: 862–865.

Tietema, A., Beier, C., de Visser, P.H.B., Emmett, B.A., Gundersen, P., Kjønaas, O.J. and Koopmans, C.J. (1997) Nitrate leaching in coniferous forest ecosystems: the European field-scale manipulation experiments NITREX (nitrogen saturation experiments) and EXMAN (experimental manipulation of forest ecosystems), *Global Biogeochemical Cycles* 11: 617–626.

Tilman, D. and Downing, J.A. (1994) Biodiversity and stability in grasslands. *Nature* 367: 363–365.

VEMAP members (1995) Vegetation/ecosystem modeling and analysis project: comparing biogeography and biogeochemistry models in a continental-scale study of terrestrial ecosystem responses to climate change and CO_2 doubling, *Global Biogeochemical Cycles* 9: 407–437.

Viner, D. and Hulme, M. (1997) The Climate Impacts LINK Project: Applying Results from the Hadley Centre's Climate Change Experiments for Climate Change Impacts Assessments, Norwich: Department of the Environment, Transport and the Regions Climatic Research Unit, 17 pp.

Vitousek, P.M., Mooney, H.A., Lubchenco, J. and Melillo, J.M. (1997) Human domination of Earth's ecosystem, *Science* 277: 494–499.

Warfvinge, P. and Sverdrup, H. (1988) Soil liming as a measure to mitigate acid runoff, *Water Resources Research* 24: 710–712.

Watson, A.J., Law, C.S., Van Scoy, K.A., Millero, F.J., Yao, W., Friederich, G., Liddicoat, M.I., Wanninkhof, R.H., Barber, R.T. and Coale, K. (1994) Implications of the 'Ironex' iron fertilisation experiment for atmospheric carbon dioxide concentrations, *Nature* 371: 143–145.

Watson, R.T., Zinyowera, M.C., Moss, R.H. and Dokken, D.J. (eds) (1998) *The Regional Impacts of Climate Change: An Assessment of Vulnerability*, a special report of IPCC Working Group II, Cambridge: Cambridge University Press.

Wigley, T.M.L. (1989) Possible climate change due to SO_2-derived cloud condensation nuclei, *Nature* 339: 365–367.

Wigley, T.M.L., Richels, R. and Edmonds, J.A. (1996) Economic and environmental choices in the stabilization of atmospheric CO_2 concentrations, *Nature* 379: 240–243.

Wilby, R.L. (1995) Greenhouse hydrology, *Progress in Physical Geography* 19: 351–369.

Wilby, R.L. (1996) Critical loads' sensitivity to climate change, *Environmental Conservation* 22: 363–365.

Wilby, R.L. (ed.) (1997) *Contemporary Hydrology: Towards Holistic Environmental Science*, Chichester: Wiley, 354 pp.

Wilby, R.L. (in press) Hydrological impacts of the National Forest, special issue of *East Midlands Geographer*.

Wilby, R.L., Cranston, L.E. and Darby, E.J. (in press) Factors governing macrophyte status in Hampshire chalk streams: implications for catchment management, *Journal of the Chartered Institute of Water and Environmental Management*.

Worrall, J.J. (1994) Relationships of acid deposition and sulfur dioxide with forest diseases, in Godbold, D.L. and Hüttermann, A. (eds) *Effects of Acid Rain on Forest Processes*, Chichester: Wiley, Chapter 5.

Wright, R.F., Cosby, B.J., Flaten, M.B. and Ruess, J.O. (1990) Evaluation of an acidification model with data from manipulated catchments in Norway, *Nature* 343: 53–55.

Wright, R.F., Lotse, E. and Semb, A. (1988) Reversibility of acidification shown by whole-catchment experiments, *Nature* 334: 670–675.

Xue, Y. and Shukla, J. (1993) The influence of land surface properties on Sahel climate. Part I: desertification, *Journal of Climate* 6: 2232–2245.

Yanai, J., Nakano, A., Kyuma, K. and Kosaki, T. (1997) Application effects of controlled-availability fertilizer on dynamics of soil solution composition, *Soil Science Society of America Journal* 61: 1781–1786.

INDEX

ABA (abscisic acid) 16
Abies *see* fir
Abrahams, A.D. 96
abscisic acid *see* ABA
absorption 19, 20–3, 283
acceleration 13
acid rain 160, 350, 353
acidity 159
acidophiles 157
acrotelm 153–5, 156
adaptation 103, 131, 134, 136, 211,
  250; biochemical 270; physiological
  88, 270; structural 270;
  submergence 133, 251; survival and
  reproduction 237
Addiscott, T.M. 308
adenosine triphosphate *see* ATP
adhesion 62
adiabatic effects 67
aeolian activity 88
aeration 132, 134, 154, 155; poor 211
aerenchyma 133, 134, 148
aerobic conditions 35, 133, 208
afforestation 81
Africa: East 103, 273; North 81, 89,
  105; South 80, 87; sub-Saharan 89;
  West 189, 204; *see also* Egypt;
  Hoggar; Libya; Niger; Sahara; Sahel;
  Senegal; Sudan
agriculture 50, 51; intensification 81;
  rain-fed and irrigated 81; small-scale
  108
agroforestry 189
air 26, 192; ambient 16, 18, 19;
  bubbles 27, 28; bulk 33; difference

in temperature between leaf and 29;
  dissolution of 28; heat capacity of
  32; passage prevented 27; saturation
  deficit 191; solubility in water 28;
  turbulence 33; vapour
pressure deficit 205; vertical
  convection 31; *see also* air humidity
  deficit
air humidity deficit: canopy
  conductance and 202; increase in
  219, 222; stomatal behaviour and
  199, 201, 215
Alaska 146
albedo 40, 43, 48, 155; changes 68;
  higher 82; modified 51
algae 88, 91, 243, 247, 270, 284;
  epiphytic 245, 252; filamentous 245,
  366; planktonic 271
allelopathy 246, 255
Allen, S.J. 193, 194, 223
allochthonous solids/inputs 146, 269
alluvial features: aquifers 107, 247;
  deposition 129, 160
altitudes 102
aluminium concentration 146
Amazonian forest 90, 186, 189, 202,
  205, 358; Ji-Paraná 203; Marabá
  203
America *see* USA
ammonium 146
amplitude fluctuations 140–2
analytical methods/solutions 332, 334
anatomical structures 133
Andersen, F.O. 149
Anderson, M.P. 307

374

INDEX

animals 204; choice of food 91; small,
    burrows 208
ANN (artificial neural network)
    approaches 58
annuals 85, 87, 135; biseasonal 88;
    opportunistic 141
anoxia 131, 133, 134, 148, 161;
    avoidance of 136
Antarctic Ocean 360
anthropic impact/effects 40, 65;
    dryland plants 106–8; interference
    349; responses to change 58
apoplastic water 26
approximation: Dupuit–Forchheimer
    321, 325; exact 335; numerical 335
aquatic habitats 244; amphibious 238;
    backwater-pool 247; in-stream 238,
    243; interstitial 251; macrophyte
    communities 245; micro- 243; river
    237; watery 127; wet, wetland
    species restricted to 137
aquifers: alluvial 107, 247; channel 238
Argentina 82; Patagonia 89
arid environments 27, 34, 78, 105
aridification 81, 82
aridity 87, 276; extreme 88
Arizona 85, 90; Sycamore Creek 284;
    Walnut Gulch 86, 96, 97, 100
Armstrong, W. 131, 132–3, 134, 149
Arrhenius, Svante 350
ash forest 201, 207
Asner, G.P. 357–8
Atlantic Ocean 102; North 58, 351
atmosphere 16, 28, 60, 69, 131, 192;
    above dry and wet surfaces 48;
    aerodynamic mixing with tree
    canopies 207; connection severed
    133; CO$_2$ concentrations 57;
    cyclonicity 58; dry 223; gaseous
    escape to 277; humidity 185;
    simulating features 65, 69; spatial
    characteristics 67; vapour flux
    convergence 43; variables 82; water
    returns as transpiration 183; see also
    SPAC; SVATs
ATP (adenosine triphosphate) 11
Aulehm deposits 95
Australia 86, 87, 210; central 79;

eastern 89; Melbourne 54; New
    South Wales 193; Queensland 203,
    209; south 80; southwest 80;
    Victoria 55; western 50
autochthonous development 269, 284,
    287
auxiliary variables 62
AVHRR (Advanced Very High
    Resolution Radiometer) 43, 62, 357
Avissar, R. 48
AWS (automatic weather station) 84,
    213, 218
Azores 80

bacteria 270; cyano- 271
badlands (marl) 86
BAHC (Biological Aspects of the
    Hydrological Cycle) 58
Bailey, W.G. 65
Baird, A.J. 101, 314, 333, 337
banks: ditch 151–2; river/stream 94,
    107, 143, 238, 246
Barbero, M. 108
Bardossy, A. 58
barley 147
Barnes, H.H. 241
base-richness 157, 158–60, 161
basins 273, 275, 276, 281; drainage
    282, 284; morphology 284–8
Bathurst, J.C. 242, 313
Beadle, C.L. 214
Beckhoven, K. van 149
Bedford, B.L. 149
bedrock 242, 284
beech forest 201, 207
Bellamy, D.J. 146
Beven, K.J. 42, 62–3, 310
Bharucha, F.R. 142
bicarbonate 146, 159, 255
biodiversity 65, 244, 350
biogeochemistry 197, 269
biological processes 99, 246, 251, 254,
    277; lowland streams 245; potential
    of land 81; 'resetting' 257; uptake
    163; utilisation 284; see also BAHC
biomass 197, 242, 256, 258, 350;
    above-ground 80; algal 284; dead
    246; extant community 272;

375

radiactively active 359; recycling
272; transfer 133; vapour 18; *see also*
greenhouse gas; IRGAs
Gash, J.H.C. 189, 204, 213
GCMs (general circulation models) 41,
58, 65, 67, 98, 106, 331, 358;
limitations 60; rainfall scenarios 111,
181; sensitivity experiments in
ecosystems 81
Gell, P.A. 54
geographical information systems *see*
GIS
geology 45, 104–6
geomorphology 57, 95, 111, 247, 284;
basin 281
geostrophic deflection 279
geostrophic vorticity 58
Germany (eastern) *see* Peene River
Gessner, F. 293
GEWEX (Global Energy and Water
Cycle Experiment) 43; Continental
Scale International Project 58
GEWS (Global Early Warning System)
43
Gibbs, R.J. 158
Giller, K.E. 145
Gillman, M. 304, 317
Gilmour, D.A. 203
Gilvear, D.J. 318
GIS (geographical information systems)
40, 44, 62
glacial activity 238, 285;
glacial–interglacial cycles 103
Gleick, J. 312–13, 329
Global Energy and Water Cycle
Experiment *see* GEWEX
global hydrological scale 41
global warming 40, 108, 350, 359
GLUE (generalised likelihood
uncertainty estimation 330–1
Godwin, H. 142
Goel, U. 246
Goodchild, M.F. 57
goodness of fit 324
Gopal, B. 246
Gorham, E. 162
Gould, S.J. 339
gradients 25, 64, 102, 240, 270;

base-poor 160; base-rich 159–60;
concentration 16, 133; gravitational
potential 23; humidity 191, 213;
hummock-hollow, *Sphagnum* in
ombrogenous bogs; 150–1 hydraulic
292; hydrochemical 257;
microtopographical 150; patterns
shaped by 65; potential 305–6;
pressure 23; resolving 191–2;
rich-fen-poor-fen 147; salinity 162;
shoreline 143, 148; temperature 191,
213, 255; underlying 65;
vegetation related 65, 158–62;
water depth 244; water level 137,
153
Graf, W.L. 94
graminoids 155
Granier, A. 33, 195–6
Granlund, E. 129, 130
grassland 81, 85, 86, 108, 209, 210;
clumps 97; desert 103; dry,
calcareous 131; invasive 109;
semi-arid 96; steppe merges into 88;
wet 137, 147
gravel 242, 251
gravitational forces 12, 13, 240, 279;
acceleration due to 23; water
movements and 281
Grayson, R.B. 44
grazing 89, 91, 204
Greece: Argolid 107; Attica 107;
Ionnina 106; Macedonia 107;
Tenaghi Philippon 104, 106
greenhouse gas 40, 348–9
'greenness' 51–2
'grey box' models 318
'grid-box average' 44
gritstone 52
Grootjans, A.P. 138
groundwater 142; advected 275;
brackish 162; calcareous 163;
chemical transport to 207; depleted
levels 257; discharge 130, 157; flow
160; incoming to lake 279;
movement 292; phyto-volatilisation,
phyto-stabilisation or
phyto-extraction of 365; quantity
and timing of water released to 181;

T - #0135 - 071024 - C0 - 234/156/23 - PB - 9780415162739 - Gloss Lamination